construction

materials, methods, careers

by

JACK M. LANDERS

Assistant Professor
Industrial Arts Technology
Central Missouri State University
Warrensburg, Missouri

South Holland, Illinois
THE GOODHEART-WILLCOX COMPANY, INC.
Publishers

Library of Congress Cataloging in Publication Data

Landers, Jack.
   Construction : materials, methods, careers.

   Includes index.
   1. Building trades--Vocational guidance. I. Title.
TH159.L36   690'.023   75—42032
ISBN 0—87006—202—6

INTRODUCTION

Using CONSTRUCTION, you will explore the many skills, materials, methods, and processes necessary to provide you with career awareness for the construction industry. You will become aware of the numerous career opportunities available in the cluster of construction trades and related jobs, and you will obtain a working knowledge of what people in construction do.

Specifically, CONSTRUCTION, Materials, Methods, Careers is devoted to four areas of construction to cover the entire range of the construction industry.

1. The Technology of Construction Management.
2. The Technology of Construction Personnel.
3. The Technology of Construction Materials and Products.
4. The Technology of Construction Processes and Production.

One feature of this text is the relation of all measurements to metrics. It is the intent that using metrics becomes a part of your activities. A series of discussion topics is included at the end of each chapter to assist you in checking your knowledge of the materials covered. Over 120 career descriptions in the construction industry are defined, enabling you to become familiar with various career opportunities.

A Laboratory Manual is available for use with CONSTRUCTION, Materials, Methods, Careers. Many of the suggested activities provide you with the necessary hands-on experiences. The numerous exercises allow you to practice and enrich the many construction concepts expressed in the text.

Jack Landers

CONTENTS

5

Fig. 1-1. Aerial view of sports complex. Skills and work of hundreds of people had to be brought together at the right time to complete it successfully. How many kinds of occupations do you think were involved? Can you name some of them?

Chapter 1

INTRODUCTION TO CONSTRUCTION

Construction is a broad subject. It keeps many people busy at many different occupations. A study of its processes and the people who work to make it an industry is interesting and well worth the effort.

The knowledge and skills of many persons go into construction projects such as the one shown in Fig. 1-1. Architects, engineers, estimators, carpenters and brick layers, as well as typists, computer technicians, truck drivers and others find their place in life by working in construction.

Homes, bridges, dams, highways, factories and airports start from ideas. They are made real by the labors of people who earn their living through building. This labor continues to serve the needs of people long after the project is done.

Construction is an activity that is very old. In fact, from the beginning of civilization, people have built shelter for themselves.

Today we construct homes, places of worship, commercial buildings, transportation systems and industrial complexes as in Fig. 1-2. But we also build electrical generating plants and hydroelectric dams. These are necessary to generate the electrical energy that lights and heats these structures.

Fig. 1-3. A very unusual structure built of steel and concrete expresses society's gratitude in a symbolic way.
(Chamber of Commerce of Metropolitan St. Louis)

Some buildings are erected to honor great persons or to express noble ideas. These are called symbols or monuments. Some monuments, such as the great pyramids of Egypt, have challenged modern builders to create wonders like the Gateway Arch at St. Louis, Missouri, Fig. 1-3. Complex interchanges, railroads and airports are other examples of efforts to answer transportation needs through building. See Fig. 1-4. Every day we depend on such structures to serve our needs.

HISTORY OF BUILDING

Technology was born when primitive people first started using tools to make life easier. *Technology is the science of*

Fig. 1-2. This construction site shows several different kinds of structures. What kind of material was most used here? What other material do you think will be used? (Northwest Engineering Co.)

Fig. 1-4. Structures such as this expressway allows easy movement of people and material.
(Missouri Highway Dept.)

using tools and techniques in their most efficient manner. With this new technology came agriculture, manufacturing, commerce and construction.

Early in history, primitive humans discovered they no longer needed to spend all their time gathering food for survival. Having tools, they no longer depended upon gathering pine boughs for shelter. They could cut larger logs for timbers to build better shelters. They also developed metals (copper 4000 BC), made better tools and began to cut stone into blocks. Their structures became larger, stronger and more permanent.

CONSTRUCTION MATERIALS

Around 2500 BC, someone discovered how to make glass. It was used for jars, for decorating pieces of clay and for making ceramic tile. Ceramic tile wore well when used on floors and walls. It added pleasing colors to structures.

Some 2000 years later, glass came into use as windows. Today, it is widely used in windows and window walls, Fig. 1-5.

CONCRETE

Concrete is an important and useful material in modern construction practices. One of its many uses is shown in Fig. 1-6. It is used in highways, bridges, buildings, foundations for homes, dams, irrigation canals and a long list of other structures.

Ever since humans first started to build, they have searched for a material to bind sand and stone into a solid, formed mass. The Assyrians, Babylonians, Egyptians, Greeks and Romans used a clay mixed with lime.

Shortly after 1800, Portland cement was discovered. It has

Fig. 1-5. Glass serves as the outside walls in this magnificent building.
(IDS)

Fig. 1-6. Concrete is a very versatile material capable of being formed into very unusual shapes.
(American Plywood Assoc.)

Fig. 1-7. Steel serves as the frame for this future building.   (Stran-Steel Corp.)

since become the most used cementing material in the construction industry. The word "concrete" has come to mean "Portland cement" concrete. You will have an opportunity to work with the mixing and forming of concrete in your study of construction.

METALS

The technology of building was moved forward greatly by the discovery and use of metals. Copper, steel, zinc, aluminum and other metals have become significant construction materials. Metal can be carefully controlled as to shape, strength, hardness and resistance to corrosion by alloying (mixing by melting) with other metals. Little wonder it is so often preferred in construction! See Fig. 1-7. From the raw material for tools and equipment to nails and structural beams, metal continues to serve as a basic building material.

WOOD

1. Wood is the only renewable natural resource.
2. It has been used from the earliest recorded events.

3. Wood continues to serve as a dependable material of construction.

Today, more wood is used in residential construction, Fig. 1-8, than any other material. Wood possesses certain artistic values, warmth and traditional charm, produced by color and grain patterns unmatched by any other material.

PLASTICS

Twentieth century technology has given the construction industry synthetic (man-made) materials. As shown in Fig. 1-9, plastics of various kinds are used in floor coverings, paints, sealants, roofing materials as well as in building decoration.

More building materials are available today than ever in the history of construction.

CONSTRUCTION ACTIVITIES

Over the centuries, some very imaginative building programs have been successfully carried on. A notable example is the Great Wall of China. The Great Wall, Fig. 1-10, has proven to be one of the greatest building enterprises ever undertaken.

Fig. 1-8. Wood is the traditional material for homes in the United States.   (American Plywood Assoc.)

Fig. 1-10. Built in the third century BC, the Great Wall of China extends 1500 miles (2414 kilometres).

The pyramids of Egypt also stand as a construction monument. One of the pyramids has 2,300,000 blocks of stone, each weighing 2 1/2 tons (2300 kilograms). It took 100,000 men 20 years to build this structure.

The Suez Canal, Panama Canal and Tennessee Valley Authority systems of dams for flood control, transportation and hydroelectric power are present-day construction marvels.

We have become known as builders out of a need to meet the demands of life. As our numbers grow, more and even greater construction demands will have to be met. Construction houses all of our activities.

DEVELOPMENT OF ARCHITECTURAL STYLES

Architecture is the art of designing and planning a building, employing the materials and techniques available. The pueblos

Fig. 1-9. Plastics, the twentieth century building material, takes many forms. Above. This telescoped section of escyanurate foam insulates pipes. (Manning, Selvage and Lee, Inc.) Below. Plastic laminate materials make beautiful and durable counter tops.   (Formica Corp.)

Fig. 1-11. The Pueblo Indians are most famous for their style of residences built completely of local materials.

of the American Southwest, Fig. 1-11, are a good example of local materials that have been converted to architectural needs. Regions and cultures have become known by their architectural style because of the materials available and the techniques employed to use those materials to best advantage.

The major functions of architecture are to:

1. Plan the building to meet human needs.
2. Provide sound construction.
3. Insure permanence through strong, durable materials.

Materials used in the architectural design of structures include:

1. Stone, a natural material widely used today as a surface veneer (covering).
2. Brick, a material of fired clay, often used for nonload-bearing walls and surface veneer.
3. Wood, a natural material, easily shaped and structurally strong. Suitable for simple structures of limited size. Excellent for floor and wall coverings.
4. Metals, a material smelted from the earth, have entirely changed the concept of wall and support techniques.
5. Concrete, a manufactured mixture of cement and water with aggregates of sand and stone. Used extensively in construction.

Fig. 1-13. The basic arch is used to span large openings.

STRUCTURAL METHODS

Structural methods include: post and lintel, arch, vault, dome, truss and frame.

THE POST AND LINTEL

The post and lintel, Fig. 1-12, is the simplest form of load and support construction. This is the basis for all openings such as doors and windows. The system is comprised of two or more upright posts of columns and a third member, the lintel, placed horizontally across their top surfaces. Traditionally, it is seen in colonnades (regularly spaced series of units).

THE ARCH

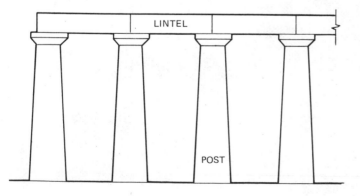

Fig. 1-12. Roman architecture made wide use of this simple post and lintel construction.

As shown in Fig. 1-13, the arch (curved lintel) can be used to span wide openings. If used in multiple, each arch tends to counteract the force of the other.

In Roman times, the arch was fully exploited in bridges, aqueducts and large scale architecture. Today, steel, concrete and laminated wood have changed the concept and mechanics of the arch.

Modern components are completely different than the wedge-shaped blocks used in the past. Today's materials provide greater freedom of design and a means of covering great spans without a massive substructure.

Fig. 1-14. The vault is another Roman architectural device often used in buildings.

THE VAULT

A vault, Fig. 1-14, is an arch deep enough to cover a three-dimensional space. The Romans discovered that openings could be made by intersecting two of the three-dimensional arches at right angles.

Fig. 1-15. Dome of the United States Capitol Building reflects an architectural form perfected by Byzantine builders.

THE DOME

Byzantine builders perfected the dome, Fig. 1-15, which also evolved from the arch. In its simplest form, the dome may be thought of as a continuous series of arches which intersect (cross) at their highest point (apex).

Geodesic domes, Fig. 1-16, were developed in the twentieth

Fig. 1-16. The geodesic dome. (Southern Illinois University)

12

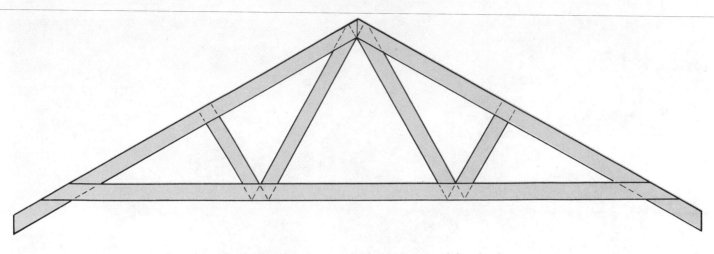

Fig. 1-17. Trusses are based upon the rigid nature of the triangle.

century by designer Buckminster Fuller. They are spherical (round) forms created by triangular facets (flat surfaces) framed by lightweight tubular struts.

THE TRUSS

The truss, Fig. 1-17, is based on the geometric law that a triangle is the only figure that cannot be changed in shape without a change in the length of its sides. The truss (when firmly fastened at the angles) cannot be deformed by its own load or by external forces such as wind pressure. When spaces become too great to span with a single triangle, the simple triangle is replaced by a series of small triangles within a frame, forming a truss.

THE FRAME

The frame, Fig. 1-18, is a skeleton that is able to stand by itself as a rigid structure without depending on floors or walls

to resist deformation (going out of shape). Materials such as wood, steel and concrete, being strong in both tension and compression, make the best members for framing.

Most homes are constructed around a light wood frame, as in Fig. 1-19. Steel framing is based on the same principle, but it is much simplified due to the greater strength of the material.

CONSTRUCTION AND SOCIETY

The construction industry is a broad and important part of the economy. Career opportunities abound, with varying amounts of education required. High school graduates will find many opportunities. Community college or two-year technical schools give sufficient preparation for technical careers in construction. The four-year college or university prepares the individual for management, architectural and engineering careers.

Basically then, construction is an activity that involves people in planning, designing, building and using the completed project. *Construction technology is the science of bringing management, personnel, materials and processes together on the building site to perform the task in the most efficient manner possible.*

PROJECT INITIATION

Before anything in our society can be done or built, an original idea or suggestion must come from an individual or a group. The idea comes after someone sees a need. It is understandable that a home answers the need of a family for shelter; a highway bridge is a result of the need to transport goods to the opposite side of the river; flood control dams are designed to prevent flooding of land. Each project is evaluated by comparing its cost to the savings which the construction guarantees.

Many times, the needs and ideas are far ahead of the technology, materials and financing available to complete the project. For instance, a tunnel under the English Channel (between the British Isles and France on the European continent) was not feasible until 1973. The idea for the tunnel

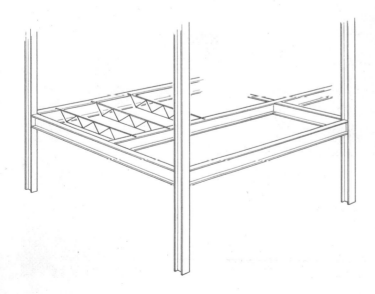

Fig. 1-18. Frame construction provides a rigid structure without floors or walls.

Fig. 1-19. Wood is the basic framing material for homes and other light construction.
(Western Wood Products Assoc.)

was proposed several centuries ago, but political and technical problems prevented its construction until recently.

Construction projects originate (have their beginnings) either in the public or private sector of our economy. Public projects are those which will serve the general public. Examples of public projects are: dams; levees and canals to insure flood control and/or inland water transportation; airports for domestic and international air travel. See Fig. 1-20. Closer to home, schools are built with public money and, therefore, are public projects.

The public project is different from the private project in the way in which the needs are established. Usually, the need is identified through public hearings. A public hearing is a means whereby individuals and groups get together with the proper governmental agencies and discuss the possibilities and worthiness of the project.

Private projects are started by an individual corporation to answer felt needs. A private project might be a home, a shopping center, a factory or an office building.

Major ways public projects differ from private projects are:

Fig. 1-20. From this airport construction project, people may travel to all parts of the earth with relative ease.
(Aviation Div., Kansas City International Airport)

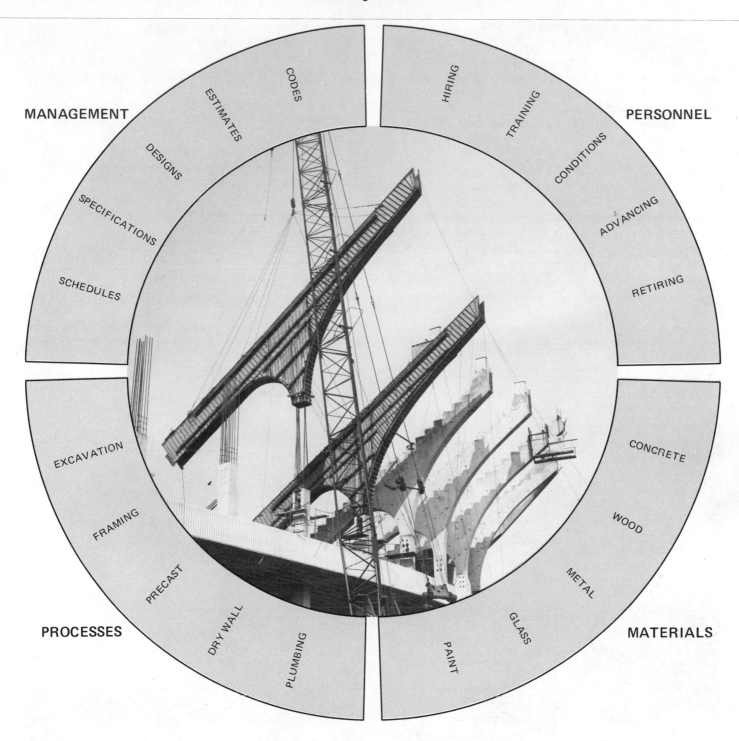

Fig. 1-21. The various parts of construction technology.

1. Public projects have no means of being financed except by a governmental agency.
2. The purpose of public projects usually is not for profit, but to provide a service.

CONSTRUCTION TECHNOLOGY

The twentieth century has seen great advances in construction activities. New techniques, improved material, improved management methods and a highly skilled labor force coupled with mechanical power have made many projects possible.

Construction, as defined, is the act of building or devising a structure by the most systematic method with the most efficient actions.

Construction may also be thought of as the process by which skilled persons assemble materials into a serviceable unit by using the most efficient techniques or processes, Fig. 1-21. In order to achieve the most efficient action, management is provided to bring the materials, processes and personnel together on the site at the proper time.

Construction

SUMMARY

Construction is one of the oldest activites in which people have engaged. During the centuries, materials and techniques have changed along with increased knowledge of the world in which we live. Architectural styles are different because of locale (availability of natural materials and methods of construction) and knowledge of how forces react through materials.

Construction in our society is important because:

1. It provides the necessary shelter for homes, industry and institutions.
2. It provides highways, bridges, dams and air terminals.
3. It provides many career opportunities and thousands of jobs so people may earn a living and get satisfaction out of life.

The remainder of this text is devoted to an understanding of the total picture of the construction industry, the many career opportunities available and the economic importance of construction to society.

DISCUSSION TOPICS

1. How does construction serve everyone?
2. List several reasons why architectural styles vary between geographical regions of the world.
3. Define the following terms:
   a. Architecture.
   b. Construction.
   c. Technology.
   d. Construction technology.
4. What are the four major factors which make up construction?
5. List several natural building materials.
6. How do each of the following structures serve society?
   a. Homes.
   b. Dams.
   c. Bridges.
   d. Commercial buildings.
   e. Highways.
   f. Schools.
7. Describe the principle of the arch.
8. List several publicly initiated construction projects in your area.
9. List several privately initiated construction projects which have been started or completed recently in your community.
10. List as many careers or jobs as you can that are related to the construction industry.

Forest areas are the nation's only renewable natural resource. More wood is used in residential construction than any other material.
(Weyerhaeuser Co.)

SECTION 1

THE TECHNOLOGY OF
CONSTRUCTION MANAGEMENT

Construction management is involved with more than supervising building operations. This section explains how management deals with site preparation, real estate laws, building codes, financial and legal aspects, geological considerations and designing and planning the project. It also covers specification writing, estimating and bidding, utilities, progressive and final inspections of materials and workmanship.

Construction management is the decision-making area involving planning and designing, financing, site selection, material procurement and building operations.

Nearly 50,000 miles (80,500 km) of superhighway have been constructed by the interstate highway system, making hundreds of thousands of potential building sites more accessible. (Asphalt Assoc.)

Chapter 2

SELECTING THE SITE

Fig. 2-1. Site preparation may include the removal of a previous construction project.

Picking the best location available for putting up a structure is very important. Success of the enterprise will, in the end, rely heavily on making a wise choice. Proper location has a direct effect on:

1. Cost of the project.
2. Accessibility. (Is the spot easy to get to?)
3. Structural techniques (construction methods used).
4. Amount of site preparation needed. See Fig. 2-1.

Clearly, a highway would not be located where the most grading is required. A hydro-electric dam is not put where the river bank is lowest and the flood plain is widest. A housing development would not be planned where there are no roads to move people to their jobs. Nor are residences built where water and other utilities are hard to get. By the same reasoning, schools are not built where children will have to travel long distances to get to them.

The national capitol was located at the site which is now Washington, D. C., because it was close to the geographical center of the original 13 colonies, Fig. 2-2. State capitols are

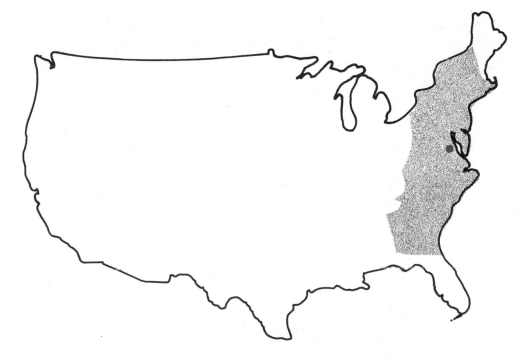

Fig. 2-2. Washington, D.C. was centrally located when it was originally selected as the site of the capitol of the 13 original colonies. Where would the geographical center of the United States be located now?

Fig. 2-3. Industry selects building sites that are near raw material sources, labor or its market area.
(American Hot Dip Galvanizers Assoc., Inc.)

Fig. 2-4. Location of the structure and its design is often keyed to the surroundings. (Koppers Co., Inc.)

generally close to the geographical center of states. County seats are normally in the center of each county.

Industry wants to be near:
1. Raw material.
2. Source of labor.
3. Markets for finished products, Fig. 2-3.

The home owner selects a building site for several different reasons, Fig. 2-4. One of these considerations is a natural setting. Other factors will vary with each individual. One may prefer nearness to daily work; another may look for schools nearby. To a lesser or greater extent, the same process of selection is followed by the prospective builder.

THE PROCESS OF SELECTING

Land is usually selected for a specific project after a feasibility study has been made. *A feasibility study is a looking into all conditions to see if the project can be carried out.* It is conducted by either a government agency, a private owner, or a hired consultant. The results of a feasibility study are submitted to the project initiator. The study will also identify

several promising sites. Each of these locations are visited and evaluated by the initiator.

Many times the selection of the site is made with the aid of a computer. However, the final choice is usually made by the individual or agency initiating the project.

BASIC REQUIREMENTS

Certain needs form the basis for judging all projects. In general, sites for all types of buildings must meet certain basic requirements:
1. Accessibility. Can the construction crews get equipment and material to the site?
2. Supply of labor. Are there sufficient men with the necessary skills available to complete the project on schedule? Is there sufficient manpower with the know-how to operate and maintain the enterprise after completion of the project?
3. Physical resources. Is there a good supply of water for drinking, heating or cooling, cleaning or processing?
4. Environmental factors. How is the climate? Are temperatures extreme? Will humidity be a problem? Is the area subject to hurricanes or other severe weather conditions?
5. Physical factors. What is the contour of the site and nature of the soil conditions?

The contour map in Fig. 2-5 was developed as part of the search for a suitable place to build a house. The client had described the kind of lot he wanted his house built on. It had to meet the following criteria:
1. Must be accessible (on or near a road).
2. Must have a south or east front exposure.
3. Must be in a wooded area.

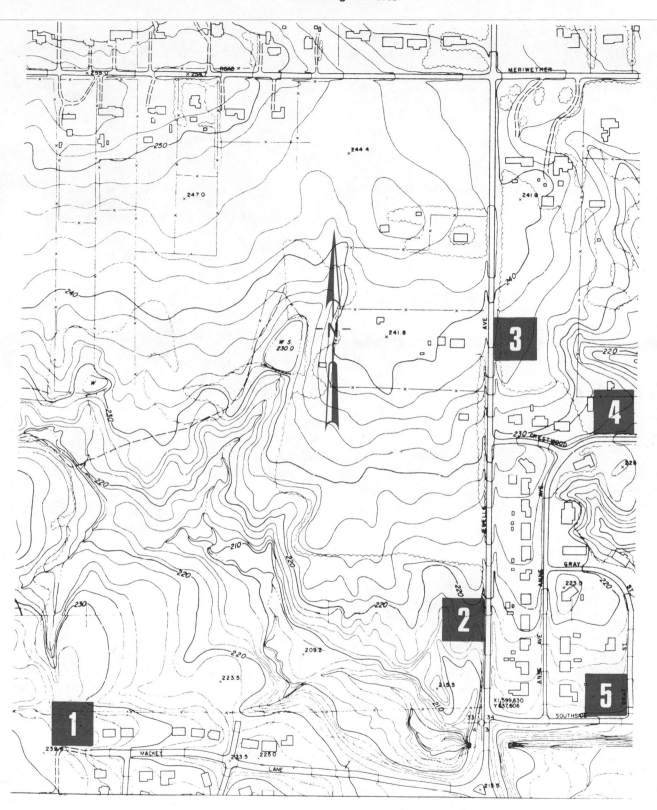

Fig. 2-5. Contour map shows elevation changes of a piece of land at 10-ft. intervals. Contour lines run through areas of like height. Scalloped lines indicate woods. Five lots are marked which have similar characteristics. How are they similar?

4. There must be no extreme elevation changes.
5. It must be high enough to provide good drainage.

By studying the map, can you tell which of the five sites meet the criteria? Which do not?

A contour map shows how level a piece of land is by showing the variation of height between the contour lines. The elevation, in feet above sea level, is given at intervals of 10 feet. Notice the drainage patterns.

Aerial photography is helpful in locating building sites. See Fig. 2-6. With aerial photographs, one can observe the physical features of the site locations. Natural and man-made landmarks are easily seen.

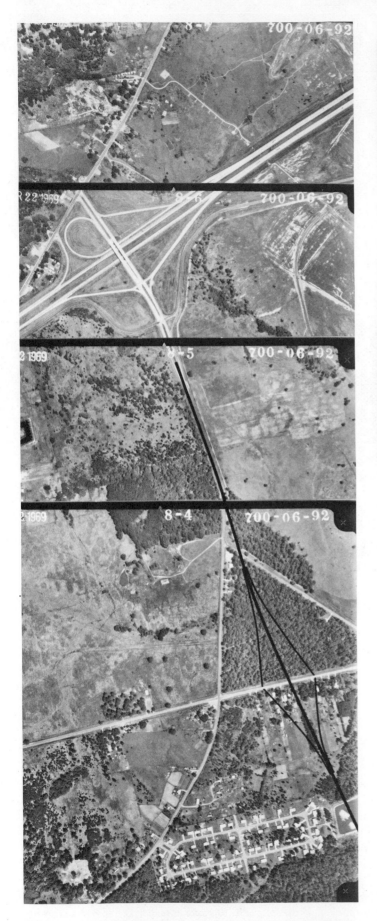

Fig. 2-6. Aerial photos are used to help select relocation site for highway interchange system. (Howard, Needles, Tammen & Bergendorf)

Fig. 2-7. Many construction projects are unsightly by nature of the activity. However, by locating industrial complexes away from housing projects, the two can co-exist successfully. (Exxon)

ZONING AND OTHER FACTORS

Political factors and land restrictions by local zoning requirments have to be considered too. Usually a city or regional governmental agency divides its area into zones. This sets aside land in an orderly fashion for residential, commercial and industrial use. It places industry and commercial activities in areas where it will not interfere with residential areas or destroy property values.

Local taxes must be considered in the selection. These will include property taxes (based upon assessed value of the site and the improvements built thereon), sales taxes and income taxes. All add to the cost of operation.

Community acceptance is still another important factor: Will the project be a benefit or an annoyance to the community? Will it damage the local ecology? Or will it be an asset by improving the general area simply because it is there? See Fig. 2-7.

A good case in point was the controversy over construction of the Alaskan oil pipeline. Ecologists feared damage to the ecological balance in the fragile tundra of Alaska. Certain industries, by nature of the raw materials used or processes required to produce the desired product, are unpleasant to have close by. They should, therefore, be located elsewhere.

Selecting the Site

REAL ESTATE CONTRACT

THIS AGREEMENT, Made and entered into this..day of ...., 19........., by

and between .... the seller;

and .... the buyer, the term,

seller or buyer, may be singular or plural according to whichever is evidenced by the signatures affixed hereto; and ....

....as agent for the seller hereinafter referred to as the agent.

WITNESSETH: That the seller, in consideration of the mutual covenants and premises hereinafter contained, hereby agrees to sell and

convey unto said buyer who agrees to purchase as herein provided the following described real estate situated in the County of.........................

....State of .... to-wit:

together with (if any) furnace, lighting and water supply apparatus, fixtures and plumbing equipment, attached linoleum,

venetian blinds, curtain rods, storm sash and awnings ....

....

Subject, however, to any restrictions, zoning laws or ordinances affecting ....of trust covering above described real

.... the rate of...............................per cent per annum, payable:

to be paid by the buyer as follows: $...

hereby acknowledge.. request then in addition to the purchase price the buyer agrees to pay all expenses

as earnest mo.. or loans.

.................................subject to the buyer's ability to obtain a loan or loans in the amount of $...........................and payable as follows:

and in the event the buyer is unable to obtain such loan or loans within .... days hereof then this contract shall be considered
null and void and the money above deposited shall be returned to the buyer.

The sale under this contract shall be closed under the GENERAL SALES CONDITIONS AND CLOSING PRACTICES as set forth
on the reverse side hereof and hereby made a part of this contract, and subject to any SPECIAL AGREEMENTS between buyer and
seller, also set forth on the reverse side hereof (if any) and hereby made a part of this contract, at the office of ....

The Seller shall pay the agent $............................. for commission at the time of the closing of this sale.

IN WITNESS WHEREOF, Said parties hereto subscribe their names.

EXECUTED IN TRIPLICATE. Note: Buyers should employ a
licensed attorney to examine title to said real property.

....

....

....
Broker.

....

....

Fig. 2-8. Real estate contract binds buyer and seller to an agreement and carefully spells out all terms of agreement.

ACQUIRING THE SITE

After a location has been found which meets all the criteria, it is important to get the right to use it. The right to build on a given site is established by:

1. Purchase.
2. Lease.

In either method, negotiation begins between the buyer and the seller. Negotiation is often conducted by a third person who acts in the best interest of each party concerned. Negotiation continues until an agreement is reached on the value of the site and on the condition for the transfer of land. When the conditions are agreed on, a contract is signed by all parties involved in the sale. See Fig. 2-8.

A real estate contract is a legal document and a general statement listing the conditions and considerations of the transfer of ownership. If the transfer agreement is reached for purchase of the site, the next consideration is to specify how much the buyer will pay the seller and at what time payment or payments are to be made.

Other considerations might include an agreement on what may be removed from the site. For example, a site may include buildings and fixtures which the seller wishes to keep. A final consideration is the date of transfer.

The second method of getting the right of use is the lease. *The lease allows the use of land without actually owning it.* The lease spells out the length of time the agreement is to last and the annual fee to be paid to the owner. A lease is usually negotiated in much the same manner as a sale. Thus, negotiation is a very important method in acquisition of the site. The negotiation method comes to a successful end when both the buyer and the seller are satisfied.

Unfortunately, the negotiation method cannot always result in agreement. If the project is of public nature, condemnation proceedings then begin. When private negotiation fails, the judicial branch of the government is asked to enter the process. The federal government, by constitutional provisions, has the power of "eminent domain" or the right to take or condemn private property for public purposes even against the objection of the owner.

Eminent domain is only invoked when the welfare of the general public is at stake. It cannot be used for the private gains of other private interests. Condemnation proceedings are unpopular and are avoided, if possible.

Sometimes several sites have identical criteria (meet stated needs equally well) and the decision to buy must be delayed. Then, an option to buy may be acquired. This will prevent the parcel of land from being sold to another interested party. *Options give the buyer more time in making his selection without the fear of losing the opportunity to purchase the site. The option is usually given for a stated period of time for a given amount of remuneration.* Remuneration is a type of payment to allow the seller a benefit in case the buyer decides not to purchase the site. The option agreement contains the basic framework for the eventual sale of the property.

SUMMARY

The selection of a construction site is determined by the purpose for which it is intended. When selecting an industrial site, for example, one important consideration is the accessibility to material, labor and markets.

Feasibility studies are conducted by individuals concerned, governmental agencies or by hired professionals to identify the most likely sites for the construction project.

Acquiring the right to build on a particular site is done by either negotiation or condemnation. Through outright purchase or lease, the builder obtains the privilege of using the land.

CAREERS RELATED TO CONSTRUCTION

CONSULTING ENGINEER is an engineer with much experience who is called in to advise on soil mechanics or site prepartion. He may be a civil engineer or he may be a specialist in structural work or soils.

SITE PLANNER, also known as a landscape architect, plans and designs development of land areas for its intended use. Compiles and analyzes data on site conditions such as geographic location, soil, vegetation and rock features; drainage and location of structures. Prepares site plans, working drawings, specifications and costs estimates for land development. Inspects construction work to insure compliance with landscape specifications.

REAL ESTATE SALESPERSON rents, buys and sells property for clients. Draws up real estate contracts and helps negotiate loans. May hold a brokerage license and then is known as a REAL ESTATE BROKER.

DISCUSSION TOPICS

1. Look at a map of your state and identify a site to locate a new hydroelectric generating plant.
2. Why was your community located on the site it is?
3. What are the zoning requirements of your community?
4. What is a feasibility study?
5. Who conducts a feasibility study?
6. How does ecology help in site selection?
7. List the two methods of obtaining the right to build on a given site?
8. How are the conditions of transfer of property established?
9. How does the concept of eminent domain affect a building project?
10. Define the following terms:
    a. Real estate contract.
    b. Option.

Chapter 3

REAL ESTATE

At some remote time in history, people asserted their exclusive right to use or occupy an identifiable part of the earth's surface. They recognized the boundaries of their living areas by relating to streams, rocks, trees and various other landmarks.

These natural landmarks, however, were not permanent, but subject to change through natural causes. For example, the early Egyptians discovered that each year after the river Nile flooded and receded, the boundary marks were obliterated. Therefore, they found it necessary to establish a system of identification and location of the boundaries.

LAND DESCRIPTIONS

The Egyptians (and the Romans and Chinese) developed the means for identifying surface spaces. Especially important was the need to reestablish boundaries after the annual flooding of the Nile.

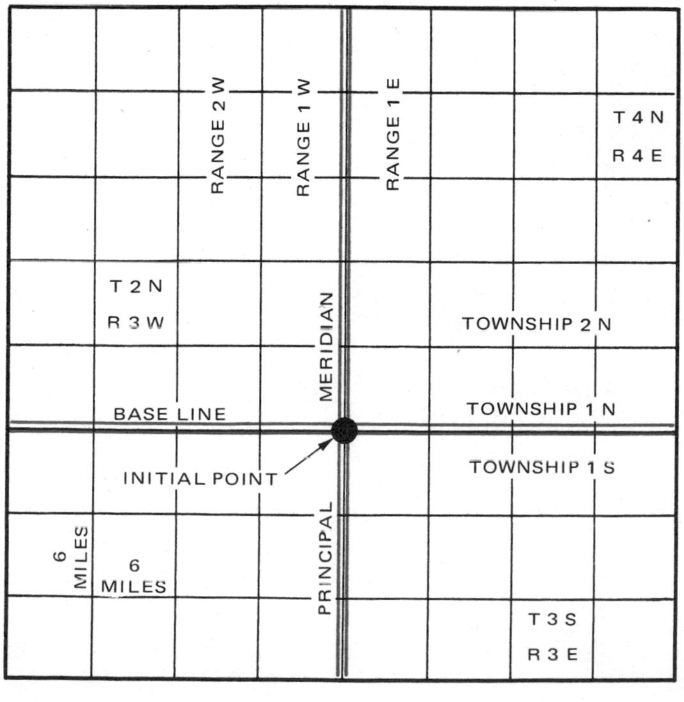

Fig. 3-1. In a land description by government subdivision: the principal meridan line runs north and south; the base line runs east and west; the point of intersection is called the initial point.

The purpose of a land description is not to identify the land but rather to furnish a means or formula by which the area or tract in question can be distinguished. A land description will not serve to identify the tract of land until it is related to known points or monuments on the earth's surface.

When a land description is coupled with available written or oral evidence, the lines which indicate and fix the boundaries of the tract or area (size, shape and location) *can* be established. The additional evidence may consist of local knowledge such as:

1. A government survey.
2. A town or city plot.
3. A fixed monument.

The principal types of land description in use in the United States today are:

1. Government subdivision.
2. Metes and bounds.
3. Reference to a map or plot.

In some cases, a land description may contain elements of all three of the above mentioned types, or all of these descriptions may, in turn, be related to the state coordinate system on the U.S. Coast and Geodetic Survey.

DESCRIPTION BY GOVERNMENT SUBDIVISION

Nearly all the land in the United States has been surveyed, mapped and described in accordance with the rectangular system established by act of Congress. The rectangular system is completed by establishing reference lines running parallel to the longitude and latitude lines of the earth. See Fig. 3-1.

Lines running north and south are referred to as principal meridan lines. Lines running east and west are called base lines. The point of intersection of these two lines is the initial point. From this point, a grid pattern is developed each six miles. This area is called a township, Figs. 3-1 and 3-2.

The township, then, consists of 36 sq. mi. or 36 sections. A section equals one square mile and each is numbered. See Fig. 3-2. Each column of townships running parallel to the meridian lines is referred to as a range. The first range to the west of the principal meridian is Range 1 West; the second is Range 2 West. See Fig. 3-1.

Beginning at the base line, the townships are numbered as follows: the first is township 1 north, the second is township 2 north and so on. Therefore, by reference to the section,

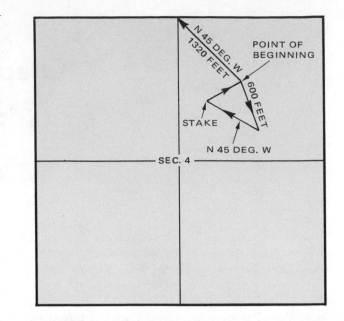

Fig. 3-4. In a land description made by metes and bounds, the location of irregularly shaped property is described by courses, distances and fixed monuments.

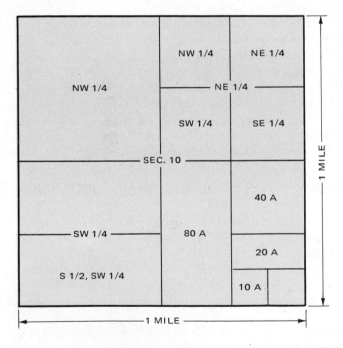

Fig. 3-2. A township covers 36 square miles, based on the rectangular system of government subdivision.

Fig. 3-3. A section is a one square mile portion of a township. Each section is numbered to simplify the land discription.

township, range and principal meridian, the location of any section can be fixed. See Fig. 3-3.

Example of a land description by government subdivision:

SE 1/4 SE 1/4 Section 2,
Township 44 North, Range 29 West of the
5th principal meridian

DESCRIPTION BY METES AND BOUNDS

Metes and bounds are used whenever areas are irregular in size and shape. The location is fixed by running out the boundaries by courses, distances and fixed monuments at the corners or angles.

A course is the direction of the line. Usually, it is given with respect to the meridian. Sometimes, a course is referenced to the magnetic north pole.

Distance is the length of a course. Ordinarily, distance is measured in chains or feet, Fig. 3-4.

A typical land description by metes and bounds follows:

Beginning at a point from which the north quarter corner of section 4, township 1 north, range 40 west of the 5th principal meridian, bears 45° west 1320 feet (370 metres), beginning at iron stake, thence south 600 feet (185 metres) to a point also marked by an iron stake; thence north 45° west 700 feet (210 metres) to iron stake thence north-easterly to point of beginning.

DESCRIPTION BY REFERENCE TO RECORDED PLOTS

Where smaller plots of area are described (residential building sites), the area usually is designated by preparing a map that divides a tract into lots, plots, blocks or squares. The map also contains intersecting streets and gives a name and number to each lot in the area. See Fig. 3-5.

A typical land description by reference to recorded plots follows:

Lots 1 and 2,
Block 10 Stiles addition to the City
Blue Ridge.

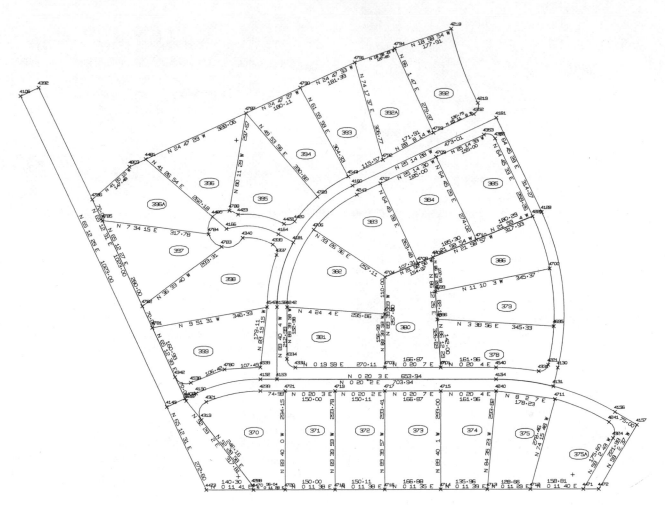

Fig. 3-5. In a land description given with reference to recorded plots (plat map), the map divides the area into lots, plots, streets and blocks. (Calcomp)

SURVEYING

Surveying is the science of large scale geometrical measurements made to establish position and size according to given requirements. A survey is necessary to the planning and execution of nearly every form of construction.

Surveying has two distinct functions:

1. Determination of relative horizontal and vertical positions, as used in mapping, Figs. 3-6 and 3-7.

2. Establishment of marks to control construction or to indicate land boundaries, Figs. 3-8 and 3-9. Instruments used for elementary surveying are quite basic:

   a. Steel tape or chain.

   b. Theodolite or transit.

   c. Compass.

   d. Level.

While the international unit of measure for surveying is the metre, the foot and yard system is used throughout the English speaking world. In surveying, a completely different set of terms is used to describe great distances:

7.92 inches	=	1 link
100 links	=	1 chain (or 66 feet)
80 chains	=	1 mile
625 square links	=	1 square pole

Fig. 3-6. A transit is a surveying instrument designed to measure horizontal distance and vertical angles. (Brunson Instruments)

Fig. 3-7. A surveyor utilizes a transit to obtain exact horizontal measurements for correct positioning of a structure. (David White)

Fig. 3-8. A level is a precision instrument used by surveyors to measure elevation. (Brunson Instruments)

16 square poles   = 1 square chain
10 square chains = 1 acre
640 acres   = 1 section (or 1 square mile)
36 sections   = 1 township

The greater the accuracy expected, the more sophisticated the equipment must become. This suggests the use of more modern technology like the laser beam and/or satellites circling the earth at great altitude, Fig. 3-10.

The surveying technique is used to chart and map the earth. A survey shows horizontal size, position and elevation of natural or man-made topographical features, Fig. 3-11. Route surveys are required for all forms of transportation: canals, railways and highways, Fig. 3-12. Nautical charts show distance and water depth.

Property boundary surveys locate property boundary lines by means of marks on the ground called monuments. The monument may consist of natural points or fixed points called bench marks. See Fig. 3-13. The bench mark is a permanent

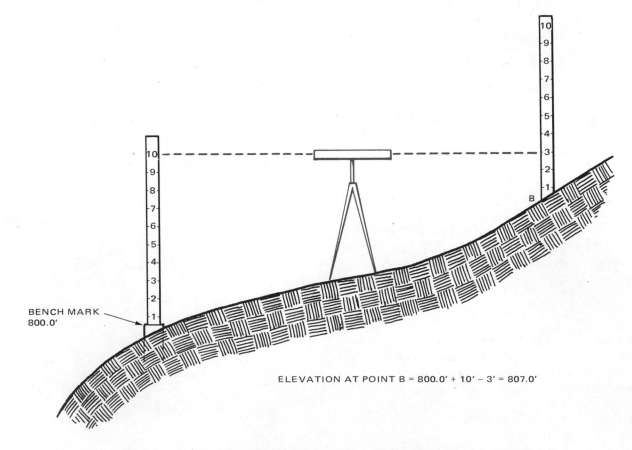

BENCH MARK 800.0'

ELEVATION AT POINT B = 800.0' + 10' − 3' = 807.0'

Fig. 3-9. A level is used to find the elevation of an unknown point from a known point.

Fig. 3-10. "Geos-A" is a 385 lb. (175 kg) geodetic explorer spacecraft. Its instrumentation system provides measurements to allow the establishment of a more precise model of the earth's coordinate system. (NASA)

Fig. 3-12. The route of this future highway was pinpointed by a survey team long before these machines were used to prepare the road bed.

Fig. 3-13. A bench mark is a permanent landmark of known position and altitude, usually installed at corners or angles of property lines.

landmark that has a known position and altitude (elevation). When used in conjunction with a surveyor's level, Fig. 3-9, a bench mark becomes the primary reference point for determining the elevation of other points within a given line of levels (series of readings).

Fig. 3-11. Varying ground level did not prevent this bridge from providing a true roadway.

Warranty Deed

KNOW ALL MEN BY THESE PRESENTS:

That _____

_____

of the County of_____, in the State of Missouri, have this day, for and in consideration of the

sum of _____ DOLLARS

to the said _____

in hand paid by _____

of the County of_____ in the State of_____, Granted, Bargained

and sold, and by these presents, do_____ Grant, Bargain and _____ said

_____ said _____

hereby covenanting to and with the said _____

_____ heirs and assigns

for _____ heirs, executors and administrators, to Warrant and Defend the title to the

said premises hereby conveyed against the claim of every person whatsoever:

_____

_____

IN WITNESS WHEREOF, _____ have hereunto subscribed _____ name _____ and affixed

_____ seal _____ this _____ day of _____, 19_____

_____ (SEAL)

_____ (SEAL)

_____ (SEAL)

_____ (SEAL)

_____ (SEAL)

Fig. 3-14. A warranty deed guarantees that the grantor has a good title, free and clear of all liens.
It protects the grantee against all claims.

OWNERSHIP OF PROPERTY

A title is a written document stating evidence of the right of ownership of a described piece of property. A deed is a written document, under seal affixed by legal authority, by which the seller transfers ownership of the land to the buyer. See Fig. 3-14.

Deeds usually consist of the legal description of the tract, amounts paid and date. A deed may also include restrictions placed upon the property. For example, a residential subdivision code may contain a restriction that the residence proposed to be built upon a lot of certain size must have at least 1000 square feet of living space.

Easements may be mentioned in the deed. Easements are necessary parts of the deed/title proof of ownership. Easement allows utilities the right to construct and maintain their services over the property.

Each county or political subdivision maintains a recorder's office, hall of records or register's office as a public service to its people. The recorder is an elected officer of the county charged with the responsibility of keeping records of the rightful ownership of property within its jurisdiction. All plots are registered with the recorder or register, and each transfer of ownership is duly recorded.

TRANSFER OF OWNERSHIP

Between the time of negotiation of the Real Estate Contract (see Chapter 2) and the actual "closing" of the transaction, certain activities are conducted by both the seller and the buyer.

The seller engages a "specialist," usually a real estate attorney, who is qualified by training and experience to give a professional opinion about the title. The "specialist" prepares an abstract, based on court house records of important information concerning the piece of land being transferred.

An abstract might be considered as a history of the piece of land. It is often referred to as a title search. The seller must give a "clear title" to the property or guarantee that no money is borrowed where the property is used as collateral. The seller also guarantees the property bounds to be correct, usually through a boundary survey.

The buyer receives the abstract and has a second real estate attorney examine the document for errors or omissions. At the same time, the seller prepares his finances for payment of the property.

At the "closing," or formal transfer of ownership, the seller delivers the deed to the buyer. The buyer provides payment and records the deed with the register at the county courthouse.

SUMMARY

The land area of the United States and the world has been identified by graph and verbal description. A land description is necessary for the owner to establish, maintain and transfer the right to occupancy and use.

Land description methods commonly used in the United States are government subdivision, metes and bounds, and reference to a plot. A survey (science of geometrical measurement) is made in any method to locate the boundaries. The survey technique is also used to chart and map the topographical features of the earth.

Right of ownership is legally established by a written document called a deed. The deed is recorded in the county register's office.

CAREERS RELATED TO CONSTRUCTION

SURVEYOR uses transit and level to chart the horizontal size, position and elevation of natural or man-made topographical features. Surveyor also establishes marks to define property boundaries and to control construction.

CARTOGRAPHER compiles data and constructs maps.

RECORDER OF DEEDS is an elected officer of the county charged with the responsibility of keeping records of rightful ownership of property within county jurisdiction.

REAL ESTATE ATTORNEY conducts a title search of county courthouse records to establish "clear title" to property being investigated. Attorney then prepares an abstract, based on findings of the search.

DISCUSSION TOPICS

1. What is a land description as used today?
2. How does a land description differ from a survey?
3. List three popular methods of land description.
4. Identify the two lines by which the rectangular system of subdivision is identified?
5. Define a section, township, range and principal meridian.
6. What is the international unit of measure for surveying?
7. What is a monument?
8. What is a bench mark?
9. What are titles and deeds?
10. Describe the process for transferring ownership of property in area.
11. What is the function of a County Recorder or Register?

Chapter 4

BUILDING CODES

Building codes are laws or ordinances that apply to the materials selected and used in building projects, Fig. 4-1. They are based on local standards which, to a great extent, govern the quality and characteristic properties of the building materials. These codes also identify procedures and techniques used in construction.

Building codes are established for the purpose of providing minimum standards to safeguard life, health and property. They protect the public welfare by regulating and controlling the design, construction, technique, quality of materials, use and occupancy, location and maintenance of all structures within a political jurisdiction (city, county or state).

The first known building code was prepared by Hammurabi, King of Ancient Babylon in 2250 BC. Its six short statements cited the wages which the architect/builder was to receive. This first code also spelled out the penalty that would be incurred for using inferior materials and employing faulty construction techniques. Simply stated:

"If the building collapsed and killed the owner, the architect/builder was to be put to death. If the owner's son should be killed by the collapsing structure, then the architect/builder's son shall be put to death."

From this clear and simple performance code have progressed nearly 18,000 local codes in the United States alone.

MODEL CODES

There are no national or universal codes for buildings. However, at present there are four model building codes, sponsored by four different organizations:
1. BASIC BUILDING CODE by Building Officials and Code Administrators International.
2. NATIONAL BUILDING CODE by American Insurance Association.
3. SOUTHERN STANDARD BUILDING CODE by Southern Building Code Congress.
4. UNIFORM BUILDING CODE by International Conference of Building Officials.

These codes are simply "models" that may be accepted through legislative action of a local municipality, county or state. See Fig. 4-2.

Some governmental agencies elect to write their own set of standards. These standards are adopted by many cities or

Fig. 4-1. For the health and safety of future occupants of this hotel under construction, many individual codes governed materials used and building practices. (Real Estate Board of Kansas City, Mo.)

Fig. 4-2. Construction codes that apply to large structures in metropolitan areas become very complicated, partly because of closeness of existing buildings.   (Northwest Engineering Co.)

counties, with modifications the authorities feel are necessary when writing a building code for their locality. In this way, they are certain that the material and construction techniques are valid and comply with proven standards.

The Office of Building Standards and Codes Services, a division of the United States Department of Commerce, aids the building industry, government and building users by assisting in the improvement of the nation's building regulatory system. Actions by the Building Standards and Codes Services reflect the needs of the building industry and society. Technical assistance in revising the codes and standards is provided by this office.

The Office of Building Standards and Code Services serves as Secretariate to the National Conference of States on Building Codes and Standards. The Conference, in turn, provides a forum for the interchange of building regulatory techniques. It provides a stimulus to needed research on technical questions, such as energy conservation in construction.

CODE SPONSORS

The four sponsoring groups for the establishment of building codes provide a testing service whereby each material or product of construction is tested in many ways, Fig. 4-3. From these tests, written research reports are prepared and made available to the architect and builder to aid in their selection of building components.

Research reports are available containing information on fireproofing, waterproofing, strength and stress values and

Fig. 4-3. Fire test of an insulated steel column places column in the center of a furnace, which is closed and sealed. Column is exposed to fire and its temperature is determined by thermocouples attached to column before the insulation is applied. During fire exposure temperature of column must not exceed certain limits.   (Underwriters Laboratory)

many other topics of concern to the designer or builder. See Fig. 4-4. *It is to the benefit of each manufacturer of building components to have each of their building products tested.* The manufacturers are responsible for submitting their building product to the sponsoring organization for testing. The sponsor, in return, conducts the evaluation test and prepares the research report.

BOCA

**BUILDING OFFICIALS & CODE ADMINISTRATORS
INTERNATIONAL, INC.**

1313 East 60th Street • Chicago, Illinois 60637 • 312 / 324–3400

APPLICATION FOR BOCA
RESEARCH AND EVALUATION

Research
Report No._____

Date
Received _____
mo.   day   year

(above for BOCA use)

Date _____

A.   Applicant _____

   Address _____ Telephone _____

   City _____ State _____ Zip Code _____

B.   Product _____

   Trade Name _____

_____

C.   Performance of product for which evaluation is requested:

☐ Structural _____ ☐ Fire/Flame resistance _____

☐ _____ ☐ Weather resistance _____

☐ _____ ☐ _____ ...cation submitted here-

☐ _____ ...upon ...e to so comply the Research

☐ _____ ...sentation by applicant will result in automatic suspension of

☐ _____ ...uation of a product does not imply any warranty or guarantee by BOCA

☐ _____ ...ce, nor any responsibility in regard to patent infringements and applicant agrees to

☐ _____ ...ployees, officers and members harmless from any litigation arising from the use or operation of

   ...and to defend and indemnify same against any loss, expense, liability or damage, including reasonable attorney's

   ...and e.) that the terms on reverse side hereof are part of this agreement.

_____
(Firm Name)

_____
(Signature of Proprietor, Partner or Authorized Officer and title)

Attest: _____
(Secretary)

Corporate Seal

Signed and Sworn to Before Me, this _____

day of _____ 19 _____

_____
(Notary Public)

Fig. 4-4. It is the manufacturer's responsibility to have company products evaluated by an independent organization.

Special interest organizations, like the American Gas Association, provide certification of gas-operated equipment, Fig. 4-5.

Electrical and plumbing components are tested and approved by special interest organizations. If evaluation is positive (the product is not dangerous to the health and safety of the user), the components are recommended for public use. All tests are based on the condition that the components have been properly installed.

Building codes for each locality may be obtained from the building inspector's office or from the office of the governmental official having jurisdiction over building codes.

Fig. 4-5. Special interest groups give certificates of approval for equipment that meets requirements fixed by their codes.

Many municipal building codes do not permit a residence to be built on a lot smaller than a given size. Also, many subdivisions have additional restrictions as to the size of residences to be built within its jurisdiction. In this case, the two separate codes would work together. For example, they would require that each lot must be in excess of 3000 sq. ft., and the residence to be built must have at least 1000 sq. ft. of living space. If this requirement were not in effect, someone might build a 500 sq. ft. residence on a 2300 sq. ft. lot, and the appearance and value of the neighborhood would be adversely affected.

SUMMARY

Sponsoring groups for model building codes provide insight into quality and characteristics of building materials so that local governing bodies can establish a set of codes for construction projects within its jurisdiction. Building codes provide laws or ordinances that require sound construction practices to protect the health and safety of the workers (during construction) and future occupants of the structure.

Under the category of health and safety, a typical requirement is that all electrical, plumbing and mechanical equipment be installed by trained and licensed workers.

CAREERS RELATED TO CONSTRUCTION

BUILDING INSPECTOR determines, before construction, whether plans for structure are suited to building site and comply with zoning laws. Inspector carefully inspects footings and foundation and structural quality of building during construction and after completion of the project. A background in the building trades usually is required.

EVALUATOR works for building codes organization, testing and reporting on the quality of materials and products of construction submitted by manufacturers.

DISCUSSION TOPICS

1. When was the first written building code established?
2. What does the building code of your community guard against?
3. Why is research evaluation of building material necessary?
4. Why are electrical codes necessary?
5. Why should there be codes which govern the sizes of lots and sizes of structures that might be built there?
6. Who in your community has jurisdiction to enforce the building codes?

Chapter 5

FINANCING

When a building project is initiated, Fig. 5-1, the first questions to be dealt with are: "How much will it cost?" and "What method will be used to get the money needed to finance the project?"

Seldom is a corporation or family able to pay "cash" for a new building or home. Therefore, a source of money other than the regular operation budget is required to finance the structure. In most cases, a loan is secured to pay for the cost of construction. With the loan, the occupant pays an annual fee called "interest" for the privilege of using these funds until the loan is repaid.

When a loan is required for a construction project, a lending institution is contacted during the early planning stages. See Fig. 5-2. If the loan is approved, most lending institutions issue a letter of commitment (legal confirmation of a loan). This written commitment guarantees the owner that the money will be available at the time and phase of construction where it is needed.

The letter of commitment usually states the loan amount, term (length of time), interest rate and the time limit for accepting the loan. In addition, the letter contains any other conditions that are agreeable to the lending institution and borrower.

Fig. 5-2. Many businesses specialize in arranging loans to make construction a reality. (Real Estate Board of Kansas City, Mo.)

THE MORTGAGE

A mortgage is a popular method of securing real estate loans. It is a contract whereby the real estate property is pledged to secure a loan, Fig. 5-3. That is: a home owner who acquired a mortgage to build a home must offer the home as security, Fig. 5-4. This security guarantees repayment of the loan without requiring the owner to give up the right to live in and use the home for personal benefit. The home remains in the possession of the owner as long as the monthly payments are made as promised in the mortgage and in the real estate

Fig. 5-1. Large building projects require huge sums of money to complete. Usually, financial assistance is required to pay for such a building program. (Business Men's Assurance)

MORTGAGE

THIS INDENTURE WITNESSETH: That the undersigned _____

_____

of the _____ County of _____, State of Illinois,
hereinafter referred to as the Mortgagor, does hereby Mortgage and Warrant to _____

a corporation organized and existing under the laws of the _____, hereinafter

referred to as the Mortgagee, the following real estate, situated in the County of _____
in the State of Illinois, to wit:

_____ ____ suit for the foreclosure hereof after the accrual of the right to foreclose, whether or not actually
commenced; or (c) preparations for the defense of or intervention in any suit or proceeding or any threatened or contemplated suit or proceeding which might affect the premises or the security hereof, whether or not actually commenced. In
the event of a foreclosure sale of said premises there shall first be paid out of the proceeds thereof all of the aforesaid items.

IN WITNESS WHEREOF, the undersigned have hereunto set their hands and seals this _____

day of _____, A. D. 19_____

_____(SEAL)   _____(SEAL)

_____(SEAL)   _____(SEAL)

STATE OF ILLINOIS   }
COUNTY OF _____} ss.

I, _____, a Notary Public in and for said county, in the State aforesaid,

DO HEREBY CERTIFY that _____

personally known to me to be the same person(s) whose name(s) (is) (are) subscribed to the foregoing instrument, appeared
before me this day in person and acknowledged that _____ signed, sealed and delivered the said instrument as
_____ free and voluntary act, for the uses and purposes therein set forth, including the release and waiver
of the right of homestead.

GIVEN under my hand and Notarial Seal, this _____ day of _____, A. D. 19_____

_____
Notary Public

My Commission Expires _____

Fig. 5-3. Printed form shown is one type of mortgage agreement.

Fig. 5-4. A mortgage allows families to enjoy a home while buying it over a period of time.
(Southern Forest Products Assoc.)

NOTE

FOR VALUE RECEIVED, the undersigned hereby promise to pay to ______________________________

____

a corporation organized and existing under the laws of the United States of America, its successors and assigns, at its office in ______________________, Illinois, or at such other place as it may designate, the principal sum of ______________________________ DOLLARS ($______________), together with interest on the unpaid balance from time to time at the rate of ______________ per centum (__________%) per annum thereafter.

Said principal and interest shall be paid in monthly installments of ______________________________ DOLLARS ($__________ on the __________ day of each month, commencing with ______________

... interest, ... per annum so long as said ... provided in this note, or in case of the ... mortgage to be performed by the undersigned, the principal sum above-mentioned, or any balance that may be unpaid thereon, together with all interest thereon as aforesaid and any advances made by the Association shall, at the option of the Association, its successors or assigns, become immediately due and payable, without notice, and all of said principal, interest and advances, together with interest thereon at the rate of   per annum, shall be collectible immediately, or at any time after such default, anything hereinbefore contained to the contrary notwithstanding.

To further secure the payment of this note, the undersigned hereby authorize, irrevocably, any attorney of any court of record to appear for the undersigned, in such court, in term time or vacation, at any time after default, and confess a judgment jointly and severally, without process, in favor of the Association, its successors or assigns, for the unpaid balance of principal and interest exclusive of other advances, together with costs and reasonable attorneys' fees and to waive and release all errors which may intervene in any such proceedings and consent to immediate execution on such judgment; hereby ratifying and confirming all that the undersigneds' said attorney may do by virtue hereof.

The makers, sureties, guarantors, and endorses of this note, jointly and severally, hereby waive notice and consent to any and all extensions of this note or any part thereof without notice, and each hereby waives demand, presentment for payment, notice of non-payment and protest, and any and all notice of whatever kind or nature and the exhaustion of legal remedies herein.

IN WITNESS WHEREOF, the undersigned have hereunto set their hands and seals this ______________

day of ______________________, A. D. 19______.

______________________(SEAL)   ______________________(SEAL)

______________________(SEAL)   ______________________(SEAL)

Fig. 5-5. The real estate mortgage note states the method of repayment.

mortgage note, Fig. 5-5.

The mortgage is recorded with the county register. The real estate mortgage note automatically becomes the second instrument whenever a mortgage is written. The mortgage note spells out the terms of the real estate loan, gives the provisions for repayment and establishes the rate of interest.

INTEREST

Interest is the price paid for the use of money. The future occupant arranges for the loan and borrows the money necessary to complete the building project. The loan is obtained from a bank, savings and loan association, insurance company or other financial institution, Fig. 5-6. Their charge for the use of the money is known as interest.

To calculate the amount of interest a given loan will cost, multiply the principal (amount of loan) times the established rate of interest.

Interest is figured by a percentage rate per annum (year). Example: A person borrowing $10,000 for one year at 8 percent interest per year would repay $10,800. The $800 is interest charged for the privilege of using the $10,000. If one agreed to repay this loan over a twelve month period of time, the monthly payments will be $900 per each month.

$$\begin{array}{r} \$10{,}000 \\ .08 \\ \hline \$800.00 \end{array} \qquad \begin{array}{r} \$10{,}000 \\ +\,800 \\ \hline \$10{,}800 \end{array} \qquad \begin{array}{r} \$900 \\ 12\overline{)10{,}800} \end{array}$$

In addition to the interest charge, money lending institutions add an "origination fee" for the new loan. This extra charge is called "points," with the point equal to one percent. Therefore, a charge of two points on a $10,000 loan would amount to $200. This is in addition to the interest charge, and it must be included in the cost of borrowing money.

Example: In looking at the original loan of $10,000, borrower must repay a total of $11,000.

$$\$10{,}000 + \$800 + \$200 = \$11{,}000$$

This includes interest and the origination fee. The monthly repayment schedule is $916.66 per month for 12 months.

Fig. 5-6. Banks profit from interest collected on loans.
(Bank of Hawaii)

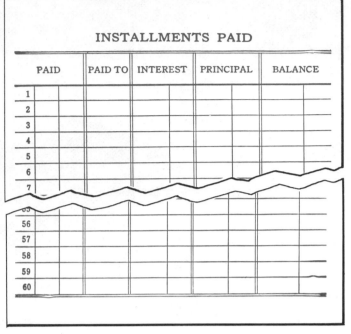

Fig. 5-8. Amortization is a planned reduction of the loan by
regular installments.

REPAYMENT OF LOAN

Most private real estate loans are made for a period of 10 to
30 years, Fig. 5-7. The loan arrangement should provide for
systematic reduction of the principal for the duration of the
mortgage term. Each installment payment is recorded, Fig.
5-8. *Remember, however, that the faster the loan is repaid, the
less the total cost of the project will be.*

In addition, a good loan should have a specific maturity
date that permits complete repayment of the principal and
interest through a regular "planned" amortization (gradual
elimination of the financial obligation by periodic payments).

Fig. 5-7. Real estate loans make expansion of industry possible.   (Stran-Steel Corp.)

Fig. 5-9. A bond is one possible method of financing public construction. (American Plywood Assoc.).

Fig. 5-10. Stocks sold by private corporations can provide money for construction. (Missouri Public Co.)

BONDS

Bonds are used by some municipalities, school districts and other public institutions to secure money for construction. Bonds are sold to individuals or lending institutions with the understanding that the seller will repurchase each at a later date for an additional amount of money. They become a guarantee for payment, Fig. 5-9. In effect, a bond is another type of loan with interest.

STOCKS

Industrial and commercial structures may be financed by the issuance of additional stock certificates by the company, Fig. 5-10. Repayment, therefore, is made through dividends paid by the company to the shareholder. The shareholder, in return, becomes owner of a certain percentage of the corporation.

APPROPRIATIONS

Federal, state and local governments finance much of the larger construction projects of a public nature that require huge sums of money, Fig. 5-11. Dams which require large amounts of money and time to construct usually are financed by government appropriation. Buildings to house state supported institutions of higher education are constructed by appropriations made by state legislators, Fig. 5-12. Many construction projects are financed by joint appropriation by the various levels of government, Fig. 5-13.

Money for appropriations comes from the working budgets of the governmental levels. The working budgets are made up of money collected from the taxpayers. Money for appropriations also may come from the sale of bonds.

Fig. 5-11. Federal legislative appropriations make taxpayers' money available for construction.
(Corps of Engineers, Kansas City District)

Fig. 5-12. State legislative appropriations finance construction of state universities.

41

Fig. 5-13. Construction for transportation requires public money.
(Missouri Highway Dept.)

SUMMARY

Any construction project that requires a loan of money before completion requires loan planning in the earliest stages of development. A letter of commitment must be secured from a bank, loan association or other lending institution to insure the availability of the loan when it will be needed.

A mortgage is negotiated and written as the debtor's obligation to repay the loan. A second written document is prepared as an agreement between both lender and borrower to establish:

1. The repayment date (monthly, quarterly, etc.).
2. Method of repayment.
3. The amount of interest paid for the use of the loan.

Interest is the price paid for using money loaned to you. Interest is figured on a percentage rate per annum on the total principal.

Generally, loans are retired by a predetermined schedule called amortization. Amortization means the gradual elimination of an obligation by periodic payments.

Means of financing construction other than a direct loan are:

1. Bonds.
2. Stocks and shares.
3. Appropriations.

CAREERS RELATED TO CONSTRUCTION

BANK OFFICER handling real estate loans must evaluate credit and collateral of individuals and businesses. Loan officer must know business operations and be able to analyze financial statements; must be familiar with economics, production, distribution, merchandising and commercial law.

LOAN COMMISSIONER prepares papers and assembles documents for builders to finance new construction; sends loan application, construction plan and credit ratings to loan company. Loan commissioner draws up closing papers showing financial transactions, conditions and restrictions. Records deeds with title company and municipal authorities.

COUNTY REGISTER is the official recorder or registrar of county records. Each mortgage entry made in record book must be in full and accurate detail.

REAL ESTATE ATTORNEY handles all legal aspects of mortgage fianancing. Some also rent and manage properties, make appraisals and develop new building projects. Besides studying law, courses in real estate, economics, business administration, appraisal, mortgage financing and property development and management are essential.

DISCUSSION TOPICS

1. What is a letter of commitment?
2. When is a letter of commitment obtained?
3. What is a mortgage?
4. Who pays the interest on a loan?
5. Calculate the interest on $100,000 for 18 months at 8 percent per annum. What would the monthly payment for interest alone be for the 18 month period?
6. What is an origination fee?
7. Define amortization.
8. How are bonds used to finance construction?
9. How are stocks used to finance construction?
10. Usually, where do appropriations originate? How are they used to finance construction? What type of construction is financed by appropriations?

Chapter 6

OTHER LEGAL ASPECTS

Fig. 6-1. Utilizing licensed contractors, most restoration projects are planned to correct, not destroy, the unique appearance of the original structure. (Bil-Jax, Inc.)

The courts state that the purpose of a contractor's (builder's) license is to protect the public against the consequences of incompetent workmanship and deception. See Fig. 6-1.

Contractors are licensed as a means by which the general public can determine the competence of a builder. A legislative body may rule that building permits should be issued only to duly licensed contractors. Therefore, it may be unlawful for any unlicensed person to act in the capacity of contractor.

The state legislature has the power to regulate the business of construction. However, the legislature may delegate this authority to local city or county governmental agencies. Then, the local governmental agency has the power and authoriza-

tion to enact legislation for the regulation of construction in their municipality. This regulation must, however, be reasonable and not discriminatory. Contractors failing to abide by these regulations may have their licenses revoked, or be prevented from filing a special claim against the property under construction for work performed. The license may be revoked for violation of a particular provision of a statute or law, such as abandonment of a construction project undertaken or failure to complete the building project for the agreed price.

LIENS

A lien is a legal claim against the property of another for the satisfaction of a debt.

Most liens arise when someone fails to do something that they are obliged to do under the provisions of the construction contract. If the property owner pays his construction bills promptly when due, there is no need for the contractor to file a lien with the courts. Liens are only filed against the property under construction when the contractor, subcontractor, workers or material supply firms are not paid and legal measures to enforce payment become necessary.

MECHANICS LIEN

Liens pertaining to the construction industry are called "mechanics liens." See Fig. 6-2. Mechanics liens involve a particular class of creditors called contractor, subcontractor, laborers or workers and the material supply firms. A mechanics lien is a special claim that the law permits to be filed against the real estate involved in the construction, when persons furnishing materials and/or labor are not paid.

The filing of a mechanics lien claim with the county clerk where the construction land is located constitutes a powerful factor in aiding the collection of past due bills for labor and material.

FORECLOSURE

Foreclosure results when a mortgage is in default (payments not being paid). In the beginning, a property owner secures money by pledging the property as payment. If the owner becomes delinquent in this debt, the lenders will take the legal action of foreclosure. The property is used to satisfy the lender as payment of the debt.

Fig. 6-2. Many people and several firms (contractor and subcontractors) have money invested in materials, labor and equipment services during the course of a building project. If construction bills are not paid, mechanics liens may be filed against the property under construction. (Allis Chalmers)

THE CONSTRUCTION CONTRACT

A contract is an agreement by two or more competent persons to do or not to do some lawful act.

Construction contracts are the same as other contracts, except where they pertain directly to the construction of that project. Conditions include:

1. Time of completion.
2. Amount of payment.
3. When payment is to be made.
4. Type and amount of insurance.
5. Workmen's compensation.
6. Any other conditions which all parties agree upon.

Fig. 6-3 gives examples of phrases used in a construction contract for a residence.

A construction contract should be in writing, and it must be signed by all parties involved. Even though the builder might be protected under a verbal contract, it would be poor business not to have the construction agreement in writing. A dispute could arise between the builder and client, and the

CONSTRUCTION CONTRACT

THIS AGREEMENT, made and entered into this _____ day of
_____, 19_____, by and between _____
_____, hereinafter called the OWNER, and
_____
hereinafter called the CONTRACTOR, WITNESSETH:

That the Contractor and the Owner in consideration of the agreements, covenants and payments hereinafter set forth, agree as follows:

1. The Contractor shall furnish all the materials and perform all the work including any and all sub-contracting as herein noted or noted in th_____ _____ove delet__ ____ ____ute which are unnecessary for ____ion herein contemplated, including the running of lines for service facilities.

21. This contract is subject to the ability of Owner to obtain a construction loan on the afore described property in the sum of _____.

OWNER _____

CONTRACTOR _____

_____

Fig. 6-3. Construction contract is shown in abbreviated form.

parties could wind up in court because the agreement was misunderstood.

The parties to a construction contract may change the contract or vary its terms. When this occurs, a new written agreement should be prepared, covering the construction changes. The new agreement would also include additional changes resulting from increased materials, labor and overhead expenses, plus a reasonable profit, whenever any changes or alterations are ordered while the building is under construction.

INSURANCE

A standard builder's risk insurance protects all parties against physical damage to the insured property during the construction period. This policy covers damage resulting from any of the perils (fire, wind, vandalism) named in the policy. It provides reimbursement based upon actual loss or damage rather than any legal liability which may be incurred.

Most builder's risk insurances are based upon the completed value concept. This method assumes that the value of a project increases at a constant rate during the course of construction. While the policy is written for the value of the completed project, the premium is based upon a reduced or average value. The monetary coverage provided is the actual work completed with the standard materials at any given time.

Within the framework of the builder's risk policy are the following perils: fire and lightning; vandalism and malicious mischief; extended coverage for windstorms, hailstorms, smoke, explosion; and riot or civil commotion. Additional perils might include collapse, landslide, water damage, breakage or theft. It might be possible to obtain endorsement or separate coverage to cover flood and earthquakes. However, this coverage may be hard to obtain.

PUBLIC LIABILITY

Public liability insurance is protection against liability for the injury or death of a person. A contractor is liable in money damages for wrongful or negligent conduct that causes injury or death to persons rightfully on the building site, or adjacent property. Therefore, the contractor must be protected with public liability insurance. Negligence invariably involves some type of wrongdoing. A claimant, other than an employee, must prove that the contractor was negligent and that this negligence was the direct cause of the injuries.

Where injury to children is involved, there is a doctrine of law called the "attractive nuisance." See Fig. 6-4. Small children who are unaware of the dangers are attracted to the construction site by the mounds of dirt, tractors and equipment. Precautionary steps should have been taken to prevent the children from being injured.

WORKMEN'S COMPENSATION

Workmen's compensation insurance protects the contractor against claims resulting from injury or death of his workmen through an industrial accident. In most states, laws prescribe that the employer must provide protection for the employee who is injured while on the job, Fig. 6-5.

Negligence is not a factor in any industrial accident when a workman is injured while on the job.

The principle of workmen's compensation has had rapid

Fig. 6-4. Construction sites are not playgrounds, yet mounds of dirt, equipment and unfinished construction attract children. For this reason, public liability insurance must be purchased by the contractor.

Fig. 6-5. Some construction jobs contain an element of danger by the very nature of the work. By exercising safety habits, however, few accidents occur. When they do occur, workmen's compensation insurance protects the contractor against claims. (Missouri Highway Dept.)

tion tasks. For example, a plumber's license is proof that the holder of the permit is competent of successfully installing the plumbing required in the building project. The license granting institution is usually a branch of the local or state government.

Liens legally allow claims against the property of another for the satisfaction of an unpaid debt. A mechanics lien involves a particular group of creditors to satisfy debts owed the contractor, subcontractor, laborers or material supply firms.

Foreclosures are the legal act of collecting payments found to be in default on a mortgage.

The construction contract is an agreement between the builder and owner. The contract states all conditions of the forthcoming construction project.

CAREERS RELATED TO CONSTRUCTION

BUILDING COMMISSIONER oversees zoning (residential, business, manufacturing, institutional) through the zoning administrator and board. Commissioner's office issues building permits; also handles complaints of building code violations through building inspectors.

CONTRACTOR is licensed to erect buildings or structures according to a written agreement between builder and client who owns the real estate being improved. SUBCONTRACTOR is hired by contractor to perform specialty jobs such as plumbing, electrical work, brick laying, floor covering and roofing.

LEGAL or CODE ENGINEER does complex, detailed legal work. Code engineer examines, authenticates and prepares legal documents such as building and electrical codes and code modifications.

INSURANCE AGENT plans insurance programs tailored to prospects' needs. Agent discusses policies, recommends coverage, makes reports and maintains records. Agent sells property/liability insurance to contractors to protect them against financial losses due to injury and death claims.

COUNTY CLERK keeps records of real estate, taxes and claims. Clerk's office generally includes election department, map department, notary public, marriage license bureau and bureau of vital statistics.

development. It has been adopted in practically all of the states. The name of the insurance may vary but the result is the same. The employer (contractor) pays the premiums to the insurance company for workmen's compensation insurance coverage. The employer's responsibility ends with the reporting of the employee's injuries to the insurance company.

The amount of compensation paid to an injured employee under workmen's compensation is never equal to his normal earnings. Most states fix the amount paid by a regular schedule that is part of its workmen's compensation laws. The compensation paid to the injured employee is not affected by the insurance company paying all medical, surgical, hospital and even burial expenses.

In many states, compensation laws are elective. The employer and employee are privileged to jointly accept the provisions of the law or ignore them. In other states, no choice is granted. Both the employer and employee are bound to the provisions of the law.

SUMMARY

Builder's licensing is the process of granting contractors or individuals the right or privilege to perform certain construc-

DISCUSSION TOPICS

1. Why are contractors (builders) licensed?
2. Who has the authority to issue licenses?
3. What is a mechanics lien? When does it apply?
4. Who forecloses on a mortgage?
5. Describe a construction contract.
6. Who are the signers of a construction contract?
7. How does insurance serve the construction industry?
8. How does public liability affect the contractor?
9. What is workmen's compensation?
10. Who benefits from workmen's compensation?

Chapter 7

GEOLOGICAL CONSIDERATIONS

Fig. 7-1. Layers of material such as clay, sand and rock, make up the earth's crust. Those materials lying beneath the layer of soil and the layer of subsoil are called the substrata. Sometimes the shell or layers of rock and other materials break causing a cleavage of the earth's crust. This cleavage or slip causes an offset with the rock structure.

The heavy weight of structures must be borne by the land surfaces upon which they are built. But the substructure of the soil and earth is not always what it seems to be from the surface, Fig. 7-1. Therefore, before building can begin, experts must take samples of soil through borings. These samples are examined carefully.

No matter what the structure, it must be supported by a sturdy foundation. This foundation links the structure to the earth. It makes it possible for the earth to support the weight of the structure. If well designed and built, the foundation will keep the structure from shifting or sagging.

When foundations are designed, it is important to understand the nature of the ground on which these foundations will rest. Geologists and soil engineers examine the soil at proposed building sites. They report what they have found to the designer. The designer takes these facts and designs the structure so that it will stand up well under the soil conditions.

Fig. 7-2. Planet earth as it appears from space. Its surface is made up of soil, rock and water. Because these materials shift, slide and have different weight-carrying ability, builders must study them carefully before putting up structures.

The earth is made up of rock, soil and water. See Fig. 7-2. *Soils are a product of mechanical and chemical action on rock. Soils are found in a wide range of particle sizes, shapes and compositions.*

ROCKS

Much of the earth's surface is rock. We do not see it all because soil overlays some of it.

There are several kinds of rock. Bedrock lies under all the earth's surface. Sometimes it is exposed. Other times it is buried deep below the surface. See Figs. 7-3 and 7-4. It is never found in solid layers over large areas. Usually it is broken into relatively small units by faults, joints and other structural weaknesses.

The faults may lie in horizontal, vertical or inclined sloping planes. Layers of rock are sometimes separated by slippery layers of clay that allow the layers of rock to slip and change position.

Fig. 7-3. Rock is difficult to remove but is very important to the stability of construction. (Allis-Chalmers)

Fig. 7-5. Texture of soil varies according to the amount of organic material it contains and the sizes of the grains of sand or gravel.

TERMS FOR CONSISTENCY OF
SOIL AND HARDNESS OF BEDROCK

SOIL		
Consistency	Estimated Unconfined Compressive Strength (Tons per square foot)	
Very soft	< 0.25	
Soft	0.25 — 0.5	
Medium	0.5 — 1.0	
Stiff	1.0 — 2.0	
Very stiff	2.0 — 4.0	
Hard	> 4.0	

BEDROCK	
SCALE OF HARDNESS	
Very soft or plastic	Can be indented easily with thumb.
Soft	Can be scratched with fingernail.
Moderately hard	Can be scratched easily with knife; cannot be scratched with fingernail.
Hard	Difficult to scratch with knife.
Very Hard	Cannot be scratched with knife.

Fig. 7-4. Terms used by engineers to describe consistency of soil and hardness of bedrock. (Corps of Engineers, Kansas City District)

Boulders may lie buried in the soil layer. Boulders are large, loose pieces of rock. They can vary in size from 8 in. to many feet in diameter. Cobbles are smaller pieces of rock from 4 to 8 in. in diameter.

SOIL

Soil is made up of rock and organic matter. See Fig. 7-3. Rocks are changed into soil through "weathering." The action of sun, water, wind, plants, gravitation and chemicals causes rocks to be broken down into pieces of different sizes.

Some of this broken-down rock is called gravel. It is made up of small pieces as shown in Fig. 7-5. Still smaller particles are called sand. Sand varies from the size of granulated sugar to pea-sized gravel. Very fine particles like powdered sugar are called silt. Clay also has a powdery texture, much like baking

flour. There are many methods of classifying soils. However, the method most commonly used by engineers is by texture and composition.

SOIL BEHAVIOR

Soils may be solid, viscous (thick but pourable like molasses), plastic or fluid. All these characteristics must be considered in designing the foundation of a highway or building.

Solid soils have a constant density and internal resistance. They are affected very little by temperature or moisture changes. However, when such soils are exposed by excavation or moving water, they may change their characteristics.

Viscous, plastic and fluid soils resist changes in volume. But such soils change their shape constantly when the slightest force is applied. Viscous, plastic and fluid soils vary only in the amount of force needed to start them moving. A plastic soil will cease moving when the force is removed. Viscous and fluid soil will continue moving until a counter force is met.

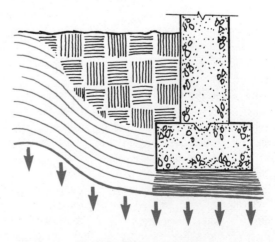

Fig. 7-6. Elastic compression.

LOAD-BEARING ABILITY

Load-bearing capacities of soils are determined by measuring the amount of weight they can support. This is stated in the greatest number of pounds that can be supported by one square foot of soil without excessive settlement or rupture (cracking) of the foundation material.

When a load (the weight of the construction project) is applied, soil tends to react in three different ways:

1. There is elastic compression, Fig. 7-6.

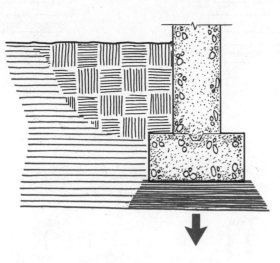

Fig. 7-7. Air and water evacuation.

2. The soil is consolidated (packed) as air and water are squeezed out, Fig. 7-7.
3. There is plastic creep or flow of the soil particles. See Fig. 7-8.

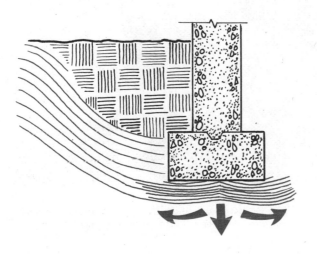

Fig. 7-8. Plastic creep.

All foundations settle when they are loaded. The settlement of a rigid structure resting on a uniform soil and subsurface of bedrock can be measured. If settling is within the design

limitation, it is not serious. If the underlying strata are not the same throughout the site, then settling may become a serious problem.

The elastic compression of a soil depends on the weight or load on it. If the heavier portion of the structure settles at a faster rate than the rest, the structure will tilt. The classic example of this is the Leaning Tower of Pisa of Italy.

LANDSLIDES

Certain geological formations located throughout the earth are made up of sedimentary deposits of rock. This rock was originally formed in a horizontal layer. Later, internal forces of the earth pushed these rock layers upward leaving them tilted as shown in Fig. 7-9. Such strata, when left exposed to water and weather, tends to slip. Sometimes it carries tons of earth downward. Exposure of such layers may be the result of indiscriminate grading for new construction.

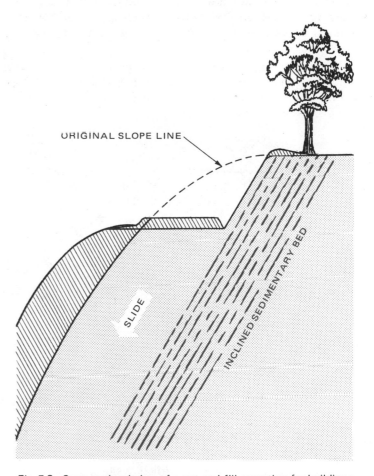

Fig. 7-9. Cross sectional view of a cut and fill operation for building a road bed. Under certain conditions, earth lying above the inclined sedimentary bed could slide, destroying the road.

EARTHQUAKES

An earthquake is a vibration or shaking of the ground. It starts deep within the earth. The vibrations begin at one point on the earth's surface called the epicenter. Then the shock waves move outward in all directions. The ground motions affect buildings and structures to varying degrees depending upon:

1. Distance from the epicenter.
2. Type of soil upon which they rest.
3. Construction methods used.

Sudden weakening of the soil during earthquakes often causes buildings and structures to settle unevenly into the ground.

Sandy, water-filled ground turns into a liquid during the strong shaking and loses strength. This loss is caused by liquefaction (turning into a liquid). It loosens the bond between grains of soil. Time is needed to reestablish stable grain contacts in a closely arranged soil state. Meanwhile, the soil structure is weak and cannot support the weight of buildings. Landslides also may result from such liquefaction of the soil.

Construction projects in areas that are likely to have earthquakes (high seismic probability) require special precautions regarding soil structures and foundations. Fig. 7-10 shows where these areas are located. There are certain known geological faults in the earth's crust along which major earthquakes are more likely to occur.

In some cases, all damage to buildings and structures occurring during seismic activity (shaking) can be attributed to the soil conditions upon which the structures rest.

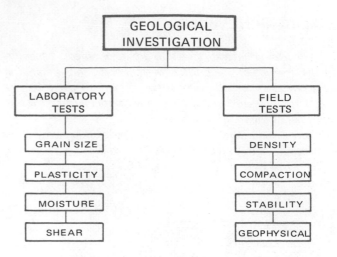

Fig. 7-11. Eight separate tests are given to soil before the soil engineer or geologist makes a report to the designer.

Fig. 7-10. Location of earthquake-prone areas in the United States.

GEOLOGICAL INVESTIGATIONS

The portion of a building or structure which is below the ground is the responsibility of the geologist or soils engineer. You will recall that this is the person who conducts a number of tests to determine the physical characteristics of the soil and its trends. Number and nature of these tests is shown in Fig. 7-11. This expert also locates and describes the rock strata. The soil report must satisfy local building codes (Chapter 4) and lending institutions (Chapter 5). Building permits are issued and loan commitments are made on the basis of such reports.

Fig. 7-12. Large machines assist the soil engineer by drilling deep holes from which samples are taken at different depths.
(Soiltest, Inc.)

SUBSURFACE EXPLORATION

Sampling of the subsurface consists of boring test holes and making core samples as in Fig. 7-12. Core sampling and tests will indicate the depth, thickness and texture of each layer.

To get the sample, an auger is turned into the ground for a

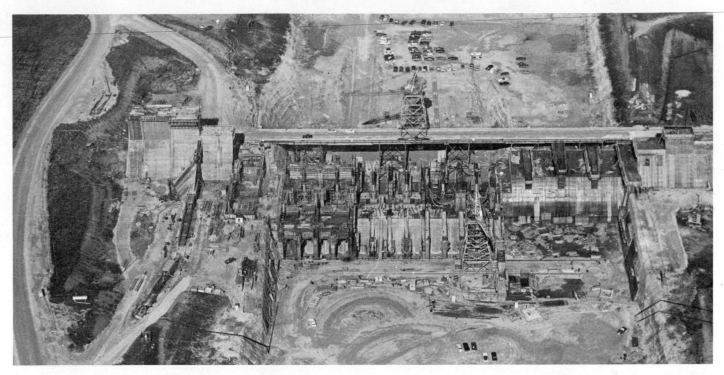

Fig. 7-13. The foundation of this dam, needed for a new hydroelectric plant, is keyed into a bedrock of limestone.
(Corps of Engineers, Kansas City District)

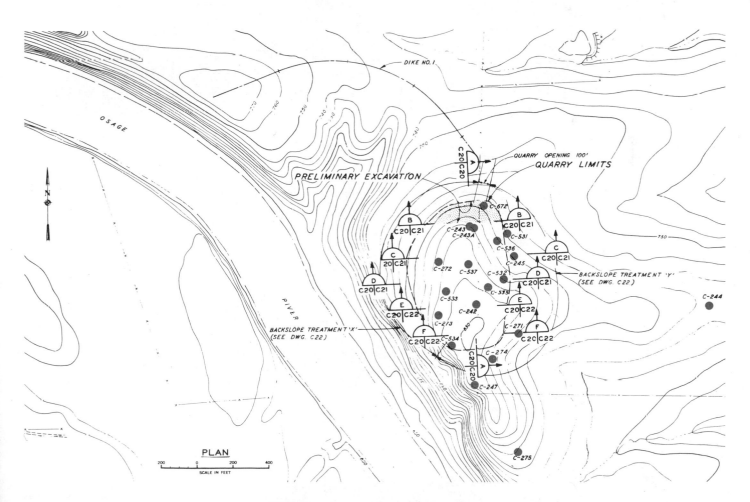

Fig. 7-14. Contour map is used to show location of test borings. Holes are shown in brown.
(Corps of Engineers, Kansas City District)

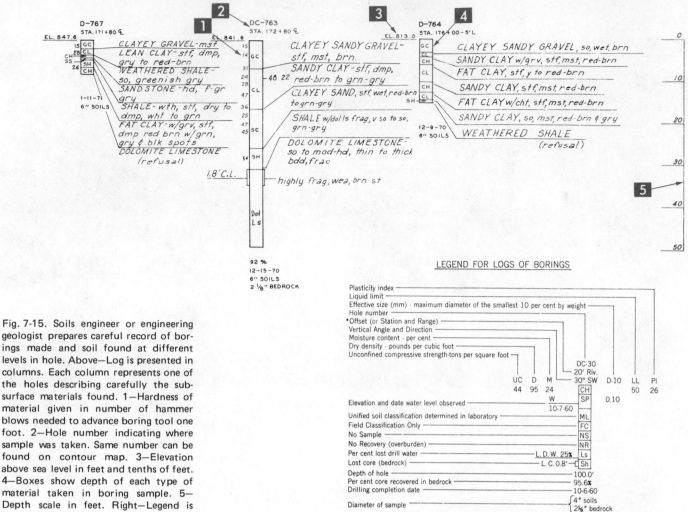

Fig. 7-15. Soils engineer or engineering geologist prepares careful record of borings made and soil found at different levels in hole. Above—Log is presented in columns. Each column represents one of the holes describing carefully the subsurface materials found. 1—Hardness of material given in number of hammer blows needed to advance boring tool one foot. 2—Hole number indicating where sample was taken. Same number can be found on contour map. 3—Elevation above sea level in feet and tenths of feet. 4—Boxes show depth of each type of material taken in boring sample. 5—Depth scale in feet. Right—Legend is prepared to explain log.

Fig. 7-16. A geological section.   (Corps of Engineers, Kansas City District)

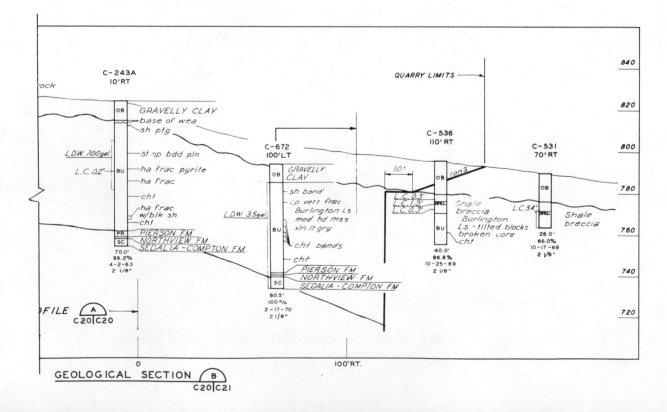

given depth and a 1 1/2 to 2 in. (38 to 51 mm) sampling tube called a "Shelby tube" is lowered into the hole. This Shelby tube is forced into the earth at the bottom of the hole by blows of a hammer of a given weight. The resistance of the earth can be determined by the number of blows necessary to sink the tube a given distance. When the soil sampling is more critical, for tall buildings, dams, and bridges, 5 in. (127 mm) diameter core samples are taken. See Fig. 7-13.

Determining the proper number and depth of borings requires judgment and experience. The number needed depends on the reason for the soil investigation. Preliminary borings on a large site may be limited to one per acre, Fig. 7-14. Borings within the proposed building site may be needed every 100 ft. (30.5 m).

KEEPING OF LOGS

Logs of test borings are prepared by the soils engineer. From the core samples collected, the kinds of materials found at the various levels of subsurface strata are recorded, as in Fig. 7-15.

Geological sections are prepared in the laboratory from data gathered and recorded by logs of borings, Fig. 7-16. (Refer also to Fig. 7-14.)

Finally a generalized geologic column is prepared, Fig. 7-17. The geologic column is a general statement of the total area and its geological history.

Plasticity of soil indicates the content and quality of silt or clay soils and predicts how they will behave. The analyses of

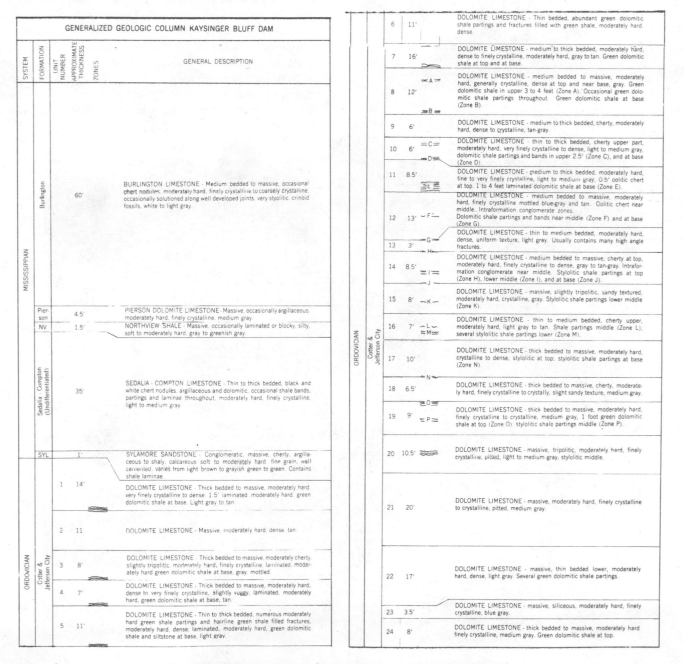

Fig. 7-17. Geologic column. (Corps of Engineers, Kansas City District)

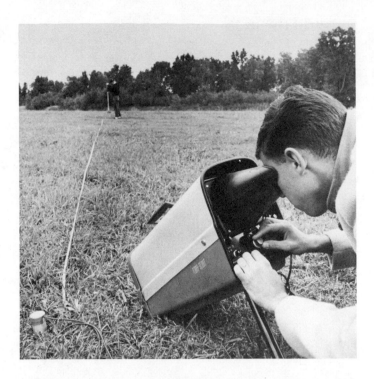

Fig. 7-18. Operating a surface refraction seismograph. (Soiltest, Inc.)

clay content are indications of the soil's cohesiveness. (This means that the grains stick together.) Some soils stick together very strongly, nearly like cement. Other clays are plastic and can be molded into different shapes.

Compaction tests are made on samples in the laboratory. These tests are designed to predict the amount of settling which might occur due to compression. Compression of soil allows the building or structure to settle: air and water are expelled (squeezed out) leaving the soil with greater density and sheer strength. *Proper compaction of soil is very important in highway construction to prevent uneven settling of the road surface.*

Other investigation procedures used by the geologist are:

1. Seismic refraction surveys which measure the time required for a shock wave to travel a known distance from an explosion point to the ground surface. Fig. 7-18 shows a geologist using a seismograph. The velocity of the waves indicates the characteristics of the underlying strata.
2. Aerial photographs of the site and surrounding terrain, Fig. 7-19, are useful for studying drainage patterns, slopes and vegetation. This information, along with geological reconnaissance, provides general knowledge of subsurface conditions. (Reconnaissance means a kind of advance survey to get information.)

Fig. 7-19. Aerial photograph is identified by series of numbers and date as shown at top of illustration. Can you tell how this land is drained?

SUMMARY

Regardless of the type of structure, method of construction, or materials used, the building is no more stable than the geological base it rests on.

Geologists and soils engineers are required to identify the soil and rock formations throughout the building site. These geological investigations require test borings into the earth and taking out of samples. From a test of samples, predictions can be made on how these formations will react under the weight of the proposed structure. The soils engineer prepares reports of these geologic formations. They are used by the architect/designer and others in planning the structure.

CAREERS RELATED TO CONSTRUCTION

ENGINEERING GEOLOGISTS study the structure, composition and history of the earth's crust and advise construction companies of suitability of site for building.

SEISMOLOGISTS study the courses and characteristics of earthquakes using special equipment such as the seismograph. They establish active fault lines or areas where it would be hazardous to build cities, dams or tall structures. Also sometimes called SEISMIC ENGINEERS.

DISCUSSION TOPICS

1. Why is the earth structure important to construction?
2. How is soil formed?
3. What is bedrock?
4. What is sedimentary rock?
5. What is the meaning of "load bearing capacity of soils?"
6. In what way does soil react to applied loads?
7. What are geological sections used for?
8. What is a geologic column?
9. Discuss the objective of compression tests.
10. What is meant by soil liquefaction?
11. What does plasticity of soil indicate to the builder?

Chapter 8

DESIGNING THE PROJECT

Fig. 8-1. Bridge designed of laminated wood is treated against decay. (South Dakota Dept. of Transportation, Div. of Highways)

Designing of buildings and structures is the act of finding the best way to solve the construction problem. Design considers many aspects of construction. For example, the site selected may dictate how the structure is designed. The clay, sand or rock found there may make certain types of construction necessary. The materials available for building will also affect the design. The problem of design facing the architect or engineer might be as big as the total project or it might be as small as the hinge of a door.

The solution to the same kinds of problems does not always result in like structures. This is demonstrated in Figs. 8-1 and 8-2. The problem was to bridge a space of equal distance. However, conditions were different and lead to different solutions. Geological strata and climate were not the same. Probably, the largest difference was people. Different people were involved in "designing" the bridges. The designers had different ideas. Both bridges serve their intended use equally well but are structurally different.

Fig. 8-2. Bridge is constructed of concrete with steel superstructure. The superstructure is hot dip galvanized to save money on long-term maintenance expenses. (American Hot Dip Galvanized Assoc., Inc.)

Fig. 8-3. The Cathedral of Notre Dame is considered a masterpiece of architecture. Study the soundness of structure. Observe what is known as the "flying buttress" support design.
(French Government Tourist Office)

Fig. 8-4. Each individual component of construction, no matter how small, is the product of somebody's ideas and thinking.

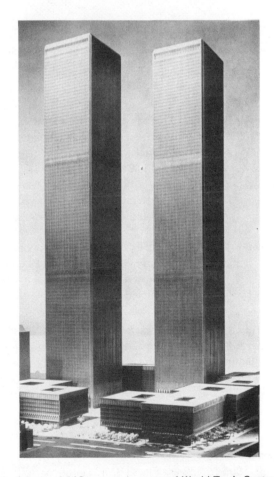

Fig. 8-5. Facade of 110-story twin tower of World Trade Center used 9 million lb. (4.1 megagrams) of aluminum.
(Cupples Div., H. H. Robertson Co.)

MEANING OF DESIGN

The terms "design, designed or designing" mean different things to different people. Design, to many, means the appeal that an object has to the eye, Fig. 8-3. Architectural masterpieces are often termed "well designed" because of their interesting shape. Time-honored structures which have proved to be well built are said to be well designed. *Designing is a process in which the architect or engineer identifies the material and techniques to be used in the construction project.*

Design is not limited entirely to monumental structures people have erected. Each building venture has many small design problems which must be solved as in Fig. 8-4. The smallest electrical insulator or metal fastener must be capable of performing its intended function without fault.

With each new design problem, there are various kinds and amounts of information the designer can draw from. Past solutions are good sources of information for new problems.

Fig. 8-6. What looks like a launch tower is a huge chamber for testing aluminum curtain walls which sheath the World Trade Center. The 55-ft. (17-metre) high structure houses four duplicate floors, the 101st to 104th, including the structural steel, insulation, heating and air conditioning and aluminum outside walls.

Fig. 8-7. The designer needs many types of information before structural problems are solved.

The designer of buildings and other structures is usually called an architect. The architect finds that certain problems require entirely new information. An example is the 110-story twin towers of the World Trade Center in New York City, Fig. 8-5. Never before had a building of this nature been constructed. The architect-designer-engineer could not be certain how the materials would react under these structural conditions.

However, such information must be obtained before the structural design could be completed. Testing structures, as shown in Fig. 8-6, are often used to test certain materials before the final solution is reached.

Tests include those which subject the materials to extreme atmospheric conditions, hurricane winds and torrential rains, and temperatures from −29 F to 122 F (−34 C to 50 C). Tests of flammability of materials provide additional information for the architect. Data is gathered and converted to specifications for the builder to use. See Fig. 8-7.

DESIGN CONSIDERATIONS

Regardless of the type of structure being developed — a bridge, an 1805-ft. tower, Fig. 8-8, a church or residence — there are basic factors common to all designs. These are: 1. Function. 2. Form. 3. Cost. 4. Materials. 5. Technology.

Fig. 8-8. In Toronto, 40,500 cu. yd. or 30,500 cubic metres (m³) of concrete has been poured around 80 miles (128 kilometres) of post-tensioned steel cable. Such concrete has greater compression strength. (Canadian National)

Fig. 8-9. The foundation is a factor of design function. However, it is seldom seen after project is completed.
(United States Fidelity & Guaranty Co.)

FUNCTION

All architectural designs are a product of each of these factors. It is hard to separate them. In individual cases it becomes easy to pick one factor which is stronger than the others.

Function is considered in most cases by most designers to have priority in design. That is to say, the purpose or intended use of the structure should have first consideration by the designer.

Function is the ability of the structure to meet the need it is built to fulfill. A foundation, Fig. 8-9, is functional if it meets the need of the total structure. If it does not meet this need, the design is bad. It does not matter how economical or attractive it may be.

A shopping center, Fig. 8-10, is functional if shoppers can move readily from one area to another without being crowded. A home, Fig. 8-11, is considered functional if it allows the family to enjoy their activities without feeling cramped for space.

The functional home contains all necessary services for family activities. It does this efficiently without waste. Yet it is economical to build and maintain.

FORM

Form is the shape or outline of the building. It has nothing to do with color or material texture of the structure. For instance, the building in Fig. 8-12 is shaped like an inverted pyramid. It was given this form to prevent heat gain due to solar heat coming through the glass. By overhanging each floor level, an insulated heat transfer barrier is created. Only 18 percent of the sun's heat energy passes through into the building. This reduces air conditioning costs. Heat loss or gain should be considered in the design of any structure that houses people.

COST AND MATERIAL

Quite often it can be seen that the designer places form ahead of other factors, Fig. 8-13. In the case of the Air Force Academy Chapel, however, the designer used the factor of materials to achieve the form. The materials and their strength are important to the project.

Materials and costs are closely related. Cut and polished stone for a facade is obviously more costly than poured

Fig. 8-10. Shopping center design brings together many different materials for appeal and functional use of space.
(California Redwood Assoc.)

Fig. 8-11. A truly functional home depends on how it meets the needs and lifestyles of those who live in it.

Fig. 8-12. Need to solve a heat build-up problem led to the basic form of this building. (City of Tempe)

concrete. (A facade is the front or any side of a building.) Here, form and costs become directly opposed. It is the architect's responsibility to make a decision or provide a solution to the problem. The decision is made after weighing the advantages of one against the advantages of the other. Function, probably, will be the deciding criteria.

Materials include texture and color of the structure. Color and texture, then, must be taken into account by the architect/designer.

TECHNOLOGY

Construction technology is a vast body of knowledge. Many agree that it takes in all the activities related to construction. These will include the planning done by the architect, the

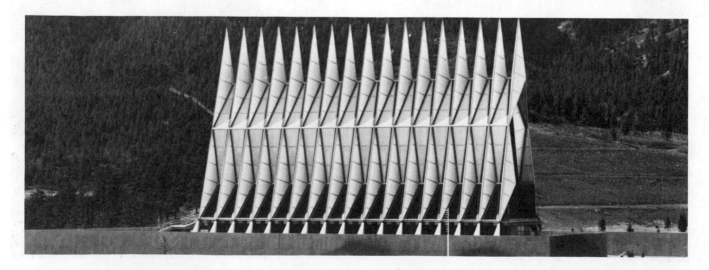

Fig. 8-13. Architects conception of the Air Force Academy Chapel is unique. Its steeply pitched shell is made from rows of 100 giant, prefabricated, aluminum tetrahedrons (four-sided forms with each side triangular in shape), the top rows of which abut to form 17 spires. Daylight enters through ribbons of glass at the joints between the big components. (Cupples Div., H. H. Robertson Co.)

materials, the work done by bricklayers, carpenters, plumbers and other skilled workers in construction. It will take in all the processes and procedures used to assemble the project.

THE DESIGN PROCESS

Design is a process. It is an organized sequence of steps which lead to a final solution in shape, material or procedure of construction. The architect's procedure is a series of steps

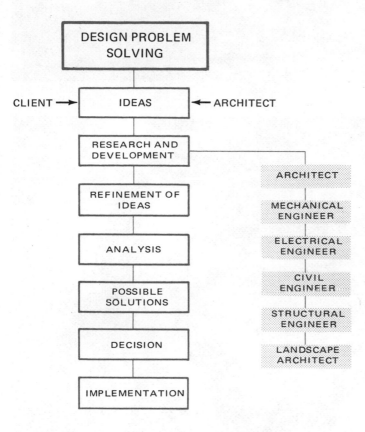

Fig. 8-14. The design process always follows an orderly movement from idea to final plan.

that arrive at a solution. The order of the steps may vary with the individual or the problem. But each approach always follows an organized pattern like the process shown in Fig. 8-14. Once the problem has been identified, the architect and his staff gather background information. They evaluate this data to understand the need more fully. Only then can they set about to solve the problem.

IDEAS

The architect considers all the possible solutions. Details, dimensions (sizes) or engineering factors receive little attention at this point. Sketches are made. Notes and ideas are jotted down. This continues until something of unusual interest begins to develop. The better ideas — those which begin to show promise — move into another stage. They are rendered into architectural sketches which client and architect evaluate. See Fig. 8-15.

RESEARCH AND DEVELOPMENT

The design process moves forward as various persons investigate structural elements of the design. Each structural element and subsystem are evaluated in their own right. A subsystem is a separate but necessary part of a structure. In a building, for example, the electrical wiring, the heating and plumbing are subsystems.

This work may involve the mechanical engineer, electrical engineer, civil engineer, structural engineer and the landscape architect. Each of these individuals makes a judgment about the tentative solutions that are offered. (Tentative means not fully worked out.)

REFINEMENT OF IDEAS

Information continues to pile up in support of various solutions. Following the research and development stage, where the subsystems have been applied to the proposed sketches, more information is brought together to make the final decision possible. These facts are needed to prove the value of one design over all others. Architectural artists produce very real looking sketches called artist's renderings to give the client a better idea of what the designs will look like when built.

ANALYSIS

Part of the design process has the architect working out how strong the related components need to be to support the design. This phase of the design process is more closely related to engineering than any other step. It becomes difficult to separate architecture and engineering. *A general rule is that architects usually design buildings while engineers' responsibility is bridges, highways and dams.* The engineer also aids the architect in those subsystems of electrical, drainage and structural (weight-carrying) problems.

POSSIBLE SELECTION

The architect or engineer, having identified two or three feasible designs, goes back to the client for a decision. Examine Fig. 8-15 again. Here are three possible solutions for a tourist's overlook. These artist's renderings all proved to be promising solutions to the problem.

THE DECISION

At this point, the client and the designer come to some agreement on the design. This must be done before the design process can go on. Remember, no work has been done at the site. Three possible decisions can be made:
1. Accept one design and proceed working out the individual details.
2. Combine features (ideas) from each design.
3. Reject all the tentative designs and start over again.

The last choice is not necessarily a bad decision. Repeating the process a second time may bring forward many more good ideas. In the case of the tourist attraction center in Fig. 8-15, the middle rendering proved to be most feasible.

Fig. 8-15. Artist's renderings of three possible solutions for an overlook to be built above the Harry S. Truman Dam and Reservoir. Which design would you select? (Corps of Engineers, Kansas City District)

IMPLEMENTATION

When the final decision is made to accept a design, the architect or engineer begins to make final drawings. Dimensions and specifications are prepared. These indicate sizes and type of material. This information is necessary before the project can be constructed.

SUMMARY

The design process demands the time and talent of many individuals. The major responsibility for the design of buildings rests with the architect. For the design of bridges and highways, the engineer has chief responsibility. However, both are aided by ideas and input from the client and the supervising engineers for each of the subsystems.

Designing is the most creative phase of construction. Ideas are fostered and developed from which the structure will eventually be built. A good design solution in any construction problem is one which is feasible to build, economical, structurally sound and is "artfully" done.

Of the original solutions proposed by the designer one may be selected because it offers the best possible solution. All may be rejected because they do not solve the problem. Or, two or more ideas may be combined. It is the responsibility of the designer to present the proposed solutions to his client. He must do this in such a way that his ideas become very clear and are understood thoroughly. If not, a very good design could be rejected without a good reason. This is why artist's renderings are prepared. This is why preliminary structural analysis and tentative working drawings and specifications are made.

CAREERS RELATED TO CONSTRUCTION

ARCHITECT plans and designs structures; identifies construction problems, gathers background information, evaluates it and creates a design that solves the problems. Architect organizes services needed for construction; prepares drawings, specifications and other contract documents for builders and craftsmen.

ARCHITECTURAL ARTIST (or ILLUSTRATOR) prepares drawings, sketches, renderings and illustrations for presentation to clients or for advertising in commercial catalogs and publications.

MECHANICAL ENGINEER designs and develops variety of machines that use power — in this case, refrigeration and air conditioning equipment and elevators.

ELECTRICAL ENGINEER designs, develops and supervises manufacture of electrical or electronic equipment; plans and supervises electrical service and wiring in construction projects; designs and assists in operation of electric power generating and distributing facilities.

CIVIL ENGINEER designs and supervises construction of buildings, roads, airfields, tunnels, bridges, water supply and sewage systems.

STRUCTURAL ENGINEER works within civil engineering field as a specialist in determining building stresses; specifies strength of structural components based upon probable stresses.

LANDSCAPE ARCHITECT plans and designs development of land areas such a subdivisions, residences, commercial or public buildings and parks; analyses site conditions, geographical location, soil, vegetation and rock features; prepares drawings and estimates costs of landscaping.

DISCUSSION TOPICS

1. What is a client?
2. What are the responsibilities of the architect?
3. List the design considerations.
4. Outline the design process.
5. What is an artist's rendering?
6. What is meant by functional designing?
7. Why does an architect or engineer use sketches?
8. How do climate conditions affect buildings?
9. Explain why all bridges are not built alike.

Chapter 9

ENGINEERING THE PROJECT

Fig. 9-1. Structural engineering makes it possible to span large spaces successfully without having structural members collapse. Above. Proper selection, location, size and shape of construction materials allows many different choices in solving structural problems. Below. Completed building would have been impossible task without engineer's knowledge and understanding of materials, loads and requirements for spanning large spaces.   (Louisiana State University)

Engineering is the process or art of understanding and managing building materials to make use of their greatest strength. See Fig. 9-1. To engineer a building or structure involves two things. The engineer must first analyze the materials and then determine the structural requirements for supporting the weight of the building.

Construction technology has given us many new facts about metals, woods and other materials. We are more aware of what stresses materials are capable of withstanding. Materials receive stress not only from the structure's own weight but also from outside forces like wind, climate and earthquakes.

The engineer may never guess at his work. He must establish without any doubt, the load-bearing capabilities of certain materials. The foremost function of construction materials is to develop strength, rigidity and durability. These qualities must be equal to the service for which they are intended. Structural requirements are defined largely in such terms as "load," "stress," "weight" and "distance."

THE LEVER

The principle of the lever is used in solving architectural and engineering problems. This is particularly true in the design of beams, girders and foundations. The lever is employed constantly in the structural systems of bridges and buildings. Probably it was the chief means by which great weights were moved in the construction of ancient monuments.

The lever principle is at work wherever there is a relationship between any three parallel forces, in the same plane, which hold a beam in place, Fig. 9-2. The lever works because

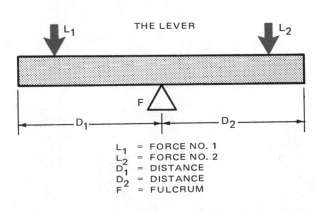

THE LEVER

L_1 = FORCE NO. 1
L_2 = FORCE NO. 2
D_1 = DISTANCE
D_2 = DISTANCE
F = FULCRUM

Fig. 9-2. Imagine that the lever above is a seesaw and that arrows representing weight or force are two children weighing the same. Fulcrum at point F is the same as supporting bar for seesaw. Like the seesaw, beam above will not tilt one way or the other because forces are equal (in equilibrium).

of the law that any force has an equal and opposing force.

Looking at Fig. 9-2, let us suppose that the load, L_1 D_1 is equal to the load, L_2 D_2. If this is true, we say the beam is in equilibrium because it is balanced over the fulcrum. (The fulcrum is the support for a lever.) It is the engineer's responsibility to use mathematics and physics to understand the weight and load forces in Fig. 9-3 created by the structural system. To keep the beam in equilibrium, (balance), 300 lb.

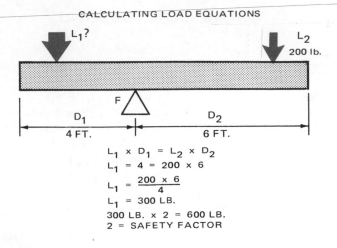

CALCULATING LOAD EQUATIONS

$L_1 \times D_1 = L_2 \times D_2$
$L_1 = 4 = 200 \times 6$
$L_1 = \dfrac{200 \times 6}{4}$
$L_1 = 300$ LB.
300 LB. $\times$ 2 = 600 LB.
2 = SAFETY FACTOR

Fig. 9-3. Engineers work out how to load beams to keep them in equilibrium. Because the distances from either end of the beam to fulcrum at F are not the same, the engineer must find out how much additional weight must be placed at L_1 to balance weight at L_2.

must be placed at point L_1 to balance the 200 lb. at L_2.

Engineers add additional force, referred to as a "safety factor" to go beyond the point of equilibrium and prevent the beam from tipping. A safety factor of 2 might be employed:
300 lb. $\times$ 2 = 600 lb. (136.2 kg $\times$ 2 = 272.4 kg)

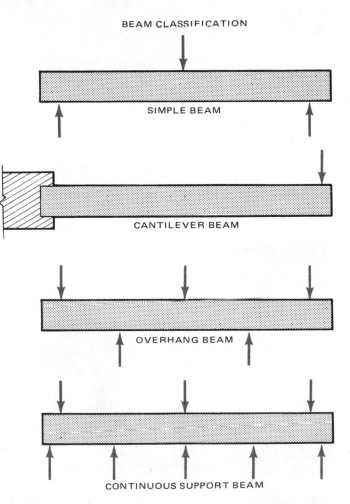

BEAM CLASSIFICATION

SIMPLE BEAM

CANTILEVER BEAM

OVERHANG BEAM

CONTINUOUS SUPPORT BEAM

Fig. 9-4. Beams are named by the way in which they are supported.

CLASSIFICATION OF BEAMS

A beam is primarily a horizontal or inclined structural member which resists bending. See Fig. 9-4. The force acting on the beam tends to bend it rather than to make it shorter or longer.

A beam resting on two supports is called a simple beam, Fig. 9-5. One which projects from a single support is a cantilever beam, Fig. 9-6. A simple beam extending beyond one or both supports is called an overhanging beam, Fig. 9-7. A continuous beam is one that rests on more than two supports, Fig. 9-8.

Fig. 9-7. This residence, constructed on natural slope, uses overhanging beam at three different levels. (Koppers Co., Inc., Forest Products Div.)

Fig. 9-5. These bridge supports are a good example of the simple beam. Beams are held up at each end by columns called piers. Another beam is about to be lifted into place atop the supports. (Northwest Engineering)

STRESSES WITHIN A BEAM

If a beam carrying a load is in equilibrium, the stresses in the fibers at any section of the beam holds in equilibrium the

Fig. 9-8. Stud-plate construction of typical frame residence is a type of continuous beam. (Western Wood Products Assoc.)

Fig. 9-6. Supported end of this cantilever is tucked back into hill formed by steep shore slope. (California Redwood Assoc.)

outside forces on each side of the section. The outside forces are the loads. The weight of the beam is one of these forces.

When a simple beam is stressed by loads, the beam tends to become concave (dish shaped) on the top and convex (dome shaped) on the bottom, Fig. 9-9. The upper fibers are in compression and tend to shorten. Those on the bottom are in tension and tend to become longer, Fig. 9-10.

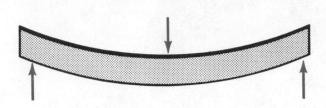

Fig. 9-9. Simple beam is deflected (bent) from stress of load placed upon it.

COMPRESSION AND TENSION IN A BEAM

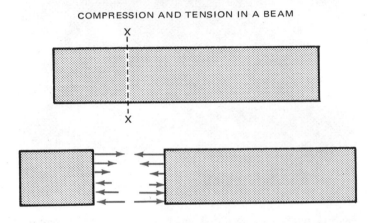

Fig. 9-10. Load placed at point X on beam causes both tension and compression in a beam. Arrows pointing towards each other show compression. Arrows pointing towards ends indicate tension (stretching).

BEAM DEFLECTION

The vertical distance moved by a point on the neutral surface during the bending of a beam is called its deflection. Deflection of beams occurs when loads are applied to them. A tendency to change shape is present in all beams under load. This change, however, does not always occur.

This resistance of the beam to deflection is called stiffness. Generally it is necessary that a beam be stiff enough as well as strong enough. A floor beam may be strong enough to carry the load if its deflection does not exceed 1/360 of the total length of span. (Span is the distance between the supports.)

Factors which contribute to a beam's deflection are:

1. Type of material.
2. Size and shape.
3. Total beam weight and load.
4. Length of span.

COLUMNS

A column is a vertical structural member subjected to compression forces in a direction parallel to its longitudinal axis. In simpler words, a column is a piece of building material that stands on end to hold up other structural members (beams). This weight tends to squeeze together the material or fibers along the column's length. Fig. 9-11 pictures steel columns used to support the floors of a building.

Fig. 9-11. Steel columns in Minneapolis building under construction give vertical support to floors. (IDS Center)

If the weight placed on a column is greater than it can hold, the column will break causing the building to collapse. Architects and building engineers must, therefore, carefully work out strengths of columns needed before the structure is built. They use a "slenderness ratio."

The slenderness ratio establishes the minimum size necessary to support the compressive force due to beam and load. In wood columns, the slenderness ratio is found by dividing the unsupported length by the smallest dimension of the column's cross section (L/d):

L = unsupported or unbraced length

d = dimension of the least side of cross section

A = area of cross section

C = allowable compression unit stress, parallel to grain for particular grade and species of wood in pounds per square inch. In SI metric measure, the answer would be given in pascals, kilopascals (1000 pascals) or megapascals (1,000,000 pascals). In Southern Yellow Pine, for example,

C = 1400 psi or 9.651 kilopascals. (A pascal is equal to 6.894 psi; 1400 x 6.894 = 9651.6 Pa or 9.651 kPa.)
F = maximum allowable (safe) load, or force, the column will support.

Total compressive force (weight) on the columns is given in pounds or tons. Compressive stress is in pounds per square inch.

When SI metric measure is used, compressive force is given in newtons (N), or meganewtons (MN). Compressive stress is expressed in pascals (Pa), kilopascals (kPa) or megapascals (MPa). Area is given in metres squared (m^2).

A column too slender for the loads placed upon it would bend under the load. The slenderness ratio of columns is found by dividing the unsupported length of a column by its smallest cross-sectional dimension:

For example, a column 11 in. by 11 in. by 120 in. long (28 cm by 28 cm by 305 cm) will support:

$$L/d = \frac{120}{11} = 10.9 \text{ or } \frac{305}{28} = 10.9$$

If the slenderness ratio is smaller than 11, the following formula is used to calculate the largest load that may be supported safely by the column:

(English)	(Metric)
F = A x C	F = A x C
F = (11 x 11) x 1400	F = (28 x 28) x 9.65
F = 121 x 1400	F = 784 x 9.65
F = 169,400 lb., safe load	F = 7565.6 kN, safe load

Columns are subject to two forces:

1. Concentric loading.
2. Eccentric loading.

A column is said to be concentrically loaded if the vertical load bearing upon it is balanced directly on top of the column. This is the case in the usual residential construction. Eccentric loading is a load bearing on a column that is unbalanced by another load opposite it, Fig. 9-12.

Columns and beams form the basic structure for many projects. In Fig. 9-13, we have an example of column and beam construction in a bridge. Between each column (pier) and beam are steel beams engineered to allow minimum deflection. The steel beams support a concrete road bed.

Fig. 9-13. Bridges are good examples of column and beam construction. Single concrete piers (a name used for columns in bridges) are reinforced with steel. Single concrete beam tops each pier.
(Corps of Engineers, Kansas City District)

SUMMARY

Engineering is the art of identifying the physical support properties of materials. The lever is a simple machine which engineers often use in construction. When so used, it provides support in spanning distances with a minimum amount of material.

The basic law of equilibrium plus a safety factor allows engineers to design beams that solve varying problems of structure. Columns, the vertical support units for beams, are subjected to forces of concentric and eccentric loading. If the slenderness ratio is too small, vertical deflection occurs.

CAREERS RELATED TO CONSTRUCTION

CIVIL ENGINEER designs and supervises construction of roads, bridges, harbors, airfields, tunnels, water supplies and sewage systems; also designs, plans, researches and inspects.

MECHANICAL ENGINEER is concerned with production, transmission and use of power; designs engines, rockets, nuclear reactors and machines that use power.

PHYSICIST studies and observes matter and energy to describe its structure and interaction in terms of mathematics; develops theories to discover laws which govern natural forces.

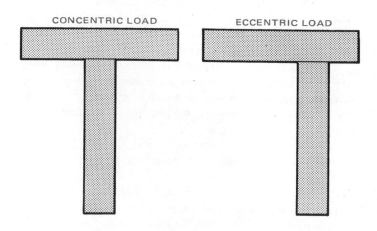

CONCENTRIC LOAD   ECCENTRIC LOAD

Fig. 9-12. Load distribution can be concentric as at left or eccentric as at right.

MATERIALS ENGINEER tests and researches construction materials; advises builders on most suitable materials; may develop new materials or find new uses for old materials.

MATERIALS TECHNICIAN assists materials engineer by testing materials, setting up tests, setting up experiments, making mathematical calculations.

DISCUSSION TOPICS

1. Define engineering as related to construction.
2. What is a load?
3. Define the principle of the lever. How does it apply to construction beams?
4. Sketch a simple beam and show how it can be deflected by a load.
5. What is a cantilever beam?
6. Sketch a continuous support beam.
7. What are the vertical support parts (components) called?
8. What is vertical deflection? What are the effects of such deflection?
9. What is concentric loading of a column?
10. How is the column and beam technique used in residential construction?

Chapter 10

PLANNING THE PROJECT

Actual planning of the project comes next in a long chain of activities needed to build a structure. At the point where this planning begins, several important steps have already been taken. One design has been chosen over the others. A structural plan has been engineered from the preliminary sketch and all of the ideas regarding the plan have been recorded.

The plans now need final drawings. These are called working drawings and will include all the necessary views, dimensions and details.

Working drawings can only be prepared after the design has been established and all engineering notes have been made. Fig. 10-1 shows the order in which work has been done.

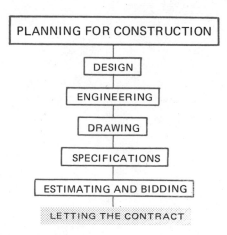

Fig. 10-1. Sequence for construction planning.

Construction involves contracts and subcontracts. Many individuals as well as individual construction companies must work together. With so many legal aspects and people involved, there is a danger that someone will misunderstand the plan. *Drawing is very important to construction. It gives the architect/engineer a means to communicate his ideas accurately to the builder.*

WORKING DRAWINGS

A set of working drawings for a small project like a residence has the following basic parts:

1. Floor plan.
2. Foundation plan.
3. Elevations.
4. Framing plan.
5. Cabinet details.
6. Wall sections.
7. Details.
8. A plot plan.

Fig. 10-2 shows a floor plan and an elevation plan. In simple projects all of the physical details are shown in one set of drawings. In large projects, there are many more drawings. Each drawing shows much more than the physical form of the project and its dimensions. Various mechanical installations such as the mechanical, electrical, structural and plumbing drawings may be included.

In large projects, the drawings are more complex. The project needs much more explaining. A set of drawings may include a site plan. This plan records the soil test boring schedule, excavation limits, paving requirements, river control and/or detour construction. When necessary, the working drawings also include:

1. Heating and cooling systems.
2. Drainage and plumbing.
3. Stairs.
4. Elevators or escalators.
5. Lighting and electrical requirements.
6. Structural details.

SCALED DRAWINGS

It is not possible to prepare working drawings full size. Instead, they are drawn to scale. This means that the drawing has been reduced proportionately from actual size so that it will fit on the drawing sheet. Small units of measure stand for large units.

The floor plan for an average size residence is prepared using a small unit — usually 1/4 in. (6 mm) to represent one foot of the completed project. This reduced scale reduces the residence 48 times, so it will fit on a 17 in. x 22 in. sheet (43 cm x 58 cm) sheet of paper.

Many persons are employed preparing the many scaled and working drawings necessary to supply construction people with plans. People working in a clean and bright atmosphere transform the architect's or engineer's ideas into scaled drawings for the builder. See Fig. 10-3.

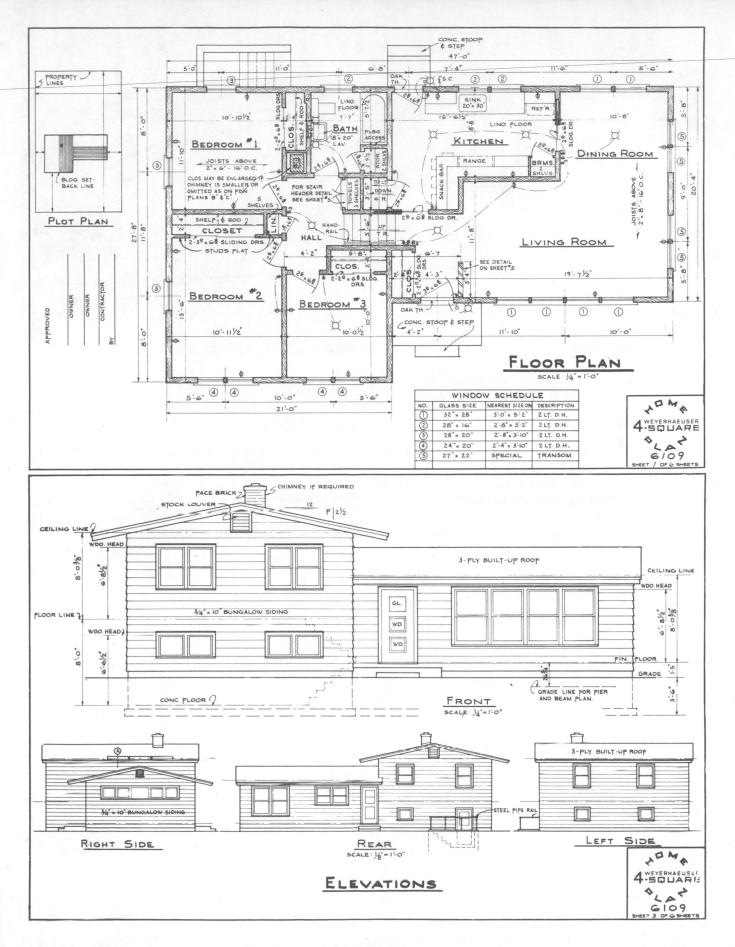

Fig. 10-2. Examples of detailed working drawings for a residence. A scaled floor plan is shown above. The elevation, which is a "straight-on" drawing of all sides of the construction, is shown below. In addition to the floor plan and elevations, a full set of drawings include a foundation plan, framing plan and detail drawings and section drawings. (Weyerhaeuser)

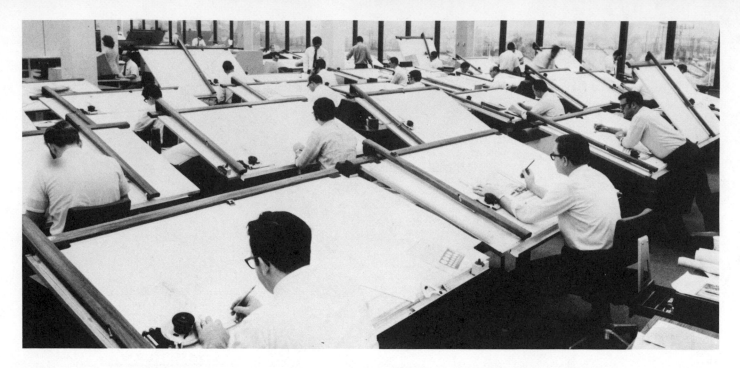

Fig. 10-3. Teams of well-trained persons continually prepare working drawings for construction. (VEMCO)

LAYOUT DRAWINGS

Through working drawings, the physical layout is clearly shown. The layout drawing of a highway interchange shown in Fig. 10-4 includes the overpasses, entry and exit ramps, location of service roads and survey station points.

ELEVATIONS

Elevation drawings are important to establish the vertical location, Fig. 10-5. A builder, studying the bridge elevation will understand pile location and height. Elevations give a general idea of how the finished construction will look.

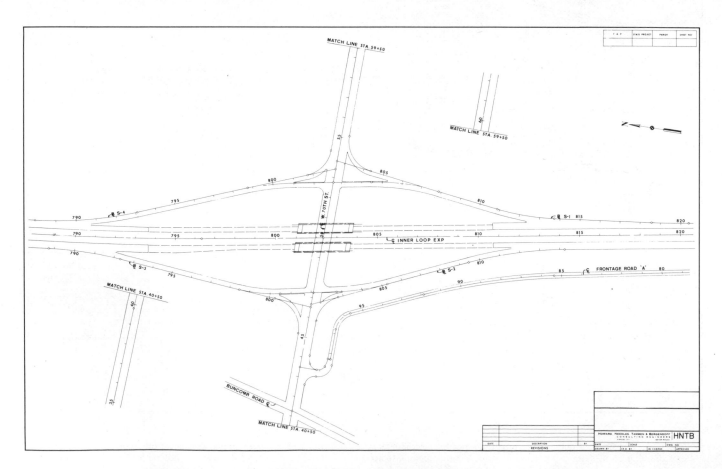

Fig. 10-4. Layout of a highway interchange. (Howard, Needles, Tammen & Bergendoff)

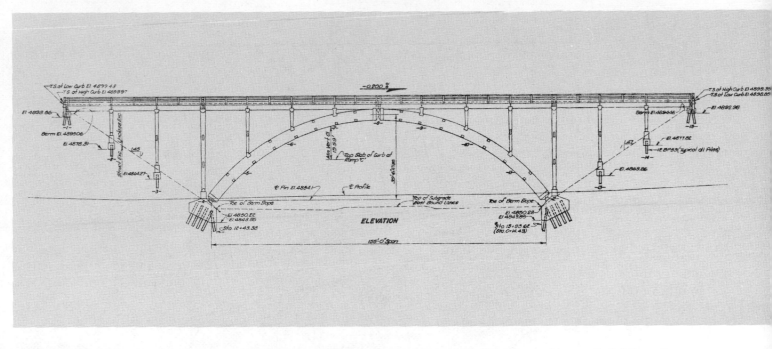

Fig. 10-5. Elevation drawings reveal a view that looks somewhat like the finished project. They also provide some basic information. (American Wood Preservers Institute)

STRUCTURAL DRAWINGS

Structural drawings show details of the structural parts of the project. Fig. 10-6 shows a cross frame for a bridge. This frame is made of steel. The drawing gives information on:
1. Cross section configuration (outline or shape) of members.
2. Size of steel members.
3. Type of welds needed to fasten these members.
4. Dimensions.

CIVIL DRAWINGS

Civil drawings have to do with the land on which a road is built, for example. They show such information as cut and fill necessary for a road bed. See Fig. 10-7. Such civil drawings or contour maps are based on survey data.

Before a contour map can be drawn, elevations must be measured in the field for various key points controlling the establishment of the contour lines. Various methods are used:

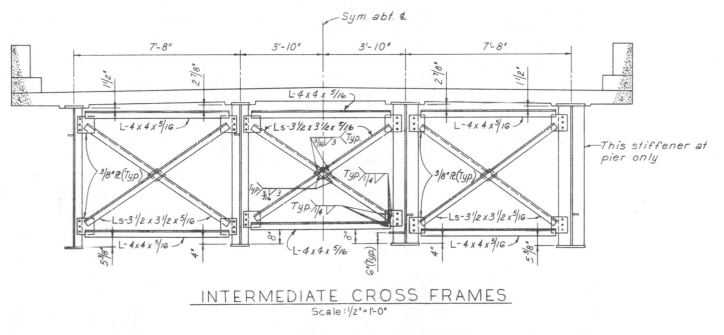

INTERMEDIATE CROSS FRAMES
Scale: 1/2" = 1'-0"

Fig. 10-6. Structural drawing of a bridge cross frame. (Corps of Engineers, Kansas City District)

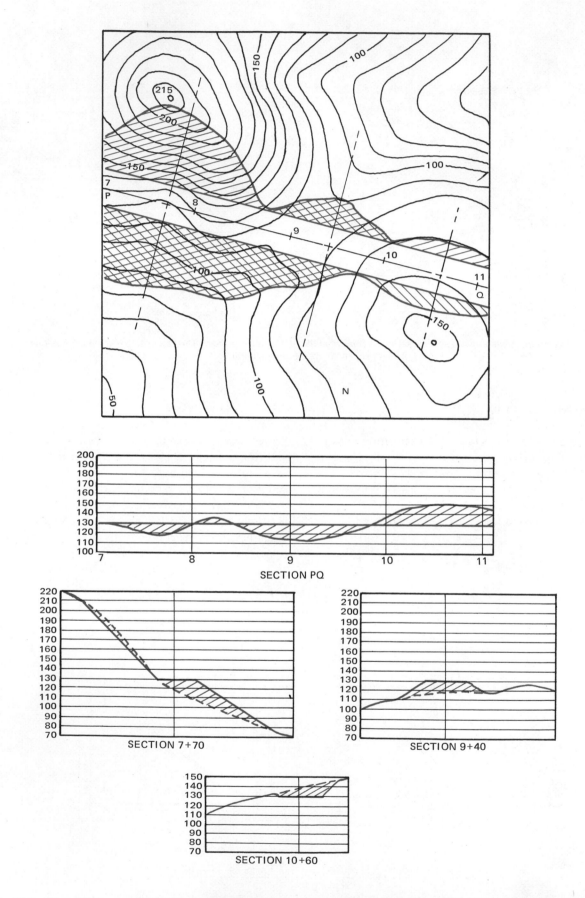

Fig. 10-7. Cut and fill for a road bed. Note that illustration at top is like a contour map. Other illustrations are cross-sectional views. Elevations above sea level are shown at left of these cross sections.

1. Grid system is used where all intersection elevations are grid lines.
2. Points located by transit and stadia rod with corresponding elevations calculated by plane table.
3. Aerial photograph surveys.

All methods require experience in surveying. This is an important function of civil engineering.

ARCHITECTURAL DRAWINGS

Architectural drawings, Fig. 10-8, show many details of the physical form of the project. Architectural drawings include elevations, floor plans, foundation plans, sections and details. Elevation and floor plans are shown in Fig. 10-2.

ELECTRICAL AND MECHANICAL DRAWINGS

Electrical and mechanical drawings show the plumbing, Fig. 10-9, heating/cooling systems, electrical requirements, ventilating systems and lighting. Usually the mechanical drawings appear in a set of working drawings as sections or detailed views. See Fig. 10-10. Any type of mechanical installation must be planned in detail to insure that space is saved for the installation.

SCALING DRAWINGS

Those who prepare and use working drawings for construction, use an architect's scale. The draftsman uses it to reduce

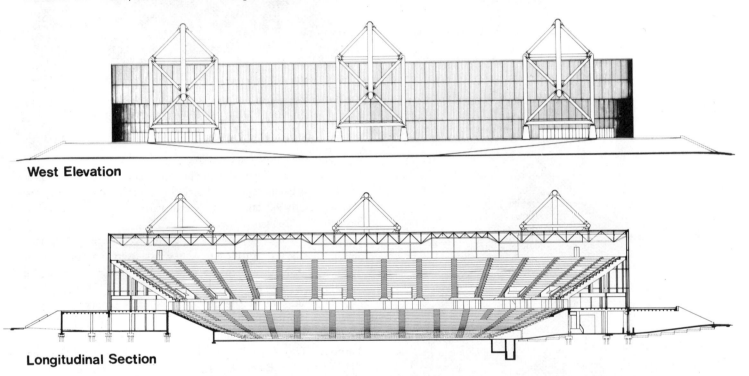

West Elevation

Longitudinal Section

Fig. 10-8. Elevation shows exterior wall appearance. The figure above has a cutaway elevation (longitudinal section) revealing the interior elevation.

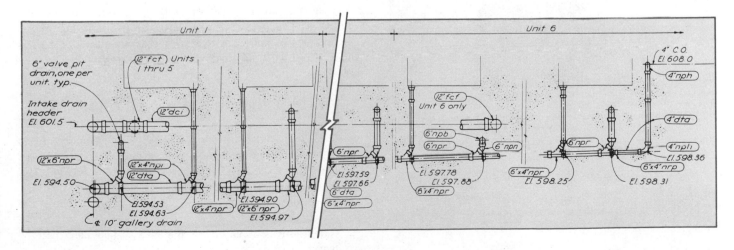

Fig. 10-9. Detail of plumbing system.

measurements from actual size to scale size. Those who must read drawings, use the scale to find actual size.

If you examine the scale carefully, you will notice that each of its three sides have two different scales. Notice also that a section of the end is divided into 12 parts. Each part represents one inch. See Fig. 10-11. You will recall that, if one were using metric, for architectural drawings, 6 millimetres represent 30.5 centimetres. (In common English measure, 1/4 in. is equal to one ft.)

On the English scale, find the 1/4 in. scale. From the 0 (zero) point at the end, move toward the center to locate the number of feet. Move toward the other end of the scale to locate the inches.

DRAFTING EQUIPMENT

Other tools and instruments valuable to the draftsman who prepares working drawings are:
1. Drafting table.
2. Templates.
3. Pencils.
4. Erasers and eraser shields.

Fig. 10-12. Draftsman working with various instruments such as the drafting arm shown, prepares plot plan.

5. T-square or drafting machine.

An elevated top makes the drafting table, Fig. 10-12, easier for the draftsman to work. Usually there is a comfortable seat. Tables are of various sizes so that various paper sizes can be used.

Templates are thin pieces of plastic with common shapes and symbols cut into them. They are used where common lines and shapes are to be drawn. For instance, plumbing fixtures usually are of the same size and shape. These templates are cut at the architectural scale sizes of 1/4 in. = 1 ft. 0 in. (6 mm = 30.5 cm).

Drafting pencils are made in varying grades of lead hardness. This degree of hardness is indicated as F, H, 2H, 3H, 4H, 5H and 6H. Being softest, F is used for sketching while the harder 2H to 6H lead is used to prepare clear, sharp and distinct lines on the working drawings.

Erasers and erasing shields are necessary for making changes in drawings.

T-squares and triangles or drafting machines have several purposes. These are the very heart of the drafting process serving as straight edges to draw either horizontal or vertical lines. They may be used also to establish inclined lines or angles.

DIMENSIONING A WORKING DRAWING

In dimensioning a working drawing, the first concern is to show the information in a way most helpful to the workers building the structure. This concern has led to development of some general rules of dimensioning:
1. Everything is dimensioned to actual size even though the drawing is in scale.
2. Dimensions are placed on the view which describes the feature best.
3. Dimensions are always placed where they can be read from

Fig. 10-10. Mechanical installations are carefully planned, as are these see-through elevators. (Hallmark's Crown Center)

Fig. 10-11. Architect's scale. Divisions to the right of zero are each representative of an inch.

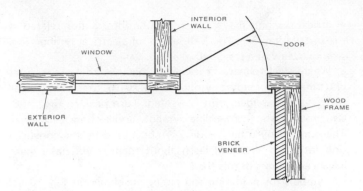

Fig. 10-13. Symbols for wall sections are easily understood.

ARCHITECTURAL SYMBOLS

Working drawings are made to a very small scale, so it is not possible to draw many features or actual parts of the structure as they appear to the eye. It is not economical to even consider drawing these. Therefore, many parts are represented by standardized symbols.

Architectural symbols are standardized to represent walls, wall materials and methods of construction. See Fig. 10-13. The different styles of windows each have a symbolic representation.

Electrical symbols, Fig. 10-14, are another good example of simple signs representing actual parts of a structure. These

ELECTRICAL SYMBOLS

○	CEILING OUTLETS FOR FIXTURES
	WALL FIXTURE OUTLET
○ₚₛ	CEILING OUTLET WITH PULL SWITCH
○ₚₛ	WALL OUTLET WITH PULL SWITCH
	DUPLEX CONVENIENCE OUTLET
	WATERPROOF CONVENIENCE OUTLET
	CONVENIENCE OUTLET 1 = SINGLE 3 = TRIPLE
	RANGE OUTLET
	CONVENIENCE OUTLET WITH SWITCH
▲	SPECIAL PURPOSE (SEE SPECS.)
⊙	FLOOR OUTLET
⊗	CEILING LIGHT FIXTURE
Ⓟ	PULL CHAIN LIGHT FIXTURE
⊕	EXTERIOR LIGHT FIXTURE

■	LIGHTING PANEL
▨	POWER PANEL
S	SINGLE-POLE SWITCH
S₂	DOUBLE-POLE SWITCH
S₃	THREE-WAY SWITCH
S₄	FOUR-WAY SWITCH
Sₚ	SWITCH WITH PILOT LIGHT
⊡	PUSH BUTTON
	BELL
◀	OUTSIDE TELEPHONE CONNECTION
TV	TELEVISION CONNECTION
S	SWITCH WIRING
▭	EXTERIOR CEILING FIXTURE
	FLUORESCENT CEILING FIXTURE
	FLUORESCENT WALL FIXTURE

Fig. 10-14. Electrical symbols used in construction.

the bottom or right side of the drawing.

4. When metric system is used, distances are recorded in metres. The English system uses feet and inches. Example 12 ft. 3 in. or 3.8 m.

Each drawing should be completely dimensioned. If it will help the builder, dimensions may be repeated. At no time should anyone have to add or subtract to find a size.

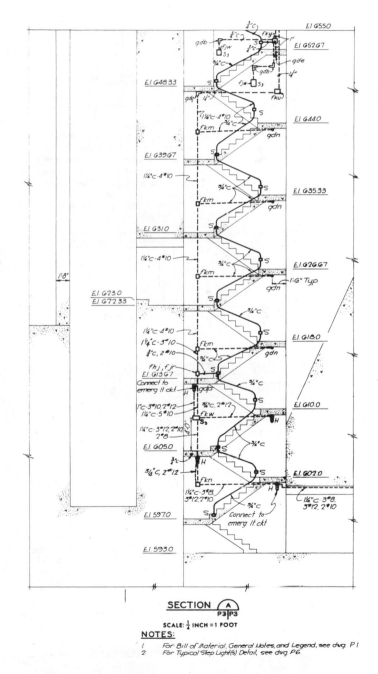

SECTION A

SCALE: ¼ INCH = 1 FOOT

NOTES:
1 For Bill of Material, General Notes, and Legend, see dwg. P1
2 For Typical Step Light(s) Detail, see dwg. P6.

Fig. 10-15. A stairwell with location of conduit, fixtures and switch. Note technical data on conductor sizes.

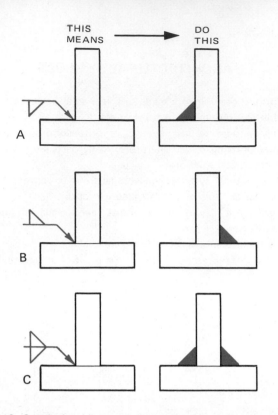

THIS MEANS → DO THIS

A

B

C

Fig. 10-16. Standard welding symbols are a way for draftsman to show what is wanted. Study symbols above. What they mean is drawn at right. A—Symbol drawn on underside of arrow means weld on same side as arrow. B—Symbol on top of arrow means weld on opposite side. C—Symbol on both sides of arrow means weld both sides.

electrical symbols may contain specific information related to the installation. Fig. 10-15 shows use of electrical symbols in a sectional view of a stairwell.

Where steel framework is welded together, the architectural draftsman uses welding symbols, Fig. 10-16, to tell the welder what type of welded joint is wanted. Turn back to Fig. 10-6. Can you locate the welding symbols and explain each one? There are many other welding symbols used by the construction industry. You will learn about them if you take more advanced courses in this field.

Symbols for plumbing and piping are shown in Fig. 10-17.

Another type of drawing used occasionally is one which resembles the aerial view. See Fig. 10-18. It is commonly prepared as a site utilization (use) study. The land use drawing is used in community development.

MODELS

Use of building models is widespread. They are made of a variety of materials — paper, illustration board and wood are but a few. In model building, all features must be kept in true scale. The advantage of a building model, Fig. 10-19, is showing the appearance of a completed building from any station point. When compared with a pictorial drawing which shows a structure from a single station point, its advantage is easy to see.

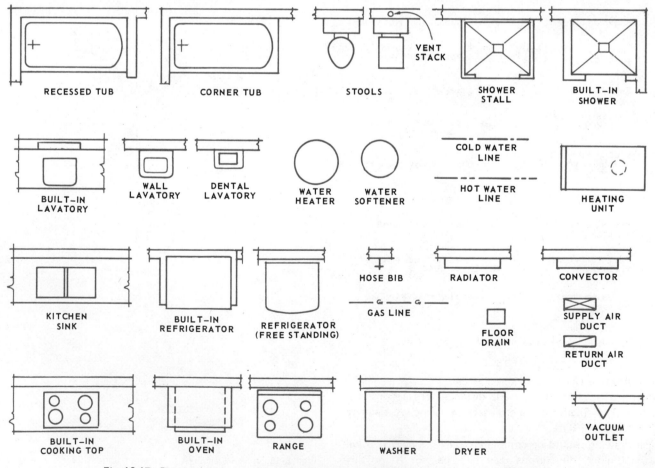

Fig. 10-17. Plumbing and piping symbols will appear on drawing for homes or public buildings.

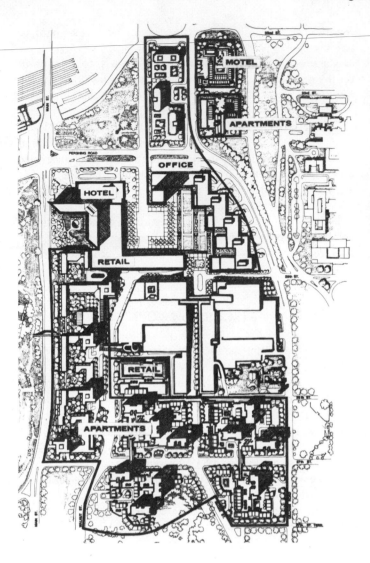

Fig. 10-18. Land use drawings give an aerial perspective of how the various components will use the plotted area. (Hallmark's Crown Center)

Fig. 10-19. Model of a 25-block development project shows true-to-life relationship of structures and traffic patterns which are expected to develop. (Hallmark's Crown Center)

Fig. 10-20. Computer memory bank analysis system. (Calcomp)

DRAFTING BY COMPUTER

Marvel of the twentieth century, the computer has its uses in the architectural design profession. See Fig. 10-20. Drawings are actually made by information fed into the computer's memory bank.

Persons who have the aptitude to program the computer and the understanding of drafting are very useful in drafting. Information feedback is plotted on the terminal plotter, Fig. 10-21, in the form of a drawing. (Feedback is what the computer gives back to its user.) For an example of a computer-assisted drawing plotted by the plotter, turn back to Fig. 3-5. This residential development plot was drawn and notations completed by the plotter.

SUMMARY

Working drawings have been used by architects and builders for centuries as a way to convey ideas related to building. Today, several types of drawings are used to complete the architectural series of drawings. Structural drawings show the frame or support system. Civil drawings describe and locate the geological characteristics at the building site so that cut and fill requirements can be worked out. The electrical and mechanical drawings show what the installation requires for lighting, ventilation, plumbing and heating/cooling.

Architectural drawings consist of floor plans, foundation plans, elevation, section and details. They are drawn to smaller

PAPER   STYLUS

Fig. 10-21. Computer plotter, note the paper and stylus (pencil). (Calcomp)

scale than the full size.

Symbols are used to represent many features of the architectural process simply because it is not economical to draw in the exact representation.

Models of architectural projects are a valuable aid to planning the project. They give the designer a perspective of the project which is not attained from any other vantage point.

Computer drafting is a valuable architectural tool. It speeds up the drawing by taking over many menial (low or repetitive) tasks while computing the answers to many problems.

CAREERS RELATED TO CONSTRUCTION

ARCHITECTURAL DRAFTSMAN draws details of a working drawing and makes tracings of drawings the designer or architect has made.

COMPUTER PROGRAMMER prepares instructions telling computer how to process data. Prepares instruction sheet for operator to follow in running computer program. Checks program for errors, makes corrections (debugs) and makes trial run.

DISCUSSION TOPICS

1. What is an architect's scale?
2. What is an architectural drawing? What does it include?
3. What is a floor plan? A foundation plan? A wall section?
4. Name one construction project where a civil drawing would be used.
5. What are structural drawings?
6. Are drawing dimensions given in scale or in actual size?
7. Why are symbols used on drawings?
8. Explain how to identify the hardness of the lead in a drafting pencil.
9. How are models used by the architect?
10. What are the duties of a draftsman?

Chapter 11

SPECIFICATION WRITING

Working drawings are complete in expressing shape and size of a construction project. It is, however, impossible for drawings to describe the quality of materials used and workmanship expected. This is even more true on projects, such as the one shown in Fig. 11-1. This information cannot be placed on the working drawing. Therefore, certain information is given through specificiations.

Construction specifications are written statements which inform the builder what is to be built, what materials are to be used and how the job is expected to be done. They are needed to prevent misunderstanding between the contractor/builder and the client/owner.

PROVISIONAL CATEGORIES

Written specifications have three categories:
1. Legal provisions.
2. General provisions.
3. Technical provisions.

Legal provisions consist of legal documents involved in the advertisement for bids. A bid form is shown in Fig. 11-2. Instructions to the bidders, bond forms and owner/contractor agreements make up the legal provision section.

General provisions state the conditions and responsibilities of all people involved in the construction project. The architect/designer, owner, contractor and all subcontractors are made to understand the conditions under which they are to function while doing their jobs.

Model sets of these general conditions are published by the American Institute of Architects. The general provisions are understood to be a necessary part of the specifications. The general provisions include items which are not covered in any part of the trades section or in any other division of the specifications.

Technical provisions, Fig. 11-3, list the kinds of materials and processes to be used in the construction. Generally, the technical provision section includes architectural, civil, structural, plumbing, electrical, heating/cooling and mechanical

Fig. 11-1. Specifications for projects large or small protect the interest of everyone involved; architect, builder and owner.

STANDARD FORM 20
JANUARY 1961 EDITION
GENERAL SERVICES ADMINISTRATION
FED. PROC. REG. (41 CFR) 1-16.401

INVITATION FOR BIDS
(CONSTRUCTION CONTRACT)

REFERENCE
Invitation No.
DACW41-71-B-0010

DATE
13 August 19 ..

NAME AND LOCATION OF PROJECT
Construction of Harry S. Truman Dam, Stage III
Harry S. Truman Reservoir, Missouri

DEPARTMENT OR AGENCY
Department of the Army
Corps of Engineers

BY *(Issuing office)*
Kansas City District

Sealed bids in one copy for the work described herein will be received until 1:00 p.m., local time at the place of bid opening, 24 September 19 .., in Room No. 140, Federal Building, 601 E. 12th Street, Kansas City, Missouri 64106, and at that time publicly opened. Hand-carried bids delivered immediately prior to bid opening shall be delivered to the above-designated room.

Information regarding bidding material, bid guarantee, and bonds

1. Bid Bonds. Each bidder shall submit with his bid a Bid Bond (Standard Form 24) with good and sufficient surety or sureties acceptable to the Government, or other security as provided in paragraph 4 of Instructions to Bidders (Standard Form 22) in the form of twenty percent (20%) of the bid price or $3,000,000 whichever is lesser. The bid bond penalty may be expressed in terms of a percentage of the bid price or may be expressed in dollars and cents. (ASPR 16-401.2)

2. Performance and Payment Bonds. Within ten (10) days after the prescribed forms are presented to the bidder to whom award is made for signature, a written contract on the form prescribed by the specifications shall be executed and two bonds, each with good and sufficient surety or sureties acceptable to the Government, furnished, namely a performance bond (Standard Form 25) and a payment bond (Standard Form 25A). The penal sums of such bonds will be as follows:

Fig. 11-2. First page of an invitation for bids. (Corps of Engineers, Kansas City District)

specifications. Good technical specifications reduce erection problems on the site, Fig. 11-4.

Each section of the technical provisions of the construction project is organized into:

1. The general scope of the work.
2. Description of the work.
3. Materials to be used.
4. Time of installation.
5. Installation procedure.
6. Testing and measurements.
7. Special conditions.

WRITING SPECIFICATIONS

Those preparing specifications must know all aspects of construction. They must understand all materials, their classification, strengths and structural capabilities. They must be able to present clearly all the information necessary to identify construction techniques and processes. At the same time, they must understand working drawings and be capable of giving complete technical specifications without duplicating information already in the drawings.

Good specifications give serious weight to the appropriateness, quality and function of material and fixtures. Similar attention is given to ease of installation and serviceability of special equipment. See Fig. 11-5.

SUMMARY

Drawings and specifications for a structure or a residence make the pre-construction planning complete. With this sequence of activities, the architect/designer and owner has a reasonable degree of certainty that the structure will be built as they have envisioned it.

Specifications are a written document containing:

1. A legal provision.
2. General provisions.
3. Technical provisions.

CAREERS RELATED TO CONSTRUCTION

SPECIFICATION WRITER prepares all written information which describes workmanship, materials, fixtures and methods to be used in structure.

3.5 Special workmanship requirements:

3.5.1 Welding procedure qualification shall be in accordance with the qualification requirements of AWS Standard Specification D2.0 and shall be submitted for approval in accordance with the requirements for shop drawings of the SPECIAL PROVISIONS. The procedures shall be such as to minimize residual stresses and distortion of the finished members of the structure. Heat treatment, if required, shall be included in the procedures specification. Should it be found that changes in any previously approved welding procedure are desirable, the Contracting Officer will direct or authorize the Contractor to make such changes. Approval of any procedure, however, will not relieve the Contractor of the sole responsibility for producing a finished structure meeting all requirements of these specifications.

3.5.2 Stress-relief annealing: Where stress-relief annealing is specified or indicated, it shall be in accordance with the requirements of Article 412(a) of the AWS Standard Specification D2.0 unless otherwise authorized or directed.

3.6 Welding of aluminum: Aluminum members specified to be welded shall be welded by the inert gas shielded metal arc-welding method following the recommendations of the manufacturer of the materials and the instructions outlined in the current edition of the American Welding Society, "Welding Handbook."

3.7 Brazing shall be in accordance with the recommendations and instructions outlined in the current edition of the American Welding Society, "Brazing Manual."

4. FLAME CUTTING: Low-carbon structural steel may be cut by machine-guided or hand-guided torches instead of by shears or saws. Flame cutting of material other than low-carbon structural steel shall be subject to approval and where proposed shall be definitely indicated on shop drawings submitted to the Contracting Officer. Where a torch is mechanically guided, no chipping or grinding will be required except as necessary to remove slag and sharp edges. Where a torch is hand guided, all cuts shall be chipped, ground, or machined to sound

Fig. 11-3. A typical specification of the technical provisions providing for workmanship requirements.

Fig. 11-4. Specifications help to insure the product uniformity. This is important to installation on the site.
(South Dakota Dept. of Transportation, Div. of Highways)

Fig. 11-5. One concern of the specification writer is the function, ease of installation and serviceability of special equipment.

DISCUSSION TOPICS

1. List some problems which might arise if specifications did not exist.
2. How do specifications affect installation of mechanical equipment?
3. Who prepares construction specifications?
4. What is included in the general provisions of the specifications?
5. What does the legal provision include?

Chapter 12

ESTIMATING AND BIDDING

Fig. 12-1. Construction estimates for this complex structure require the knowledge of many people trained to determine the cost of excavation, material, labor, equipment and profits. (Hallmark's Crown Center)

Estimating the cost of a construction project must be done very carefully because there are so many variables. Each can affect the actual cost of constructing the project. Fig. 12-1 illustrates a large project requiring considerable care in estimating costs.

Many of the variable costs in construction are unpredictable. Weather, material availability, subcontractors charges and transportation are such variables.

However, many others are predictable to a high degree. Material cost, labor wages, equipment cost and administration costs are in this group. Regardless of the variables involved, the construction estimator must prepare a cost analysis as accurately as possible. The cost estimate can spell the difference between profit or loss on a construction contract.

Building construction estimating is the careful determination of probable construction cost of a given project. Many items influence and contribute to the cost of a project. Each

must be compiled and analyzed. Since the estimate is prepared before construction is started, the estimator spends a great deal of time studying the working drawings and specifications.

Many consider estimating the most important function in contracting. It is from this estimate that the contractor enters into competitive bidding to win the contract. Competition in construction bidding is keen. Sometimes many firms bid for the same project. In order to compete and stay in business, a contractor must be a low bidder. He must offer to build the structure for less money than anyone else. However, the estimate must reflect a profit. Otherwise, it becomes impossible to stay in business.

In a small company, the contractor will do the estimating himself. In larger companies a specialist is hired to do it. He works from drawings supplied by the architect or designer.

Working drawings contain the information needed on design, location, dimensions, material and general construction

Fig. 12-2. This residence is used in estimating and bidding project in text. (Western Wood Products Assoc.)

of the project. Specifications are a written supplement to the drawings. They include information having to do with workmanship and quality of materials. Since the estimate is prepared from these documents, the ability of the estimator to visualize all the different phases of construction becomes a critical element in the success of the company.

The successful estimator organizes the estimate into specialty divisions of:

1. Earthwork costs.
2. Materials costs.
3. Labor costs.
4. Mechanical equipment cost.
5. Equipment depreciation.
6. Overhead and contingencies.
7. Profit.

The rest of this chapter will show how to organize an estimate. The project will be the residence in Fig. 12-2.

EARTHWORK

One of the first items to consider in estimating construction is the type of soil found at the building site. The estimator may begin by studying the soil borings provided in the drawings. (Review Chapter 7.) This would help him determine the difficulty in removing the soil. If a great deal of rock were found, for example, more time would be allowed for removal.

Then the quantity of rock or soil to be removed, Fig. 12-3, would be estimated. You will recall from your math courses that volume is measured in cubic yards or cubic metres. It is calculated by multiplying the length times the height times the width. Fig. 12-4 shows how to find volume and area for various shapes of excavations.

The estimator must bear in mind that excavated earth takes up more space than settled (compacted) earth. Thus, it will take up more room in the dump truck, and the storage space for excavated earth must be larger than excavated space.

Fig. 12-3. Cost of moving excavated earth from construction site is part of building cost. Cost is based upon volume, not area. (Caterpillar Tractor Co.)

Cost of excavation is based on the number of cubic yards or cubic metres of soil that must be removed. A cubic yard is equal to 0.764 cubic metre. Cubic yard is usually abbreviated to cu. yd. in computation. Cubic metre is shortened to the symbol, m^3.

From Fig. 12-5, 375.6 cu. yd. or 288 m^3 of earth need to be excavated for the basement of the house. The estimator must calculate the time required to complete the excavation. Rate of removal is 25 cu. yd. (19 m^3) per hour. There are 375.6 cu. yd. of dirt, so we divide that quantity by the amount that can be moved in an hour:

375.6 ÷ 25 = 15 hours; or

287 ÷ 19 = 15 hours

Hourly cost of operating equipment ($9.78) plus hourly cost of operating crew ($13.65) is multiplied by time needed to complete job (15 hrs.).

($9.78 + $13.65) x 15 = $351.45

The estimated cost of excavating the basement is $351.45.

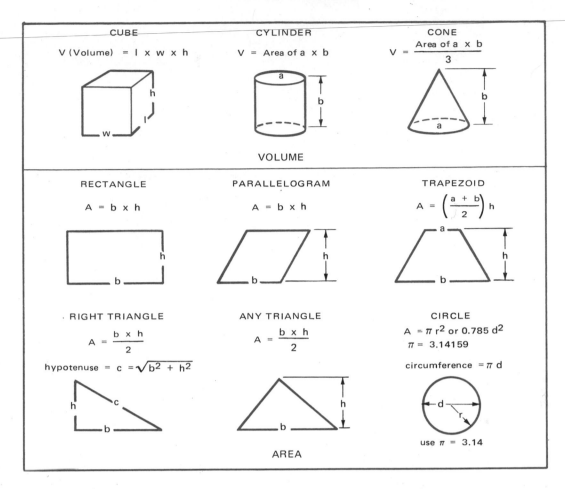

Fig. 12-4. Methods used to calculate volume and area on basic shapes.

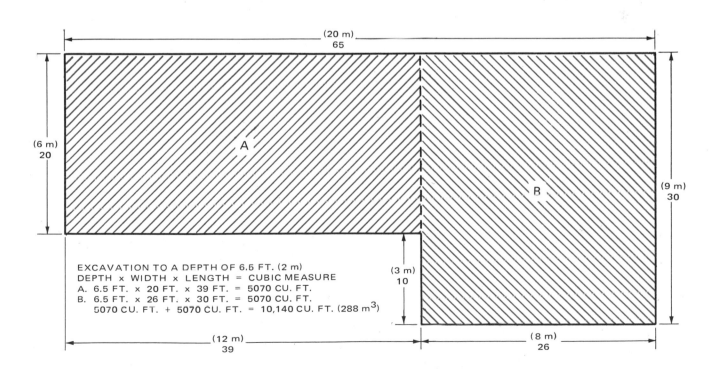

EXCAVATION TO A DEPTH OF 6.5 FT. (2 m)
DEPTH x WIDTH x LENGTH = CUBIC MEASURE
A. 6.5 FT. x 20 FT. x 39 FT. = 5070 CU. FT.
B. 6.5 FT. x 26 FT. x 30 FT. = 5070 CU. FT.
 5070 CU. FT. + 5070 CU. FT. = 10,140 CU. FT. (288 m³)

Fig. 12-5. Calculations for basement excavation.

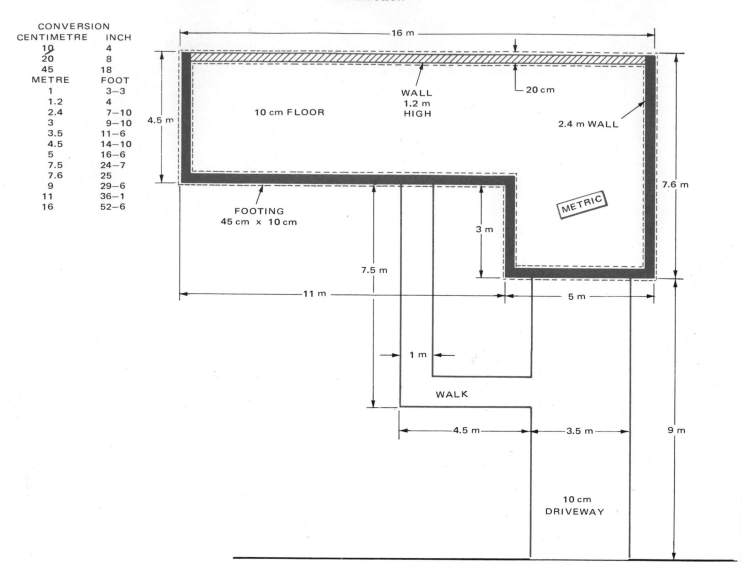

CONVERSION	
CENTIMETRE	INCH
10	4
20	8
45	18
METRE	FOOT
1	3—3
1.2	4
2.4	7—10
3	9—10
3.5	11—6
4.5	14—10
5	16—6
7.5	24—7
7.6	25
9	29—6
11	36—1
16	52—6

Fig. 12-6. Calculations for volume of concrete needed for a residence.

MATERIALS COSTS

Material is the single largest cost item of most construction projects. An estimate includes the cost of raw materials, mechanical equipment and all labor necessary to complete the construction job.

Amounts of material are calculated by the "takeoff" method. The estimator studies the plans and specifications counting quantities of each kind of material. Prices of materials are found in catalogs or by calling suppliers.

Most often, quantities must be computed after studying the working drawings. Quantities of each kind of material are determined and cost is computed.

CONCRETE

Concrete is estimated by the cubic yard or the cubic metre. The footing, foundation, basement floor, drive and sidewalk of our house are of poured concrete, Fig. 12-6. Using the same project we used in the excavation problem we will determine the cost of concrete needed to fill the form.

In English measure:

Footing	.11 yd. x	.49 yd. x	51.3 yd. =	2.8
Wall	.22 yd. x	2.6 yd. x	33.9 yd. =	19.4
	.22 yd. x	1.3 yd. x	17.4 yd. =	5.0
Floor	.11 yd. x	4.9 yd. x	12.0 yd. =	6.5
	.11 yd. x	8.3 yd. x	5.5 yd. =	5.0
Driveway	.11 yd. x	9.8 yd. x	3.8 yd. =	4.1
Sidewalk	.11 yd. x	1.1 yd. x	12.0 yd. =	1.5
			TOTAL	44.3 cu. yd.

In SI metric measure:

Footing	.10 m x	.45 m x	47.1 m =	2.1
Wall	.20 m x	2.4 m x	31.1 m =	14.9
	.20 m x	1.2 m x	16 m =	3.8
Floor	.10 m x	4.5 m x	11 m =	4.9
	.10 m x	7.6 m x	5 m =	3.8
Driveway	.10 m x	9 m x	3.5 m =	3.2
Sidewalk	.10 m x	1 m x	11 m =	1.1
			TOTAL	33.8 m³

Ready mixed concrete delivered to the site costs the contractor $25.00 per cu. yd.

44.3 cu. yd. x $25.00 = $1107.50

Fig. 12-7. Steel reinforcement used for floors and columns must be calculated in estimating. (American Hot Dipped Galvanized Assoc.)

STEEL

Reinforcing steel for the concrete may be either reinforcing bars or wire mesh, Fig. 12-7. Reinforcing bars are sold by the lineal foot. Price follows size. Our driveway is reinforced by No. 4 (1/2 in. diameter) reinforcement bars spaced at 2 ft. intervals running in alternate directions, with a total length of 360 lineal ft. (110 m).

Structural steel is handled by the contractor in one of two ways. Either he purchases the steel fabricated and erects it with his construction crew, or he has the steel company fabricate and erect the steel.

When estimating structural steel, you should add each item such as column bases, columns, trusses, beams and lintels separately. Structural steel is purchased by the ton, and the cost per ton varies depending on the type and shape of steel required.

The estimate of the field cost of erecting structural steel will vary depending on weather conditions, equipment available, size and fastening technique.

The beams for the basement in Fig. 12-8 weigh 18 lb. per

foot. A total of 69 lineal (running) feet are needed. The cost is $.29 per pound.

In English measure:

18 lb. per ft. x 69 ft. x $.29 per lb. = $360.18

In metric:

26.8 kg/m x 21 m x $.64 per kg = $360.18

This does not include the cost of erecting or fastening sections of the beam together on the building site.

LUMBER

Wood is a traditional material of construction. It is used widely in light construction and residential buildings. On large projects it is used for temporary work and in concrete form work.

Lumber commonly used for framing is called yard lumber. This is the classification given dimensional stock 2 to 5 in. (3 cm to 13 cm) thick in any width. The dimension given refers to rough lumber. It is measured before it is dressed or planed smooth. A 2 x 4 dressed actually measures 1 1/2 x 3 1/2 in. (3.8 cm x 9 cm). However, the number of board feet is calculated at 2 in. by 4 in. This is called the nominal size.

When any quantity of lumber is purchased, it is priced and sold by the thousand board measure. The estimator must calculate the number of board feet required on a job.

One board foot is equal in volume to a piece of wood 1 in. thick and 1 ft. square, Fig. 12-9. By using the following formula, the number of board feet can be determined.

$$\text{Board feet (N)} = P \times \frac{(T \times W)}{12} \times L$$

N = Number of board feet
P = Number of pieces
T = Thickness of lumber in inches
W = Width of lumber in inches
L = Length of lumber in feet

To calculate the board foot measure of 10 floor joists 2 x 10 in., 16 1/2 ft. in length:

$$N = 10 \times \frac{(2 \times 10)}{12} \times 16.5 = 275 \text{ board feet}$$

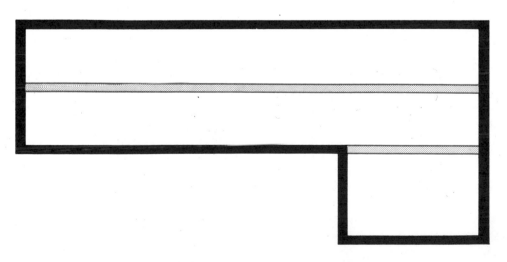

Fig. 12-8. Cost of steel beam placed on foundation wall, as shown in drawing, includes cost of materials as well as installation cost.

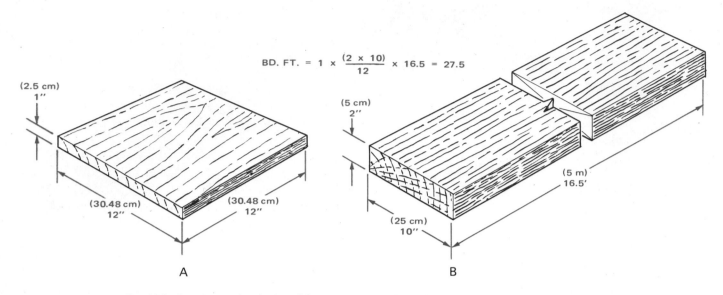

$$\text{BD. FT.} = 1 \times \frac{(2 \times 10)}{12} \times 16.5 = 27.5$$

Fig. 12-9. Calculations for the board foot measure. A—Board 1 in. thick by 1 ft. wide by 1 ft. long equals 1 board measure. B—Calculating board measure of yard lumber.

In SI metric:

$$N = 10 \times \frac{(5 \times 25)}{100} \times 5 = 62.5 \text{ BMM}$$

(Board Metre Measure)

Continue in metric:

To estimate the cost of 10 floor joists 5 cm by 25 cm by 5 m multiply the cost times the board metre measure. If the cost is $160 per 100 BMM:

62.5 BMM x $160 ÷ 100 = $100.

To determine the number of pieces required for a project, the space is divided by the distance used for spacing the individual pieces. For example, studs are commonly spaced on 16 in. centers. Therefore, the total length of partitions are divided by the spacing. To this add one more stud for each intersection of walls, two studs for each wall opening, and two extra studs for each corner. About five percent is added for waste. Rafters are calculated in basically the same manner. Plates are triple the lineal footage of walls because one runs along the bottom of the wall and two run along the top.

PLYWOOD

Cost of plywood for sheathing, subfloors and boxing is figured by the sheet or the square foot. A standard 4 by 8 ft. panel of plywood contains 32 sq. ft. In the metric system, the surface metre measure (SMM) is a measure of 1 metre by 1 metre disregarding the thickness. Cost is based on thickness and kind of material.

While dimension lumber is determined by the board foot, plywood quantities are usually given in the number of sheets needed. To determine the number of sheets needed find the area of all surfaces that are to be covered by the plywood. Find area by multiplying length by the width. Then divide the area by 32 since this is the number of square feet in each 4 by 8 ft. sheet of plywood.

Example: How much plywood will be needed for sub-flooring if the area of the floor is 1120 sq. ft.?

Sq. footage ÷ sq. footage of 1 sheet =
    No. of sheets needed
1120 sq. ft. ÷ (4 x 8 ft.) = No. of sheets needed
1120 sq. ft. ÷ 32 sq. ft. = 35 sheets

A sheet of 5/8 in. plywood underlayment costs $11.52. Thus: 35 sheets at $11.52 per sheet will cost $403.20.

If you were to use metric measure in the same problem, determine the area in square metres and then divide by 2.88. This is the number of square metres in a standard 1.2 by 2.4 m sheet of plywood.

If estimates are given in terms of square feet, the cost of each square foot of plywood would have to be determined. For example, plywood at $11.52 per sheet would cost $.36 a sq. ft.:

$11.52 ÷ 32 (sq. ft. per sheet) = $.36.

Insulation, roofing, drywall and siding are also estimated by square foot method.

LABOR COSTS

Labor cost is time multiplied by the hourly wage. This eliminates the effects of work days and work weeks that are longer or shorter than normal.

To complete the estimator's task, many variables must be considered. To be considered are the skills and mental attitude of the worker which may affect the length of time required to complete a certain piece of work.

Allowances must be made for variation in wages during the construction of the job, working conditions, availability of skilled and unskilled workers, climatic conditions and supervision. A worker seldom works continuously for 60 minutes of each hour.

The actual time will range from 30 to 50 minutes per hour. Time must be taken to start up the operation, for clean-up after the operation, coffee breaks and trips to the restroom and stops for a drink of water. These are some of the many variables which interrupt work. Waiting for materials and the scheduling of work contribute to the difficulty of preparing the construction estimate.

Estimators first identify the job or operation to be performed. They estimate the man-hours required to do the unit of work. They include all trades involved in the job — masons, carpenters, machine operators and laborers.

All workers will probably not spend exactly the same number of hours on the work. They may only be doing one part of the job. Once the hours worked by each trade have been estimated, they are multiplied by the wages per hour and totaled.

This is how it might be totaled:

Trade	Hour Rate	Hours Worked	Cost
Form setters	$ 5.25	10	$ 52.50
Concrete finishers	6.65	5	33.25
Carpenters	8.90	160	1424.00
Masons	10.05	25	251.25
		TOTAL	$1761.00

The estimator is responsible for deciding the number of hours required to perform a given unit of work. This can be done only through experience. If labor costs are uncertain, past project records are the most accurate source for preparing an estimate.

A second method of figuring labor costs is called the square foot of surface area. Consider brick for instance. The first four feet in height on a wall will require 15 mason hours per 500 sq. ft. of wall surface. The mason's helper will work 18 hours to mix mortar and bring mortar and brick to the mason for the same 500 sq. ft. The mason will lay the brick, square it and clean up at the end of the day.

The hourly wage is multiplied by the total hours, per worker to calculate cost per 500 sq. ft.

	Rate	Time	Cost
Mason	$10.05	15 hr.	$150.75
Mason's helper	4.50	18 hr.	81.00
Total labor cost for 500 sq. ft.			$231.75

MECHANICAL EQUIPMENT

Estimates of mechanical equipment such as electrical, plumbing, heating/cooling and other special fixtures are usually submitted by a single subcontractor. To work up the estimate, go through the plans and specifications systematically, item-by-item counting the quantity of each.

Identify each piece of equipment and obtain the manufacturers price. Again, experience will be the best guide to the hours of labor required to install each piece.

EQUIPMENT DEPRECIATION

As soon as a piece of equipment has been purchased, it begins to lose some of its value, Fig. 12-10. When this equipment is used on the project, it begins to wear out and one day will have become completely worn out or obsolete. If an allowance for depreciation is not included in the estimate, then when the piece is worn out, there will be no money set aside to replace it.

Depreciation can be figured on a yearly basis. For estimat-

Fig. 12-10. Cost estimates reflect expenses of owning and operating equipment needed. Sometimes equipment is rented. (South Dakota Dept. of Transportation, Div. of Highways)

ing the depreciation costs, you should assign the equipment a useful life. This is given in years or units of production.

If a unit of equipment had an original cost of $20,000 and an estimated life of five years, the depreciation value would be $4000. Depreciation of each piece of equipment must be estimated as part of the cost of completing the project.

OVERHEAD AND CONTINGENCIES

Overhead constitutes a large percentage of costs on the job. Failure to allow enough money for overhead has caused many firms to lose money and even go out of business.

GENERAL OVERHEAD

General overhead costs are not readily chargeable to any one project. They represent the cost of doing business and the fixed expenses that must be paid by the contractor. Fixed expenses include rent on office space, electricity, heat, office supplies, postage, insurance, taxes and telephone services. Salaries for office employees are also a fixed expense.

JOB OVERHEAD

Job overhead expenses are all the costs that can be readily charged to a project. Expenses incurred at the spot where construction is taking place are such costs. Construction workers wages, building permits, photographs, surveys, cleanup and winter construction expenses of temporary enclosures are good examples of job overhead.

CONTINGENCIES

Contingencies are those items of expense which are left out, not foreseen or forgotten on the original estimate. Almost every construction project has these items. See Fig. 12-11. In some

Fig. 12-11. On a complex project, it is possible to overlook certain unforeseen costs. Contractors allow for such additional costs in their bids. (American Plywood Assoc.)

cases, the items left out could not have been anticipated at the time the estimate was prepared.

Contingencies should not be the excuse for a poor estimate. The proper approach is to be as careful and thorough as possible in listing all items from the working drawings and specifications.

PROFIT

Profit is the amount of money added to the total estimated cost of the project. This amount of money should be clear profit. It compensates the company and its owners for the use of their money and for the risk they have taken in financing the company and managing the construction job.

Profit can be figured in several ways. Following are two acceptable methods:

1. A percentage is added to each item as it is estimated. Varying amounts are allowed for the different items from 8 to 15 percent for concrete, 3 to 5 percent for subcontracted work.
2. A percentage is added to the total estimated price of materials, labor, overhead and equipment. Twenty to 25 percent is normal on small projects; 5 to 10 percent on a larger project.

BIDDING

Bidding is part of a competitive method of selecting the builder for a project. Bids are prepared by the contractors according to the specifications submitted by the owner. Bid forms, like the one shown in Fig. 12-12, may be furnished by the owner.

Before bidding can begin, word must be spread to contractors that a project is ready for bidding. There are several ways to let this be known. Public advertising is required for many public contracts. The advertisement is generally placed in newspapers and trade magazines. Often, notice is also posted in public places.

Private owners often advertise in the same way to attract a larger cross section of bidders. Included in the advertisement is a description of the nature, size and location of the project, the owner, availability of bidding documents and bond requirements. Also given is the time, manner and place that the bids will be received.

SUMMARY

Building construction estimating is a method of adding up the probable construction costs of a given project.

The estimate is organized into several areas of cost:

1. Earthwork.
2. Materials.
3. Labor.
4. Mechanical equipment.
5. Equipment depreciations.
6. Overhead and contingencies.
7. Profit.

Bidding is the act of preparing a detailed estimate of building cost and submitting the estimate to the owner. Estimators prepare bid proposals. These bid proposals are evidence of the estimators finding the most economical way of doing a job. After considering all the costs, the contractor makes final decisions on what price goes into the bid proposal.

STANDARD FORM 21
DECEMBER 1965 EDITION
GENERAL SERVICES ADMINISTRATION
FED. PROC. REG. (41 CFR) 1-16.401

BID FORM
(CONSTRUCTION CONTRACT)

REFERENCE
Invitation No.
DACW41-71-B-0010

Read the Instructions to Bidders (Standard Form 22)
This form to be submitted in

DATE OF INVITATION
13 August 19

NAME AND LOCATION OF PROJECT

Construction of Harry S. Truman
Dam, Stage III
Harry S. Truman Reservoir, Missouri

NAME OF BIDDER *(Type or print)*

(Date)

TO: DEPARTMENT OF THE ARMY
Kansas City District, Corps of Engineers
700 Federal Building
Kansas City, Missouri 64106

In compliance with the above-dated invitation for bids, the undersigned hereby proposes to perform all work for construction of Harry S. Truman Dam, Stage III, Harry S. Truman Reservoir, Missouri,

in strict accordance with the General Provisions (Standard Form 23-A), Labor Standards Provisions Applicable to Contracts in Excess of $2,000 (Standard Form 19-A), specifications, schedules, drawings, and conditions, for lump sum and unit prices set forth in the attached Bidding Schedule.

The undersigned agrees that, upon written acceptance of this bid, mailed or otherwise furnished within 60 calendar days XXXXXXXXXXXXXXXXXXXXXXXXXXXXXXXXXXXXXXXXXXXXXXXX after the date of opening of bids, he will within 10 calendar days XXXXXXXXXXXXXXXXXXXXXXXXXXXXX after receipt of the prescribed forms, execute Standard Form 23, Construction Contract, and give performance and payment bonds on Government standard forms with good and sufficient surety.

The undersigned agrees, if awarded the contract, to commence the work within XXXXXXXXXXXXXXXXX 10 calendar days after the date of receipt of notice to proceed, and to complete the work within the completion dates XXXXXXXXXXXXXXXX after the date of receipt of notice to proceed. XXXXXX specified in the SPECIAL PROVISIONS.

RECEIPT OF AMENDMENT: *The undersigned acknowledges receipt of the following amendments of the invitation for bids, drawings, and/or specifications, etc. (Give number and date of each):*

The representations and certifications on the accompanying STANDARD FORM 19-B are made a part of this bid.

ENCLOSED IS BID GUARANTEE, CONSISTING OF | IN THE AMOUNT OF

NAME OF BIDDER *(Type or print)* | FULL NAME OF ALL PARTNERS *(Type or print)*

BUSINESS ADDRESS *(Type or print) (Include "ZIP Code")*

County of ____________________

BY *(Signature in ink. Type or print name under signature)*

TITLE *(Type or print)*

DIRECTIONS FOR SUBMITTING BIDS: *Envelopes containing bids, guarantee, etc., must be sealed, marked, and addressed as follows:*

MARK:
Bid under DACW41-71-B-0010
To be opened 24 September 19
Receipt of Amendments No. ____
______________acknowledged.

ADDRESS: DEPARTMENT OF THE ARM
Kansas City District,
Corps of Engineers
700 Federal Building
Kansas City, Missouri 64106

Fig. 12-12. Both sides of a bid form used in bidding on a public project.

CAREERS RELATED TO CONSTRUCTION

ESTIMATOR or PROJECT ESTIMATOR studies working drawings of proposed structure to determine costs of materials, labor, workmanship and overhead charged to a building project.

CONSTRUCTION COST ENGINEER or COST ANALYSIS ENGINEER is a specialist in estimating materials and labor costs especially on large construction projects. May also analyze changes in materials, construction methods, etc., to learn their effects on construction costs. Makes recommendations on methods and materials to be used.

CONTRACTOR makes contract to build for others. Owns machines and equipment and supervises construction workers. Purchases materials and may prepare estimates.

LISTER computes lumber, material and labor costs for all types of millwork manufactured by woodworking firm.

DISCUSSION TOPICS

1. What does a project estimator do?
2. How do weather and climate affect the estimate?
3. How would rock affect the cost of excavation?
4. What is meant by competitive bidding?
5. What is profit?
6. Why is equipment overhead added to the estimate?
7. What is a contingency?
8. Figure the cubic yards of concrete in a floor 56 x 75 ft. poured 6 in. thick.
9. What is the cost of 100 2 x 4's 8 ft. long priced at $180 per thousand board foot measure?
10. What is a subcontractor?

Chapter 13

SCHEDULING

Work scheduling is the responsibility of the contractor. Upon notice of the contract award, preparation of a job progression chart begins. Schedule charts are detailed studies of how the contractor will use his labor force, equipment, materials and money to do the work outlined in the working drawings, specifications and estimate. Fig. 13-1 shows lumber packaged and ready for delivery "on schedule."

Scheduling is having the materials and skills necessary to complete a particular job at the site, when it is time to perform the job. Nearly every unit of work on a project depends upon the completion of the previous unit of work.

ANALYZING THE PROJECT

Regardless of the project's size, planning the work is a necessity. There is no use in receiving the roofing material on the construction site before the foundation is poured. The mechanical equipment should not arrive on the site before it is to be installed. Weather or construction damage could impair the use of that equipment in the meantime. By the same token, if the job is delayed because material, equipment or men do not arrive on the site at the scheduled time, then the job will be slowed. Cost to the contractor and owner will thereby be higher.

SCHEDULING TOOLS

Scheduling a project is possible through understanding the individual units of work required. A project is made up of major tasks like excavation, building of foundation, erection of frame, applying exterior walls, roof, electrical, mechanical, and finishing the interior. The scheduler estimates how long it will take to complete each of these tasks. He works out a time table and prepares a work schedule. A schedule can be organized by one of several methods:
1. The Critical Path Method.
2. The bar chart.
3. The progressive chart.

CRITICAL PATH METHOD

The Critical Path Method (CPM) is the most economical way to plan all operations to meet desirable completion dates. It provides a means of assessing the effect of all variation —

Fig. 13-1. Proper scheduling assures movement of materials to the site at the time of need. (Allis-Chalmers)

changes, extra work, or deductions — upon the time of completion and upon the cost of the work. It is an open-ended process that permits different degrees of involvement by management to suit their various needs and objectives.

In applying CPM to construction planning, it is necessary to have accurate estimates of time and costs for each operation. (See Chapter 12.) The breakdown of the project into its individual operations may be simple or detailed, as desired. The essential requirement is that the direct cost for each operation be estimated separately.

In the network or arrow diagram, Fig. 13-2, each line or arrow represents one activity. The relationship between activities is shown by the arrows. Each circle represents an event. The event depends on the completion of all activities entering the circle. Length of the line between circles has no meaning except to show passage of time.

THE BAR CHART

This method prevents delays. The contractor makes a complete analysis of the work to be done. Then he prepares a bar chart as shown in Fig. 13-3. This chart was prepared for major concrete work on the Harry S. Truman dam. In a

Construction

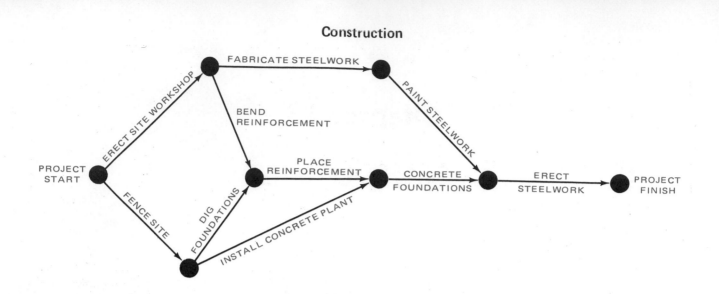

Fig. 13-2. Critical Path Method of job analysis traces major events in construction of small project.

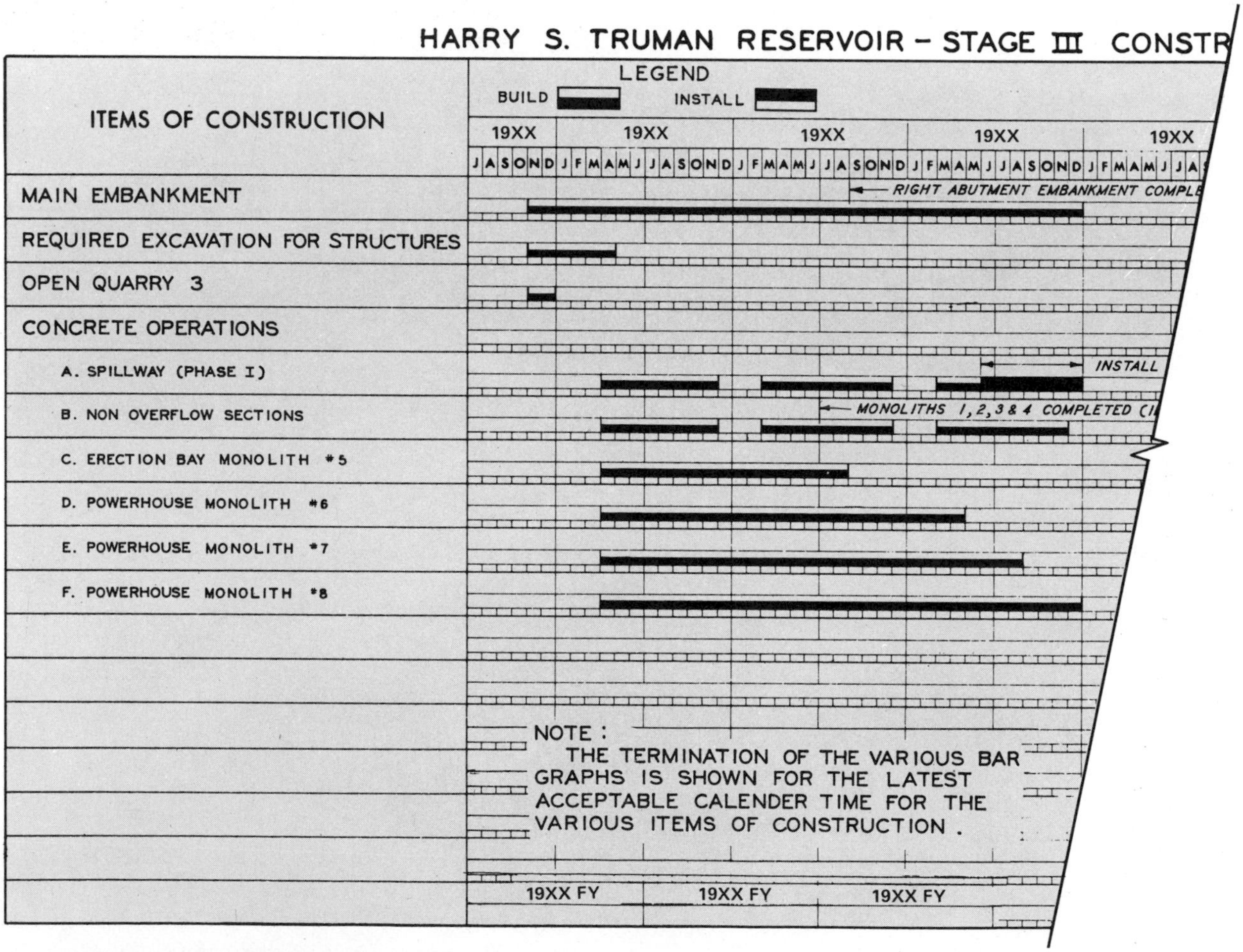

Fig. 13-3. Bar chart. Black bars are time modules for building of parts of power house. Note from legend at top that placement of heavy inked section of bar indicates either building or installing.

project of such great size, Fig. 13-4, time is scheduled in much longer periods than would be necessary in a smaller project. The bar chart is capable of providing sufficient lead time to stock pile materials, in advance of need. Fig. 13-5 shows need for large stock pile. A double bar, Fig. 13-3, shows whether the work is construction or installation of equipment or pre-assembled units.

The bar chart is really a time chart. The contractor may also break down large parts of the work into smaller tasks. He will then estimate the cost and percentage of the whole represented by each item in the total job. Fig. 13-6 shows how this break down might appear.

Fig. 13-4. Exact scheduling and timing is essential to large, complex projects.

Fig. 13-5. Large spaces are necessary for stock piling materials of construction.   (Corps of Engineers, Kansas City District)

INSTALLATION OF TAINTER GATES

ITEM	COST	PERCENTAGE OF JOB	TIME (IN DAYS)
DELIVERY FROM SUB-CONTRACTOR	$50,000	50	216
STORE AT SITE	1,000	1	10
WELD FITTING	10,000	10	14
MOVE TO ERECTION SITE	2,000	2	4
HOIST TO POSITION	5,000	5	4
PLACE AND FIT	12,000	12	14
PRIME COAT	3,000	3	6
THREE COATS, PAINT	17,000	17	36

Fig. 13-6. Builder may prepare list of work to be done, cost and time required to build or install. This list breaks down tasks for installing tainter gates for a dam and power plant.

PROGRESS CHART

Cost of the job, rather than time elapsed or amount of work actually done at the site is the concern of the progress chart shown in Fig. 13-7.

The heavy line represents progress the contractor had planned to make before work started. The thin, broken line is progress actually made.

To plot the graph on the progress chart, first place a fairly heavy dot directly above the time elapsed and directly across from the percentage of dollars spent at that time. Place similar dots representing other periods of time and percentage of cost.

Then draw a series of straight lines connecting each dot to the next. This is called the graph line.

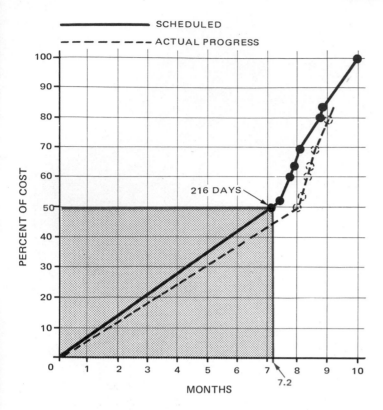

Fig. 13-7. Progress chart above was set up by contractor to keep track of money and time spent on each part of the job. Heavy black line was prepared to estimate cost and time. Broken line plots actual progress. Brown lines show how location of points is determined by intersection of vertical and horizontal lines extending from coordinate numbers.

Speaking in technical language, the contractor may call the numbers on the chart "coordinates." Coordinates are simply a set of numbers that lead to a point. Looking at the graph, you can see that the brown vertical line from number 7.2 intersects with the brown horizontal line coming from 50. Numbers 7.2 and 50 are coordinates because they determine where a point will be placed on the chart.

The progress chart represents a subcontractor who is assembling tainter (radial) gates for the construction of the

Truman dam. He was to deliver to the construction site in 216 days. In reality he took 246 days. This is shown by the second broken line. The contractor had to adjust progress to bring the graph lines back together. The adjustment was an increase in the labor force, thereby decreasing the time necessary to install the tainter gates.

Progress charts may need to be revised and updated several times during construction. It is often better to set up a new chart after the job is under way to keep the schedule accurate.

SUMMARY

Scheduling is the result of a thorough study of the construction project. This schedule is based on the contractor's idea of the best way to use men, money, materials and equipment. The contractor lists each part of the work along with an estimate of time and costs. Next comes preparation of bar charts and progress charts. These plot and keep a record of how work is moving toward completion.

CAREERS RELATED TO CONSTRUCTION

CONTRACTOR contracts to perform construction work. Estimates costs and time. Purchases materials for construction. Supervises work directly or through supervisors.

CONSTRUCTION SCHEDULING ANALYST estimates time needed to complete parts of jobs and coordinates delivery of materials. Also may coordinate completion of subcontracted work.

DISCUSSION TOPICS

1. What information does a bar chart contain?
2. What is scheduling?
3. What can the contractor do if the time schedule begins to lag behind estimated time?
4. How is a construction project analyzed for scheduling?

Chapter 14

UTILITIES

Utilities are services useful to the public. Especially necessary are electric power, gas, water, communication and sewer. Every residence and building where people live, work and enjoy their leisure time, is expected to have these services.

Utilities are a vital part of the basic construction of many types of structures. Bridges, streets and highways need lighting and traffic control to function safely. Highways need drainage control. City parks need water for fountains.

In many building projects, much of the total cost can be attributed to the installation of utilities. Yet, in almost every case and in spite of the cost, their installation is demanded by modern living. From the standpoint of community living, a city obviously cannot operate without the services of utilities.

Providing utility services is a construction project in its own right, Fig. 14-1. Electric power must be transmitted great distances over land to the user. Water and gas is transported by pipelines. Communications may be beamed through the air to receivers many miles away, or transmitted by wire over land or underground.

In our larger cities, utilities are an unseen servant, Fig. 14-2, continuing to serve any who request their use. These utilities come from passageways deep underneath the streets out of the way and out of view of the people who accept these services.

WATER

A community obtains its supply of water from a surface reservoir, deep drilled wells which tap the subterranean water tables, or from a lake or river, Fig. 14-3. The stored water is purified, then pumped through underground pipelines or water mains and is distributed by a branching system to each home, store, factory, school and office in the community.

Fig. 14-1. Providing utility services has become a major construction project in its own right.
(American Hot Dipped Galvanizers Assoc.)

Fig. 14-2. Utilities of the city are located beneath the street, out of view and out of danger of being interrupted.   (IDS)

Fig. 14-3. Reservoir for water, for neighboring area. The stored water is purified, then pumped to the consumer.
(Howard, Needles, Tammen & Bergendoff)

Fig. 14-4. Concrete pipe is layed to serve as a sanitary sewer from a housing project. (Northwest Engineering)

Water is distributed throughout the construction project to provide for drinking, washing services, fire protection and to operate the sewerage system. The water company or municipal utility service supplies the water to the project. From each property line, the contractor installs a system of smaller lines to connect each plumbing facility.

The craftsman who works to install the water distribution system is called a plumber. At the property line, or just inside the new structure, a meter is installed by the water company to measure the amount of water flowing into the new system. The measured amount is used to assess the owner for this service, based on an established rate of cost per gallon.

SEWERAGE

Sewerage is a system installed to carry off excess water or waste water. Sewerage is of two basic types:
1. Sanitary sewers.
2. Storm sewers.

The sanitary sewers carry waste water away from plumbing fixtures to a far away sanitation treatment plant which disposes of the solid waste material and dispels clean purified water into the natural drainage systems, Fig. 14-4.

Storm sewers are a system of drainage ducts which collect rain water and discharge it into the natural drainage system untreated. The storm and sanitary systems are two completely separate water carry-off systems. Both operate by gravity, or by gravity and pump. Pumping is provided by lift stations spaced at the lowest gravity points.

ELECTRICITY

Electrical energy is the service of power. It provides: power to operate lighting, dishwashers, clothes dryers, heating units, air conditioners and a multitude of other conveniences; power to operate traffic control systems, elevators and escalators.

Fig. 14-5. Huge generating plants are constructed to convert fossil fuels into electrical energy.
(Missouri Public Service Co.)

Fig. 14-6. Steel towers support high voltage conductors which transmit power cross-country from source to consumer. (American Hot Dipped Galvanizers Assoc.)

Electrical energy is distributed throughout the construction project by a network of copper wire, aluminum wire or other conductors coated with insulating material.

Electricity is produced by generators that convert other sources of energy into electrical energy, Fig. 14-5. Power generating plants produce an electrical current by converting falling water, coal, oil, gas or atomic fusion into electrical energy.

Electrical energy is easy to transport to the consumer through a series of conductors supported by steel towers or wood poles, Fig. 14-6. Electrical energy is transmitted from generating plants to consumers at a high voltage. When the electricity reaches the consumer in the city, its voltage (electrical force in the line) is reduced by a transformer. The line is branched to many locations by a substation near the city, Fig. 14-7. From the substation, the conductors may be strung on poles or buried underground. Each branch of the system carries electricity to a local area within the community, where it is reduced again. At this time, the transformer reduces the voltage to 240 volts and branches to each house. At the residence, a meter measures the amount of energy used by each consumer. From the meter, attached to the exterior of the home, it enters a distribution panel. From this panel, a third system of branching conducts the electrical energy to each service device.

Workers employed by the electrical subcontractor install the interior utility service to distribute the power to each light fixture, appliance, air conditioning, heating and all outlets.

GAS

Gas is a fuel which heats and cools our homes, cooks our meals and serves in many ways in the home and industry where heat is necessary. Some gas fuel is manufactured from coal. However, the major portion exists in the form of natural gas from underground sources.

Natural gas is produced from wells drilled deep into the earth. From the gas field, high pressure (either from the natural source or by pumps) forces the gas through pipelines

Fig. 14-7. Well planned and durably constructed substations help distribute electrical energy to consumers. (American Hot Dipped Galvanizers Assoc.)

Fig. 14-8. Sections of pipe are welded together before burial; one of many jobs necessary to bring gas company services to users.   (Exxon)

to the consumer, Fig. 14-8. In the community, the pressure is reduced and gas is distributed throughout the community by gas mains located beneath the streets. At each building, the gas pressure is further reduced and metered. A subcontractor for plumbing usually installs the distribution system to pipe gas to each gas burning appliance.

COMMUNICATION

Communication systems consist of the telephone, telegraph, television and radio. These systems may be divided into two classifications:
1. Personal.
2. Mass.

Personal communication systems, primarily the telephone permit one to make contact with any other person and carry on a conversation, throughout the United States and most of the world. The telephone system consists of several means of transmitting voice contact: lines over land, underground, beneath the ocean; beamed by radio or bounced off a satellite located high above the earth. Mass communication systems, television and radio are media capable of getting messages to the majority of the people. A one-way communication exists.

From the telephone in your home, wires lead to a central location where a bank of switches automatically connect you with the telephone having the number you are calling.

Construction of the communication systems is completed by a joint effort of the subcontractor and the utility company to install all of the wires and receivers necessary.

SUMMARY

Utilities, water, sewerage, gas, electricity and communications provide services essential to maintain buildings and structures. Utilities make living more comfortable and take the drudgery out of many menial tasks. Utilities attribute to our standard of living. Installation of these services contribute a large part of the cost of the project.

CAREERS RELATED TO CONSTRUCTION

ELECTRICIAN follows blueprints and specifications to assemble, install and wire electrical systems for heat, light, power, air conditioning and refrigeration components. Some also install electrical machinery, equipment, controls, signal and communications systems. Maintenance includes troubleshooting, test and repair.

PLUMBER "roughs in" pipe systems that carry water, steam, air and other liquids and gases, then installs and connects plumbing fixtures during final stages of construction. Installations also include heating and refrigeration units, appliances and waste disposal systems.

DISCUSSION TOPICS

1. List each of the utility services you have in your home, school and community.
2. Describe any local utility construction projects you have noticed in your community.
3. List the mass communication systems in your community.
4. Identify a structure in your area which has none of the utilities.
5. What is a pipeline?
6. What are conductors?
7. How does the cost of installing utilities compare to the total cost of structure?
8. List some services, other than those noted in this chapter, which may be installed in a building (for instance, a hospital).
9. What is a storm sewer?
10. How is gas obtained?

Chapter 15

INSPECTIONS

Fig. 15-1. Inspection is the responsibility of construction management. From drawing to completion, inspections are sequenced with the progress of the building project. (Fruin-Colnon)

Inspection of a structure under construction involves a thorough investigation and careful evaluation of the technique, materials and workmanship being carried out by the contractor.

Inspections are a management function, featuring the examination of construction projects at various intervals, Fig. 15-1. A progressive inspection timetable should be a part of the construction work schedule, planned and conducted by people with knowledge and authority. Construction management uses the inspections as a method of controlling, monitoring, reporting and correcting the construction process.

As a result of an inspection report, the finished project is either accepted or rejected by the owner. Rejected work is corrected and a final inspection is made to determine if the quality of materials used and the craftsmanship are equal or superior to those stated in the specifications.

Inspection schedules start as early as the planning stages and continue until the project is completed.

THE INSPECTOR

People who serve as inspectors for the construction project fall into three general groups:
1. Those who serve the architect or engineer.
2. Those employed by the contractor or subcontractor.
3. Those who conduct inspections for outside agencies.

During the planning stages when drawings and specifications are prepared, the architect or engineer carefully reviews the work to be done. If the project is private, the plans are submitted to the city, county or state building department. Here the building engineer reviews the proposed project carefully to see if the construction plan is in keeping with all building codes.

The owner may inspect the project as it progresses, or employ an inspector to make periodic inspections. Often, the architect who designed the building is hired to inspect the

Fig. 15-2. All phases of construction are inspected by various inspectors for a variety of purposes related to construction, materials, technique and workmanship.   (Wire Reinforcement Institute)

Fig. 15-4. This intricate aluminum panel for the exterior wall of a high rise building is given a close inspection.   (Cupples Products)

work. *In any case, inspections are made to insure that the specificiations are being followed to the letter of the contract by the contractor.*

The construction superintendent is, in effect, an inspector employed by the contractor and subcontractor, Fig. 15-2. The superintendent will make certain that the floor joists and beams are level, the walls are plumb and inspect all details of a project to insure that the crews continue to turn out high quality work.

The individual worker on the project is an inspector, too. Anyone who takes pride in the finished product is an inspector, Fig. 15-3.

Agencies and organizations other than the architect and contractor also provide inspection service. Insurance companies may send their inspector to the construction site to monitor the progress and see if the work is properly and safely done. If an insurance company is to provide insurance after the structure is complete, the company will insist that proper materials are used and correct installation procedures are followed.

Inspections are made by various governmental agencies. U.S. Government inspections are concerned with health and safety precautions. The city building inspector inspects the

Fig. 15-3. All of these workers are proud of their contribution to the project when it is completed.
(Caterpillar)

structure to see if local building codes are being met. For example, a local building code specifically calls for certain plumbing practices and materials; the inspector checks the subcontractor's work for code violations.

The loan agency, bank or other lending institution usually sends an inspector to the constructing site to observe the progress being made on the project.

The manufacturer of construction materials and components makes periodic inspections before shipping them to the construction site, Fig. 15-4. Even *before* the manufacture of certain building materials and components, destructive tests are conducted to better understand the materials from which each is manufactured, Fig. 15-5. From the results of these tests, the inspectors can predict the limitations of each material or component.

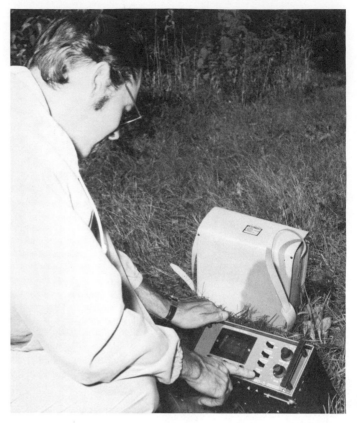

Fig. 15-6. Soil inspection is made by means of electronic equipment designed for quick and accurate subsurface exploration. Inspection results give depth of overburden, soil types and elastic qualities. (Soiltest, Inc.)

Fig. 15-5. Fire test apparatus used to determine fire resistance of roof-covering material. (Underwriters Laboratories)

Each inspector makes inspections for a specific purpose. The worker inspects for pride of satisfaction; the lending agency inspects to protect its loan; the contractor inspects to protect its employees and the owner; the city inspector inspects to make sure all construction practices comply with the local building codes. The inspector, in each case, must be completely familiar with the activity or area of construction being inspected.

QUALITY CONTROL

Quality control, to most people, means the manufacture of a product of high quality. To the manufacturer of construction components, it means producing components of consistent quality. To a contractor or subcontractor, it means building the best structure for the cost.

Actually, the process of inspection is a means of locating inferior products, material and craftsmanship. Inspection, therefore, serves the purpose of quality control by saving time

and further effort by discovering the fault early. Inspection is not only a remedial measure, it also is preventive.

The inspector's job is to find out *why* the job is substandard or inferior. In the construction field, it takes effort on everybody's part to insure that the job is completed in a craftsmanlike manner.

THE INSPECTION

Inspection covers nearly every phase of construction:
1. Materials.
2. Techniques of construction.
3. Workmanship.

Materials to be inspected include soil tests, Fig. 15-6, size and composition of aggregates, cement, glass, tile, lumber, metals, roofing material, flooring and every material thing that goes into the construction job.

Inspection calls for testing and evaluating structural techniques, Fig. 15-7. It also includes testing and evaluation of all heating and cooling equipment, and all plumbing and electrical equipment.

Workmanship is inspected in concrete finishing, reinforcement bar placement, framing carpentry, electrical and plumbing installation and all other work done on the structure. *Every phase of construction comes under the watchful eye of an inspector.*

Fig. 15-7. A concrete floor panel is lifted from pylons where it has been air drying for two weeks. The panel is about to be tested for strength. (Underwriters Laboratories)

SUMMARY

Inspection is a necessary part of construction, to provide the highest quality of construction techniques and workmanship. Inspections are necessary to insure the health and safety for employees during construction, and for the occupants upon completion of the structure. Inspections are progressive. From drawing and planning until the structure is finished, inspections are sequenced with the progress of the building. Inspections are conducted by various people and agencies for a variety of reasons.

CAREERS RELATED TO CONSTRUCTION

BUILDING INSPECTOR works for the city or village, inspecting new structures to see that all construction practices comply with local building codes. Since inspections are made in all areas of construction, the inspector must be knowledgeable in all phases of the building trades. Therefore, a trade background usually is required.

DISCUSSION TOPICS

1. Who inspects?
2. What is inspected?
3. What is quality control?
4. When are inspections made?
5. Why are inspections made?
6. What is a progressive inspection?
7. What is the professional background of a building inspector?

SECTION 2

THE TECHNOLOGY OF CONSTRUCTION PERSONNEL

Construction personnel has to do with on-the-job manpower requirements. This section explains how workers, materials and processes get the job done right and on time. The chapters that immediately follow cover recruiting, interviewing and hiring practices; the need for education and training both before and after entering the field; working conditions and union activity; economic rewards, job advancement and retirement considerations.

Company personnel builds sections of houses in production department of firm specializing in modular houses built in-plant for delivery to construction sites.

Bricks and other forms of masonry serve as basic building materials on major construction project. Note catwalk for bricklayers. (Bil-Jax)

Chapter 16

HIRING

Employee relations is a means of attracting good construction workers and getting them to perform their tasks well. Hiring new employees is one of the most important areas of employee relations.

First of all, prospective employees need to be assured of a safe and efficient environment (work site). Next, they must be made to feel that they will receive fair treatment (working conditions, wages). Finally, they need help in finding which particular job (or jobs) they like and do well.

Employee relations takes into consideration all practices that contribute to employee satisfaction, while getting construction jobs done right and completed on time. Employee relations can be divided into four areas:

1. Hiring.
2. Education and training.
3. Working conditions.
4. Job advancement.

Hiring will be discussed in this chapter. Other areas will be covered in succeeding chapters.

Fig. 16-1. Job description spells out skills and knowledge needed by worker to do the job. Prospective worker is hired if skills match job description.   (United States Steel)

RECRUITING AND HIRING

Recruiting construction workers means actively seeking new employees. It is an important early step in the hiring phase of employee relations. Each construction worker applying for a job has been hired at least once. Some are hired each season.

Before any recruiting strategy can be set up, management must first assemble data concerning the type of work newly hired employees would be expected to accomplish. From this "job analysis," a "job description" is written. Job descriptions include statements relating to general duties, Fig. 16-1, and specific tasks the worker must perform. From these given tasks, management may then preselect a labor source most likely to yield a prospect with the desired skills.

Once the job is analyzed and a likely labor source is identified, a number of recruiting methods may be used to locate a prospective employee. Some common recruiting methods include: advertising in trade magazines and newspapers; working with school, public or commercial employment agencies; doing extensive field recruiting; making an internal search within the present employment force; going to trade meetings; recruiting from a related company; securing direct applications from prospects.

WANT ADS

Placing an advertisement in appropriate trade publications is a relatively inexpensive, yet effective manner to communicate with a select portion of the population. A more popular recruiting method among construction firms is to use local newspapers to advertise for help. Newspaper want ads, Fig. 16-2, cover a large, general labor market.

PLACEMENT AGENCIES

School, public and commercial placement agencies are popular means of locating certain kinds of construction workers. One frequently voiced complaint is that referrals (prospects provided) do not match the requirements of the company. Usually, this is a problem of lack of communication as an inadequate or incomplete job description is submitted.

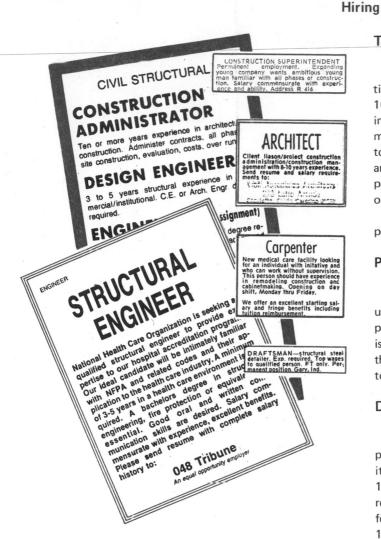

Fig. 16-2. Want ads carry brief job descriptions designed to attract strong response from prospective workers.

THE INTERNAL SEARCH

Probably one of the most overlooked sources of construction workers is within the contractor's organization itself, Fig. 16-3. The internal search for a specific type of worker has one important advantage. It creates and sustains high company morale and effort. If present employees know they are likely to move up through promotion to an existing vacancy, they are likely to put forth greater effort. However, hiring outside personnel without considering internal promotions can have an opposite effect on morale.

Employees are likely to have a "care less" attitude if positions are always filled from outside of existing personnel.

PROFESSIONAL MEETINGS AND CONVENTIONS

Professional trade meetings and conventions sometimes are used for recruiting purposes, although they do provide a ready pool of prospective employees. The obvious shortcoming here is that participants of the meetings probably are satisfied with their present positions. They are top prospects, but not likely to shift jobs.

DIRECT APPLICATION

Direct application may not qualify as an active recruiting practice, yet a construction company may exert control over it. Common methods of control are an application form, Fig. 16-4, appropriate written test and a personal interview. When releasing application forms, an employer should receive the following basic information from it:
1. Whether the applicant has the mental aptitude to perform the work.
2. If he or she is willing to do the work.
3. If, after additional training, he or she has the aptitude to do

Fig. 16-3. Many higher paying jobs in construction trades are filled from the lower ranks of the work force. The possibility of advancement is a good work incentive.

FIELD RECRUITING

Field recruiting is done by a representative of the construction firm. This recruiter covers college campuses in search of appropriate graduates, and visits military separation centers in search of men or women with construction experience.

Fig. 16-4. To obtain this job, construction worker applied directly to employer, took a series of tests, presented the recommendations of previous employers and gave evidence of having mastered the trade.
(Exxon)

the work without close supervision.

4. If the applicant is able to work in harmony with fellow employees.

In addition, the employer may want to know why the applicant left the previous job. A sampling of the applicant's attitudes is also helpful. An application form covering most construction jobs would be difficult to produce. Yet, it is in the company's best interest to do so.

SELECTING A WORKER

Selecting the most likely individual for a given job is not an easy task. It is especially difficult if many of the applicants have similar skills. In this situation, the applicant's desire to work and ability to get along with fellow workers must have great bearing on the final selection.

TESTS

Test results can be helpful to the construction firm. By comparing test scores of the applicants to actual job performance, the company will be able to predict how well the applicant will perform. Sometimes, however, job performance depends on qualities that are hard to identify. Tests are expensive and difficult to construct and administer. Yet, even with these shortcomings, tests are valuable when used in conjunction with other methods of selection.

PERSONAL INTERVIEW

Possibly the most important employee selection tool is the personal interview. Much can be learned by an experienced interviewer. Often the decision to hire or pass over an applicant is made during the interview. The applicant may give undesirable answers, or lie in an attempt to please the interviewer. A trained interviewer is able to and is expected to detect this type of deception.

SUMMARY

Recruiting and selecting the right candidate for the job is often a long, laborious task. The more complex the skill requirements, the greater the task of recruiting. The greater the demand for certain skills, the greater the task of selecting the most skillful applicant. Hiring a person to fill the position is the last activity. *Hiring is the result of a successful program of recruiting and selecting.*

CAREERS RELATED TO CONSTRUCTION

RECRUITER assembles data concerning job opportunities; compiles job analysis and writes job descriptions. Recruiter then considers labor sources, places want ads, attends trade meetings and conventions. FIELD RECRUITER covers college campuses and military separation centers looking for prospective employees among graduates and veterans being discharged.

INTERVIEWER works closely with DIRECTOR OF PERSONNEL, conferring about information contained in job application forms before conducting in-depth personal interviews with good prospective employees.

DIRECTOR OF PERSONNEL handles most employee relations activities and maintains employee records, furnishes application forms and arranges personal interviews for job applicants. Director of personnel generally directs duties of RECRUITER and INTERVIEWER; usually is in charge of training and recreation programs.

VOCATIONAL COUNSELOR advises employees on work-related problems; recommends two to five year apprentice training program, supplemented by classroom instruction. Counselor may work on joint apprenticeship committees to establish minimum standards of education, experience and training.

DIRECTOR OF PLACEMENT works for school, public or commercial placement agency; studies job descriptions furnished by employers, then attempts to "place" qualified individuals in jobs that fill the needs of both parties.

DISCUSSION TOPICS

1. List several methods that a construction contractor may use to recruit job applicants.
2. Describe an interview. Who conducts the interview?
3. What is meant by direct application?
4. Why are test results helpful?
5. What is a job description?
6. Why is a positive attitude important to any job?
7. How would you go about applying for a construction job?
8. What are "recommendations?"
9. Who is responsible for hiring personnel?
10. If you were a construction superintendent, how would you set out to locate prospective workers?

Chapter 17

EDUCATION AND TRAINING

Fig. 17-1. Workers place their safety and well being in each others' hands, confident that everyone has been trained to do the job safely. (A. B. Chance)

Each morning as the nation's construction work force gears up for the day, thousands of workers begin their tasks with confidence knowing that they are good at their jobs. They have gained this confidence by:
1. Several years of training.
2. Lengthy practice in their craft, Fig. 17-1.

This skill may be practiced at a top managerial position or as a custodian. Every worker has a similar basis for his confidence and competency. It is the amount and variety of education that establishes each person in his chosen position.

In today's world of work, education and training must come before work experience, Fig. 17-2. These training requirements are prime concerns of management. While an individual may profit from his mistakes (if he lives through them), it is not likely that construction management will realize any benefit from them. Through education and

training, both the individual and his company's management can experience success.

Notice that we refer to "education" and "training" as separate things. Employers view them as separate functions with different methods and different goals. Education, for the individual, is concerned with a change in thinking. Through education we learn more about our job. We will change our attitudes or develop better problem-solving ability.

Fig. 17-2. At each level of construction — architect/designer, engineer, fabricator, and installer — certain degrees of education and training are required to guarantee success of the project. (The Port of New York Authority)

Training, on the other hand, is concerned with improving job skill and increasing manual dexterity. These are separated for economy and efficiency of operation. Unlike the public school systems, company schools have to prove themselves. They must add to the company's profit margin or be closed.

PROFESSIONALS

Professional occupations are those requiring specialized education with four or more years of college preparation. In construction the architect, engineer, some technicians and some managers are considered professional.

To gain the knowledge and skill necessary to do designing and engineering work they must attend a college or university for four and five years. During this time they study mathematics, science (in the areas related to architecture or engineering) drafting, electrical circuitry, and the design and construction of foundations.

To graduate from an architectural or engineering school, students usually must have a "C" average or better. A successful architect or engineering student must have intelligence and a desire to learn. A good high school background is needed with good grades in mathematics and science. Engineers study the strength of materials and how they are going to react on given load stresses, Fig. 17-3.

Fig. 17-3. Here an engineer operates precision test instrument to test strength of new adhesive. (Franklin Glue Co.)

Construction management is a relatively new occupation. Many architects and engineers also work as managers or contractors and many experienced in the trades become supervisors. Some managers study professional construction management courses in college. They are neither engineers nor tradespeople. They plan, organize and control the actions of others in the construction industry.

PRODUCTION WORKERS

On-the-site construction workers are recruited from four sources:

1. Walk-on workers who join the construction activities when work is booming. They are usually unskilled.

Fig. 17-4. The apprenticeship program provides the trainee with basic job entry skills. (Dallas Independent School District)

2. Young workers who join the trade and work as helpers or laborers.
3. Apprenticeship trainees who wish to become skilled in one of the construction trades.
4. Skilled workers who are already trained and qualified.

The apprenticeship program is responsible for supplying the construction trades with skilled workers. During their apprenticeship, Fig. 17-4, the young trainees get 6000 to 10,000 hours of training on-the-job. In addition, they will receive 144 hours per year of classroom instruction every year. See Fig. 17-5.

Labor unions and contractors work together in establishing an apprenticeship-trainee program. The programs are set up to train young people between the ages of 17 and 25. High school education in mathematics and science is required for the successful completion of an apprenticeship program. The apprentice should have mechanical aptitude and should enjoy working with tools.

The apprenticeship program is usually supervised by the Joint Apprenticeship Committee consisting of representatives of local employers and unions. The committee has several responsibilities:

1. Deciding how many apprentice-trainees are needed.
2. Determining what standard of education, training and experience is necessary.
3. Locating employers who have the equipment and facilities to provide training.
4. Settling disputes between employers and apprentices.
5. Certifying the apprentice as a journeyman upon completion of program.

Fig. 17-5. Classroom instruction is a very important part of apprenticeship programs. (American Iron and Steel Institute)

This committee also registers the programs with the state apprenticeship agency and with the U.S. Department of Labor, Bureau of Apprenticeship and Training.

A young person, wishing to shorten the period of apprenticeship before joining the building trade, may take pre-apprenticeship instruction in a high school, trade school or junior college. This does not apply to all trades. The joint committee or his employer may allow credit for this pre-apprenticeship work if the apprentice does well in job-related subjects. This pre-apprenticeship instruction usually enables the trainee to qualify for a higher rate of pay during a shorter period of apprenticeship. If you are interested in learning more about the apprenticeship program, contact your guidance or vocational counselor. Fig. 17-6 shows educational needs of all types of construction workers.

SUPPORT PERSONNEL

Persons who work in the office or supervise the job are known as support personnel. Those who do office work, billing, purchasing, answer the telephone and serve as secretaries are a necessary part of the construction team. See Fig. 17-7. Receptionists, payroll clerks and file clerks serve with

Fig. 17-7. Not all who work for the construction industry are found at the project site. Secretaries and receptionists work in office buildings and receive their training outside the industry.

equal importance. Such workers keep records and provide information needed to keep the construction operation moving smoothly. Office workers usually receive their training outside the construction industry.

A high school education is all that is required for many skills in the office. Office personnel at the management level need to be highly trained. Many have four-year degrees from colleges and universities in many different programs including business and engineering.

SUMMARY

Training for the construction industry is important. It enables the worker to perform this job efficiently. It provides the worker with an opportunity for advancement to higher pay and a better job.

EDUCATION NEEDS OF CONSTRUCTION PERSONNEL

MANAGERS	TRADESPEOPLE	SUPPORT PERSONNEL
High School	Elementary School	High School
Technical Training	High School	Junior College
Four-year College	Apprenticeship	Associate Degree
Degree	Program	Four-year College
Graduate School	(Journeyman)	Degree
	Trade School	
	Technical School	
	Trade Extension	
	Junior College	
	Associate Degree	

Fig. 17-6. Construction workers include three distinct types. Many need special types of education or training as indicated above.

Construction

There are three kinds of personnel in the construction industry:
1. The professional.
2. The tradesman.
3. The support personnel.

Professional careers include the architect, engineer, estimator and contractor. They are the people responsible for designing, planning and carrying out the project. They are usually trained at the four-year institution of higher education.

The tradesperson or production worker is responsible for the actual construction work. He is trained usually through the apprenticeship program requiring three to five years of on-the-job training. Production workers make up the largest group in the construction field.

Office support personnel usually receive their training outside the industry. In many cases, graduation from high school is satisfactory to perform the duties demanded of them.

DISCUSSION TOPICS

1. What is training?
2. Who is trained?
3. Where can construction experience be gained?
4. What is an apprenticeship program?
5. What group of construction personnel require four or five years of college training?
6. Describe a skilled craftsman.
7. What on-the-job training is required for a construction manager?
8. What are the job requirements for an electrician, plumber and bricklayer?
9. What are the educational needs of an office typist, billing clerk and a receptionist?
10. What are the educational requirements for an architect or engineer?

An operating engineer usually serves an apprenticeship before becoming qualified to operate power construction equipment such as this crane. (Recruitment and Training Program, Inc.)

Chapter 18

WORKING CONDITIONS AND UNION ACTIVITY

Our work surroundings are very important to us. If we are too hot or too cold we are uncomfortable. If we find ourselves among other workers who are unfriendly or unpredictable we are unhappy. In either situation, we must adapt to the working condition, Fig. 18-1, or change jobs.

Ever since people started to leave their homes to work, they have had to contend with these two problems. The emotional effect of work surroundings are such that they may influence our life habits. Many are the instances in which a person's success or turmoil at work created a like situation at home.

So important are our work conditions that improving them has created new jobs and opened up whole areas of study. A large company may invest heavily in industrial psychology, safety engineering and systems analysis to improve morale, safety and efficiency.

We say, then, that working conditions are an important area of concern both to the worker and to management. We call working conditions an environment and distinguish between two elements. *Actual work conditions such as dust, heat or cold are called physical environment. The interactions of people at work is called social environment.* Both management and unions become involved in bettering these environments.

Fig. 18-1. Construction workers must learn to accept their work surroundings. They must also learn to work with others. (Corps of Engineers, Kansas City District)

Fig. 18-2. Construction workers are often required to be on the job during all kinds of weather. (The Port of New York Authority)

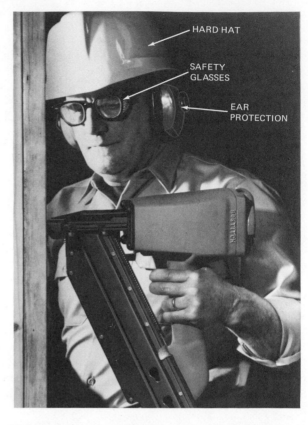

Fig. 18-4. Personal safety equipment protects the construction worker from noise, flying or dropped materials. (Bostitch)

HARD HAT

SAFETY GLASSES

EAR PROTECTION

Fig. 18-3. Many physical conditions are not controllable and must be endured. Safety equipment is provided as a matter of company policy. (A. B. Chance Co.)

PHYSICAL ENVIRONMENT

If our work place is in an enclosed area, the company can control such things as dust, heat and cold. Rooms can be air conditioned and miners may have air conditioned suits to cope with hot mining conditions. Not much can be done about dust and temperatures on the construction site, Fig. 18-2. However, safety clothing and safe equipment can be provided for those who work in dangerous surroundings or high places, Fig. 18-3.

Strikes or work stoppages are sometimes called by labor because of physical conditions. Labor and management work together to improve such things as safety equipment, Fig. 18-4, lighting, dust and noise control. Certain conditions cannot be eliminated or made less disagreeable. These must be endured as part of the work condition.

SOCIAL ENVIRONMENT

The social working environment refers to the actions and interaction of one working group and all other people who work with and around them. If the worker is happy with these people, the work will generally be of better quality. Social environment includes:

1. The general attitude workers have toward each other.
2. What after-hours get-togethers they have.

The work place should have a centrally located place where the workers can congregate and interact in their leisure time. This might be a cafeteria or lounge to be used during breaks and lunch hours. These places are important to the morale of the workers. It would also be helpful to have after-work activities. Companies may sponsor bowling leagues, softball teams, Little League teams, picnics, outings and other company functions. See Fig. 18-5.

ECONOMIC REWARDS

Economic rewards involve more than the money the construction worker receives for performing certain tasks. Fringe benefits such as an insurance program, life and hospital

Fig. 18-5. Usually, great ceremony attends the placing of material which signals the reaching of a special point in construction. These ceremonies are staged for company morale and for public relations. Above. Topping out ceremony marks the placing of the highest piece of steel in every major structure. (Missouri Highway Dept.) Below. Public ceremony attends the placing of the keystone section of a public monument. Workers and dignitaries share in the celebration. (U.S. Department of Interior)

insurance, social security, workmen's compensation and a paid vacation are other important economic rewards.

Construction trades offer especially rewarding career opportunities for those who are not planning to go on to college, but who are willing to spend seven years learning with pay. Well-trained construction workers can find jobs in any part of the country. Generally, their hourly wage rates are much higher than those for most other manual workers. Those who have above-average business ability have a greater opportunity to establish their own business.

COLLECTIVE BARGAINING

One person alone has little bargaining power. When people with common interests join together, they can jointly elect one person to speak for them. This person has more authority than any other single member in securing pay raises or benefits for all.

When this leader seeks benefits for the group from the employer, it is called collective bargaining. The group of organized workers who wish to bargain collectively is called a labor union.

Working together through a union, all employees can speak with one voice at the bargaining table just as management does. Unions negotiate for benefits such as better wages, better working conditions and a comfortable retirement program.

LABOR UNION HISTORY

During the early industrial revolution, labor did not become organized for several reasons:

1. Free movement of families from one area to another and rapid growth of the frontier found workers drifting in and out of industrial occupations.
2. The agricultural nature of society kept public support away from organized labor.
3. Laborers accepted the idea that they were like servants.

Labor did not become a significant part of the American industrial scene until mass production placed many employees under one management.

Some local attempts at strikes and work stoppage began as early as 1800. However, not until the middle of the 1800's did these groups gain numbers. They were local groups very loosely organized.

Generally, the courts did not favor labor unions. Also, during this period, immigration to the United States was heavy. Employers were able to hire cheap labor. Although many attempts at organization were made during this period, unions generally were short lived.

In 1869, a group called the "Noble Order of the Knights of Labor" organized in Philadelphia. They grew to over a million members. However, after unsuccessful attempts at striking, they lost many members.

In the 1870's a cigarmaker, Samuel Gompers, began to organize local unions. In 1881 he and a few leftovers of the Knights of Labor organized a national union called the "Federation of Organized Trades and Labor Unions of the United States and Canada." This appeared to be too loose an organization and five years later, Gompers seized total control. This was the beginning of the American Federation of Labor (AFL).

At very best, labor/management relations in those days were strained. Labor organized strikes, boycotts, pickets and civil disruption. Management fought back with injunctions, lockouts, blacklists and more civil disruption. Labor unions became more militant.

This pattern continued until well into the 1900's. The labor unions were doing more than just creating civil disorder; they were pushing for legislation to give them recognition and support.

The depression of the 1930's hurt labor unions. No one escaped the effects of the economic down swing. Labor had been against the government-sponsored work programs. However, they were unable to provide any relief of their own for out-of-work people and the labor union found they needed help.

The United States government responded with legislation. Though short term in scope, measures such as the National Industrial Recovery Act, and its successor, the National Labor Relations Act (Wagner) had long range effects. These pieces of legislation gave legal recognition to the fact that employers had responsibilities to laborers. When these were declared unconstitutional Congress responded with the Walsh-Healey Act and the Fair Labor Standards Bill, which provided maximum work hours and minimum wage standards.

It must be understood that many benefits have been made possible by labor unions. They have fought for and gotten shorter hours, higher pay and better working conditions. Often, improvement in work conditions has also made construction operation more efficient. See Fig. 18-6.

Fig. 18-6. Concrete workers pour a form, using a crane and bucket to move materials from truck to form. Power vibrator removes air pockets from poured concrete. These improvements in technology increase efficiency and take the drudgery out of concrete work, eliminating the need for wheelbarrows and hand "rodding" of fresh concrete.
(Corps of Engineers, Kansas City District)

LABOR AGREEMENTS

What management and organized labor are committed to do for each other is governed by the terms and conditions of the labor agreement. Labor agreements have varied widely from one industry to another since Samuel Gompers first organized the AFL.

A labor agreement is the total relationship between a union and management. Agreements are of two types, simple and supplementary.

SIMPLE AGREEMENT

Simple agreements set down a few elementary rights and duties of the employer and employee. Simple agreements are confined to a few fundamental and procedural steps for disposing of problems as they arise from day to day. This document seeks to prescribe a general method of handling grievances.

SUPPLEMENTAL AGREEMENTS

Supplemental (added-on) agreements may be classified as either major or minor. Major supplemental agreements are those issues which either labor or management consider vital to their operation.

These issues usually center around policies and rules that have been accepted for a long period of time. One party usually is trying to get the other party to change or modify the contract. Both parties will take a militant position all the way to the arbitration.

The party that loses the decision will keep that issue in mind until collective bargaining begins for a new contract. It will be used as a bargaining issue for a wage increase, or perhaps a comprehensive health and welfare plan.

Minor supplemental agreements are the result of grievances that are settled locally and without arbitration. In this way, the words of a contract come to have a specific meaning in a variety of circumstances. Records are kept of all grievances handled and how they were settled. These records constitute a system of precedents which fill in the general provisions of a simple agreement. (A precedent is an act that serves to authorize or justify another act of the same kind.)

In some labor/industrial management relations, minor supplemental agreements are reached through the practice of exchanging letters setting forth agreements on problems that arise during the contract period. These letters deal with issues that are not spelled out in the contract. They are useful to both parties involved.

GRIEVANCES

A grievance is a complaint by a worker that he has not been treated fairly in some aspect of his work or pay. Usually the grievance is in an area specifically covered by the collective agreement.

Seniority grievances are common. So are grievances where workers feel their seniority has not been reckoned with properly in cases of promotions or transfers.

When unions bargain with small employers, the grievance system is usually very simple. The workers' complaints are delivered verbally to their immediate supervisor. If satisfaction is not given, they go to the union president or its business agent. That person talks to the owner or manager of the firm.

Such informal procedure is clearly unworkable in large companies. Therefore, large formal systems of handling grievances have been worked out over the years. In large companies the worker would go first to the shop steward. If the grievance appears to be valid, the steward will help the worker put it in writing.

If the steward cannot settle the grievance with the foreman, it will travel upward through a well-defined series of steps. Each step involves increasingly higher levels of authority within the union and company management.

In a large firm, the second step might be a conference between an officer of the local union and management's personnel department. A third step might be a meeting between the grievance committee of the local union and a committee from management.

The grievance procedure helps to keep unions in close touch with their members. It is an important function that maintains life and interest in the local union.

ARBITRATION

Suppose labor and management get into an argument over the contract and cannot come to an agreement. What then?

They decide that they will put the facts before a third party who has nothing to gain whichever way the argument is decided. This is called arbitration.

Both labor and management agree in advance to live by the decision made by the arbitrator. There can be no appeal to the decision later.

Like a judge, the arbitrator listens to arguments and reviews evidence offered by both sides. Then he or she decides on the basis of the evidence what is fair. Arbitrators are, in effect, both judge and jury. However, it is not part of our system of courts. Rather, it is designed to keep the argument out of the courts. The legal process is slow and expensive.

Labor arbitration is not collective bargaining. It may become a part of it. Its purpose is to arrive at a decision. Interpretation of a collective bargain contract may be arbitrated.

The arbitrator's fee is usually divided equally between the disputing parties. Services of an attorney are not necessary but may be helpful at times. The process is quick, lasting from two days to two months.

Two kinds of disputes are handled by arbitration:

1. A difference of opinion over what a collective bargaining agreement means. This is called a rights dispute.
2. An argument over the terms and conditions of a collective bargaining agreement. This is called an interest dispute.

Voluntary labor arbitration provides a process for an orderly solution to disputes. It also gives a solid basis for good labor/management relations. Nearly all issues to be arbitrated go through a grievance procedure set up by both labor and management.

MEDIATION

Mediation places an impartial person into the role of advisor in collective bargaining. There is a difference between mediation and arbitration. Arbitrators have the responsibility and authority to decide one or more disputed issues. The decision is binding on all parties. Mediators have no such authority. The parties make all the decisions by agreement. Mediators must rely on persuasion. They may suggest or they may recommend, but the individual parties may choose not to accept the advice.

Mediators are servants of the public. In a very direct sense, they are a servant of the parties involved in the labor dispute.

STRIKES

In construction, as well as in other industry and business areas, it takes people to operate the various pieces of equipment and perform the various tasks. Unlike machines, people have a voice in their destiny. Consequently, people organize to bring about a change.

A strike is a means of doing this. *The term strike means a concerted withdrawal from work by a group of workers employed in the same economic enterprise.* Its purpose is to force the employer to be aware of their demands. The work stoppage is understood to be temporary and subject to laws.

WHEN WORKERS STRIKE

Strikes occur most often in those nations that have a large industrial base and a relatively free, democratic political system. The right to strike is normally guaranteed in the Western democracies. Such rights are less frequently found in those nations that are governed by dictatorships. Strikes occur more often when employment is rising and when economic conditions are favorable. At such times, employees are less concerned about economic security and employers are more sensitive to work stoppage.

The public is usually unaware of reasons behind a particular strike. A few reasons are:

1. An employee group strikes to impose its will on an employer in the broad area of wages, hours and working conditions.
2. Many strikes result from conflicting interpretation of the terms of a collective bargaining agreement.
3. Jurisdictional strikes may be caused by a disagreement between two unions over the assignment of work or jobs.
4. In the recognition or organizational strike, employees try to force the employer to deal with the specific union as the collective bargaining representative.
5. A strike may be called to change the collective bargaining to national bargaining or to change the effective date on contracts so that they expire on dates more favorable to labor.

There may be other reasons not mentioned, but most strikes fall somewhere in these categories. After a strike has begun and run its course, it usually comes quickly to a point of settlement.

SETTLEMENTS

Most strikes are settled fairly soon by mediation, fact finding or arbitration. Governmental or private third parties enable management and labor to narrow the issues and bridge differences while saving face with their constituencies (those they represent).

People are human beings capable of making mistakes. But only people can rationally make decisions to live by. Strikes, mediation and arbitration are tools whereby people can come to a mutual understanding and control change.

CONTRACTS

Labor contracts are the result of mediation, arbitration and collective bargaining. They are statements of the provisions agreed on by both parties. Labor and management have come to an agreement on hours, wages and working conditions. Labor contracts usually are written for a given period of time. At the end of the contract period, collective bargaining is repeated and a new contract is bargained for.

SUMMARY

Most working conditions are established by negotiation between labor organizations and management. The worker and the employer reach a compromise so that both can be satisfied.

Certain rules are established for the settling of labor disputes. Mediation and arbitration allow for third-person help in reaching a just and honorable settlement.

CAREERS RELATED TO CONSTRUCTION

UNION STEWARD is a union representative who acts as treasurer for union; collects dues and takes up grievances with management.

MEDIATOR hear arguments on both sides of labor dispute; suggests solutions.

ARBITRATOR hear arguments on both sides of labor dispute; weighs arguments and determines just settlement of differences.

BUSINESS AGENT manages functions of labor unions such as public relations, membership building, rents halls, hires speakers, helps in negotiations.

LABOR RELATIONS SPECIALIST represents either labor or management in contract negotiations; studies and interprets collective bargaining agreements and labor market conditions to assist in establishing policy; helps resolve grievances.

DISCUSSION TOPICS

1. What is meant by physical environment? What is meant by social environment?

2. How does economic reward affect working conditions?
3. List some fringe benefits.
4. What is collective bargaining?
5. How is collective bargaining conducted?
6. What is the responsibility of the mediator?
7. What is arbitration?
8. How does arbitration differ from mediation?
9. How do strikes improve labor's position in the collective bargaining process?
10. Who benefits from a labor contract? How?

The work enviroment often requires workers to protect themselves against weather conditions. Here masons protect themselves against cold while on the job.   (Recruitment and Training Program, Inc.)

Chapter 19

JOB ADVANCEMENT

Fig. 19-1. Construction jobs demand widely different responsibility and offer widely different advancement opportunities. (Corps of Engineers, Kansas City District)

Construction advancement practices affect the kinds of employees who are attracted to work in construction, Fig. 19-1. The ambitious person will choose a company with the best advancement chances. These people will probably be the best workers.

Advancement practices also affect the employers. An employee who is unhappy has low morale and may not be working up to ability. An unhappy worker will probably quit. When a great number of employees are unhappy and quit their job, a poorer quality of work will probably result. Then too, it is expensive to hire and train new employees.

CAREER PROGRESS

If employees have worked for a company long enough to exhibit desirable work habits, they have a good change for advancement. As new positions or job classifications open up in the company their record will be reviewed. Job classifications in the middle or upper part of the job scale, Fig. 19-2, are quite often filled by transferring people from lesser jobs.

In the construction industry, as in other industries, there are four directions that a career may take:

1. Positive progression, upward.
2. Negative progression, downward.
3. Lateral progression, across.
4. Separation, out.

We usually think of career progression as positive, upward.

NEGATIVE PROGRESSION

Negative progression or demotion is being moved to a job with less responsibility or to one that requires less knowledge or skill. This may or may not mean less pay.

If a company has a lay-off, there is a good deal of demoting especially in a company that acknowledges seniority. (Seniority is established by the number of years spent with the employer. Those with less seniority are laid off first.) This type of demotion is usually temporary until the company's work load or production is back to normal again.

Another type of demoting occurs when a person is found to be incapable of handling a certain job. The employee is moved back to a lesser job while another person is tried in the more responsible position.

LATERAL PROGRESSION

Lateral progression is usually called a transfer. To transfer is to move to another job of the same level and pay. This may happen for several reasons:

1. Employee may have a bad attitude or cannot get along with others. If reliable in other ways, the worker may be transferred to another job or to another department.
2. Personality clashes between supervisor and worker. Management may want to keep this situation from getting worse or affecting a whole department.
3. Lack of advancement opportunity. Employee may have advanced as far as possible without realizing true worth to the company. In this case, the worker may choose to transfer into a job of equal pay but with a chance for further advancement and higher pay.

SEPARATION

Separating is the process of discharging, relocating, laying off or retiring of personnel. Discharging means firing or removing an employee from the work force.

Many reasons may be given for discharging: lack of skill, an

Fig. 19-2. Ambitious workers move on to more responsible jobs in construction as they prove their reliability and readiness to accept responsibility. Left. Either a crane crew foreman or a rigger is responsible for directing the movements of the crane operator who may not be able to see the huge steel section the crane is going to lift. Right. Crane operator carefully controls lifting of huge sections of building framework by voice instructions from the ground. Operator may never see the load until it is topside. (The Port of New York Authority)

undesirable attitude, or simply inability to get along with his co-workers. In any case, the construction crew can function better without that member.

Laying off is the temporary release of an employee during a cutback in the number of workers. Laying off may be the result of a work slowdown, strikes of other building trades, weather and possibly material shortages.

RETIREMENT

Retirement is the last form of separation. Retirement has become a very important part of the American scene. We take retirement at a time in life which many look forward to. Retirement gives workers time to do the things they never had the time to do, Fig. 19-3. However, retirement is a relatively new concept. Never before could individuals quit earning their livelihood with so many productive years ahead of them. Today, company retirement plans, Social Security, life savings, Medicare and investments make retirement possible from an economic point of view.

PLANNING FOR RETIREMENT

Retirement is something to be planned for from a very early period of your young career. The best way to enjoy your

Fig. 19-3. Modern retirement programs are based on a savings program and social security which enable older workers to stop working while healthy and still capable of productive work.

retirement is to plan for it while you are still young enough to shape your future.

Seek a company with a progressive pension plan, or save money that will be invested and not used until you retire. The federal government offers a kind of retirement fund called Social Security.

Social Security is not a total retirement system and you should not place your retirement dependence on this source. It is only meant to supplement your savings.

Investments are another way to earn money for retirement. Real estate, mutual funds and insurance companies all pay a good rate of return, assuming you invest wisely.

Medical and hospital insurance are protection against expensive medical bills, which will increase during the retirement years. Medicare, another government fund, is designed to aid in off-setting medical bills for retired people.

SUMMARY

Advancement practices are very important to the construction industry. Poor advancement policies lead to poor morale of the work force. Many persons looking for a job will not consider working for an employer who has no advancement program. Promotion usually means better salaries.

A career must go in one of four directions:
1. Positive progression, upward.
2. Negative progression, downward.
3. Lateral progression, across.
4. Separation, out.

Advancement depends upon skill, attitude and education.

DISCUSSION TOPICS

1. How does advancement practice affect the work force?
2. Describe the difference between positive and negative progression.
3. What does lateral job progression mean?
4. List the types of separations. Describe each.
5. Prepare a good retirement plan for yourself.
6. What is Social Security?
7. How do mutual funds aid one's retirement?
8. What is Medicare?
9. Describe a temporary separation?
10. On what factors are advancement based?

The prospect of advancement in the construction field will serve as an incentive for workers to move up to better jobs and better salaries. Through work experience and study, for example, a laborer can become a foreman or supervisor.

SECTION 3

THE TECHNOLOGY OF
CONSTRUCTION MATERIALS AND PRODUCTS

Construction materials and products cover the broad areas of building supplies and equipment. This section details the use of concrete, brick, stone, clay tile, wood, steel and related products. It describes the construction of doors and windows, building hardware, electrical and mechanical equipment. It explains the need for sealants and adhesives, insulation, exterior wall and roof coverings, interior wall and floor coverings, trims, paints and finishes.

Construction materials range from stone and sand to concrete blocks and chemical compounds, and from paper and plastics to wood and steel.

Workers erect sectional forms of lightweight aluminum made to hold semi-liquid concrete until it solidifies and forms a continuous foundation wall. (Precise Mfg. Co.)

Chapter 20

CONCRETE

Fig. 20-1. Concrete is a long-lasting and dependable material widely used in construction.   (American Hot Dipped Galvanizer's Assoc.)

Concrete has long been used as a construction material. It first became popular with the Romans and is widely used today. See Fig. 20-1.

Around 100 BC, Roman builders developed an excellent concrete which allowed them to erect vast structures. Some of these buildings endure even to this day though in various states of ruin.

The main ingredient of their concrete was a special volcanic ash. It was mined on the slopes of Mt. Vesuvius near the village of Pozzuoli.

When this ash was mixed with limestone, burned and combined with water, it became a very strong cement. It was then mixed with small pieces of rock to form a strong and lasting concrete. Roman builders also found that this cement would harden under water. We call this hydraulic cement.

Today, the characteristics of concrete vary widely. Qualities of the mix can be changed depending on the composition of the aggregates (sand and stones), chemical composition and physical properties of the cement. But in every instance concrete is a mixture of cement, water, sand, gravel and stones of various sizes. See Fig. 20-2.

The term cement is given to any material that will bond two or more nonadhesive (nonsticking) substances together. In concrete construction, the term means portland cement.

PORTLAND CEMENT

Knowledge of how to make hydraulic cement was lost in the middle ages. It was rediscovered in 1750 by an Englishman, John Smeaton. In 1824 in England Joseph Aspdin

Fig. 20-2. Concrete may be thought of as a plastic material which takes the shape of the container or form which holds it. Construction workers formed these huge stadium supports by pouring concrete into huge wooden forms. Usually they are poured on the site; however, it is possible to pour them elsewhere and then move the finished pieces to the building site. (Fruin-Colnon)

developed and patented a hydraulic cement that was superior to any other of the time. He called it portland cement because it looked like a grayish limestone mined on the isle of Portland. He found that if a carefully controlled mixture of limestone and clay was burned at a higher temperature than had been used before, the resulting cement had better hydraulic qualities. At this higher temperature, the limestone and clay fused together forming a new material.

Portland cement today has the following basic composition:

Lime   60—65 percent
Silica   10—25 percent
Iron oxide   2— 4 percent
Aluminum   5—10 percent.

Most of the ingredients in portland cement are found in nature. They cannot always be used in their natural state.

MANUFACTURE OF PORTLAND CEMENT

Portland cement is manufactured by mixing and pulverizing the materials and then feeding them into a kiln (furnace). These kilns are steel cylinders 20 ft. (6 m) in diameter and 300

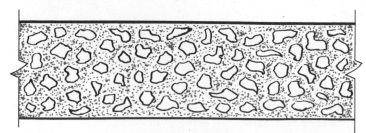

Fig. 20-3. Section through concrete showing distribution of aggregates and cement bond.

to 700 ft. (91 to 213 m) long. Fuel is fed into one end and ignited. As the raw material rolls in the rotating and slightly sloping cylinder, the heat fuses the raw material into green-black clinkers. The clinkers have specific chemical and physical properties which give cement its hydraulic characteristics. The clinkers, direct from kiln, are ground into fine powder. Some of it is shipped in bulk to ready-mix concrete plants. The remainder is packed into paper bags of 94 lb. (44 kg) and shipped to building supply dealers.

HARDENING AND SETTING

When portland cement is mixed with enough water and left undisturbed, the mixture loses its fluid state and becomes a solid. Cement does not harden by drying. It hardens because of a chemical reaction.

Water and cement combine chemically to form a new compound. This process is called setting.

Initial setting may take place after only a few minutes or it may take several hours. When the cement has been combined with aggregates and has set, it will continue to harden (gain strength) for several months or even years.

WATER RATIO

If too little water is used in the mix, some particles of cement will not be chemically changed. If too much water is used, the excess water may be trapped in the mix. In either case, the concrete will be weakened. The temperature of the materials and moisture in the air (humidity) affect the exact amount of water the mixture needs.

A 94-lb. (44 kg) bag of cement requires 2.5 to 3 gal. (9 to 11 l) of water for a complete chemical combination of materials. The use of exactly this amount is not practical for field conditions. Usually 4 to 8 gal. (15 to 30 l) must be used for each bag of cement. The extra water serves as a lubricant to carry the cement into small pores of the aggregate.

DESIGN OF MIX

The water-cement ratio of a given concrete determines its strength. However, to produce concrete that is economical as well as strong, the proper aggregate type, grade, size and proportions must be used.

Proper portions of the various sizes of aggregates can be determined by trial mixes, either in the field or in the laboratory. A concrete with too much coarse aggregate may contain too many voids. A mixture with too much sand may be smooth but is not economical. It will not attain its greatest strength. See Fig. 20-3.

PROPORTIONS OF THE MIX

In the past, the proportion of cement, sand and aggregate was thought to control the various strengths of concrete, Fig. 20-3. Little or no importance was placed on the water-cement ratio. Water-cement has been found to govern the strength of finished concrete. A 1:3:4 mix contains, by volume, one part

Fig. 20-4. Concrete mixing on the site. Equipment of this type is portable and may easily be moved to another site.   (Ross Co.)

cement, three parts sand and four parts rock. Enough water is added to this dry mix to make it flow into the forms.

MIXING OF CONCRETE

Another factor in the workability and strength of concrete is the method used to mix the ingredients. It is essential to mix thoroughly. Sometimes concrete is mixed on the site using equipment like that shown in Fig. 20-4. In other cases, the concrete is mixed at a central batching plant and is hauled to the site in a ready-mix truck. See Fig. 20-5.

Fig. 20-5. Concrete may be mixed in transit to construction site using a mobile mixing unit.   (Jay H. Maish Co.)

The time required to mix the ingredients varies. It depends upon the size and efficiency of the mixer. Mixing time should not be less than one minute for concrete of medium consistency. A concrete mix will not remain useable as long during hot weather.

Mixing is calculated from the time all solid materials are mixed together. Prolonged mixing will not greatly add to the strength of concrete as long as additional water is not added to increase the slump.

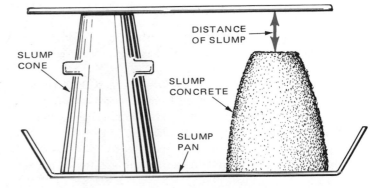

Fig. 20-6. Setup for slump test of concrete. Rodding tool or straight-edge is laid across top of cone. Slump is measured with a rule.

SLUMP TEST

Consistency of fresh concrete is measured in terms of its slump. Slump is the distance the wet concrete will settle under its own weight. See Fig. 20-6.

To test the slump of concrete, samples are taken directly from the mixer and placed in a cone-shaped container. The cone is inverted (turned upside down) in a pan to release the concrete. No longer supported by the cone, the wet concrete settles or slumps. If the concrete sample does not have the specified slump, the mix must be redesigned.

Water should never be added to the slump. It will reduce the resulting strength.

The minimum slump recommended for various types of work are:
1. Mass concrete for slabs on soil (not reinforced), 2 to 5 in. (5 to 13 cm).
2. Reinforced concrete other than thin walls, 3 to 6 in. (8 to 15 cm).
3. Thin reinforced concrete walls, 3 to 6 in. (8 to 15 cm).

AGGREGATES

Aggregates are sand and larger particles of stone, rock or other material which make up about 65 to 80 percent of the total volume of concrete. They do not react chemically with cement and water. Sizes range from fine sand pieces to 1 1/2 in. (3.5 cm) in diameter or larger. The quality of the concrete is affected in several ways by the aggregate:

1. Strength of concrete is no greater than the strength of the aggregate.
2. The size and shape of aggregate affects the flowability of the aggregate.

STANDARD WEIGHT AGGREGATES

Natural aggregates, such as sand and crushed gravel or gravel taken directly from the river, may be used. Natural aggregates must be free from dirt or other foreign materials.

By-product aggregates are the inert materials such as blast furnace slag or cinders. Cinder concrete has a low strength factor and is not suitable for structural concrete.

LIGHTWEIGHT AGGREGATES

Certain materials can be processed by heat to produce a lightweight aggregate. During the heating process gas that forms inside the material expands it.

Vermiculite, a type of mica, may expand 20 times its original volume. In its expanded form it weighs only 6 to 20 lb. per cu. ft. (98 to 330 kg/m^3). Natural stone weighs 135 to 160 lb. per cu. ft. (2100 to 2500 kg/m^3). Vermiculite will form a nonstructural concrete fill that weighs 20 to 40 lb. per cu. ft. (330 to 660 kg/m^3) instead of the 150 lb. per cu. ft. (2400 kg/m^3) of standard fill.

Perlite, a type of volcanic rock, traps water during mixing. When it is heated, the water changes to steam which forces the material into a lightweight fluffy form. Perlite weighs approximately 85 lb. per cu. ft. (1350 kg/m^3) before expansion and 6 to 12 lb. per cu. ft. (98 to 196 kg/m^3) after expansion.

Concrete made of perlite or vermiculite aggregates has excellent insulating properties. It is often used on the ceilings of large buildings. See Fig. 20-7. Perlite concrete is easy to place and work, Fig. 20-8. However, because of its low strength, it is not suitable for structural concrete.

Fig. 20-7. Perlite concrete will lighten the load of the structure, thereby reducing the size of structural support members needed.
(Perlite Institute)

Fig. 20-8. Lightweight concrete is easy to handle and finish. It can be used where it will not have to support heavy weights.

ADMIXTURES

Admixtures are additives to concrete. They include any materials other than water, cement, or aggregates added to concrete before or during mixing. Admixtures are used to modify the properties of the concrete mix in one of the following ways:

1. To speed or retard setting (hardening).
2. To improve workability.
3. To reduce separation of aggregates.
4. To make a stiff mix more fluid without addition of water.
5. To resist freezing.
6. To resist abrasion (wear).
7. To add air to mix.
8. To add color to concrete.
9. To minimize expansion and contraction.

REINFORCEMENT

Concrete has great compression strength (squeezing action). Each square inch of concrete can be designed to support loads of 10,000 lb. (4500 kg) or more.

However, concrete has little tensile strength (resistance to pulling action). In reinforced concrete, steel and concrete are combined to take advantage of the high compressive strength of concrete and the high tensile strength of steel. Sometimes, to reduce the size of the concrete member, steel is expected to assist in accepting part of the compression load. Concrete is cast around reinforcement steel, Fig. 20-9. As it hardens, it grips the steel to form a bond with it. This bond becomes stronger as the concrete hardens.

TYPES OF REINFORCEMENT

Steel reinforcement is manufactured in round bars or wire mesh. Bars are used in structural framing members; wire mesh is used in concrete slabs.

Deformed bars, rolled with small projections on the surface, are more often used than smooth bars because of the added

Fig. 20-9. Wire reinforcement is being placed around a steel beam. Later, concrete will be cast around the beam.
(Wire Reinforcement Institute)

Fig. 20-10. Deformed reinforcement bars protrude from precast construction components waiting transportation to building site.
(American Hot Dipped Galvanizer's Assoc.)

and wire sizes may be equal in both directions or they may vary in size and spacing to form what is called a one-way mesh.

PLACING THE REINFORCEMENT

Before the concrete is placed, the steel must be cleaned of loose scale which would reduce the bonding strength. Reinforcement steel must be placed and secured before any concrete is poured. This is important because flowing concrete will cause unsecured reinforcement to drift or relocate.

Fig. 20-11. Welded wire mesh sheets are placed on layer of freshly poured concrete. Wire reinforcement will be buried in the concrete, where it will increase strength of the slab.

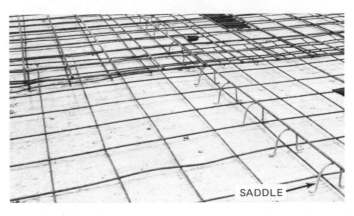

Fig. 20-12. Saddles support wire mesh reinforcement above form floor. Suspending wire above floor allows concrete to flow underneath.

bonding strength, Fig. 20-10. Bars are made in different sizes and are numbered from 3 to 8, depending on diameter. There is 1/8 in. difference from one size to the next. Thus, a No. 3 bar is 3/8 in. while a No. 7 is 7/8 in. diameter.

Deformed bars larger than an inch are numbered differently. A No. 9 is 1.128 in. in diameter; a No. 10 is 1.27 in. and a No. 11 is 1.41 in. in diameter.

Bars are usually bent or hooked at the end. This improves their holding power under load conditions.

Several types of wire mesh are used to reinforce concrete slabs. The most common is a right-angle mesh. See Fig. 20-11.

Wire mesh reinforcement is manufactured in various sizes of wire diameters and in various sizes of spacings. The spacing

Saddles, Fig. 20-12, are designed to hold wire mesh up away from the bottom of concrete slabs. Bars are usually spaced at prescribed distances according to the size of bar and strength required. Concrete walls require both vertical and horizontal reinforcement bars. Fig. 20-13 shows workers placing bars. Each intersection is fastened by either a 16 gauge soft wire tie or a weld.

PRECAST CONCRETE

If concrete members for construction can be cast on the ground and then lifted or tilted up into position, cost of concrete work will be greatly reduced. Work done at ground

Fig. 20-13. Deformed bar has been placed in vertical wall. Horizontal bars are being fastened to vertical members and properly spaced.
(Corps of Engineers, Kansas City District)

level is simpler and less expensive.

Intricate shapes with various textures and colors may be cast in wood, metal, plastic or plaster forms. The best quality concrete is achieved in a precast concrete manufacturing yard. See Fig. 20-14. These precast units are transported to the

Fig. 20-14. Precast concrete wall panel receives final surface finishing touches before shipment to construction site.
(American Hot Dipped Galvanizer's Assoc.)

construction site and hoisted into place, as in Fig. 20-15.

Slabs can also be precast for wall sections, then tilted up into position as shown in Fig. 20-16. Floors and roof slabs are cast at ground level and then lifted into position.

PRESTRESSED CONCRETE

Prestressed concrete products are usually precast. A prestressed concrete beam, for instance, is a structure of concrete and reinforced steel. The steel is placed in tension by applying hydraulic pressure, Fig. 20-17, before concrete is poured. After the concrete is cured, tension is released causing the concrete to be squeezed together.

A prestressed structural unit may be designed to use only half the concrete and a quarter of the steel used in the conventional structural unit.

SUMMARY

Concrete has been used as a building material since the Romans developed cement from volcanic ash. Today, portland cement has become the leading bonding agent. Cement produces a hard, strong building material through a chemical reaction with water.

Aggregates of sand, gravel, or lightweight material of vermiculite and perlite are used to control concrete weight, insulative and strength properties.

Various admixtures are added to give concrete special properties. Some, for instance, prevent freezing and expansion of the concrete during the setting stage. The freezing would cause the concrete to fracture (break).

Steel is used to reinforce concrete. Bars and wire have been found to give concrete additional strength under compressive and tension loads.

While most concrete is cast in place on the job site, it has

Fig. 20-15. Precast beams are being hoisted into position on bridge piers. (Northwest Engineering)

Fig. 20-16. Precast slabs can be finished at the time of casting while a formed wall cannot because of the forms.
(Wire Reinforcement Institute)

Fig. 20-17. Precast section of floor is given additional compressive strength by placing reinforcement steel under tension pressure before concrete is poured. Tension is created by pulling reinforcement material very tight. When section is released from form, tension compression loads result in stronger concrete.

been proved that many building components can be precast and transported to the site and erected. Precasting eliminates many forming problems and does away with ugly form marks on the finished product.

CAREERS RELATED TO CONSTRUCTION

CEMENT MASON smooths and finishes surfaces of poured concrete floors, walls and sidewalks according to specifications.

CONTROL CHEMIST makes periodic chemical analysis of concrete mix to make sure it is uniform in content and meets specifications; keeps records and makes reports.

CONCRETE MIXER OPERATOR or BATCH PLANT OPERATOR tends mixing machine to mix sand, gravel, cement and water to make concrete; cleans and maintains mixer.

CONCRETE MIX DESIGNER determines proper quantities of aggregates to meet specifications for the concrete.

CONCRETE MIXING TRUCK OPERATOR drives truck equipped with concrete mixer; transports concrete from batching plant to construction site; cleans and lubricates mixer.

CRUSHER OPERATOR tends a crusher that sizes rock for construction use; uses hand tools; adjusts screens, conveyors and fans; cleans and oils equipment.

KILN OPERATOR (FIREMAN) controls equipment that superheats lime and clays or rock to make portland cement.

TRUCK DRIVER uses heavy duty truck to transport concrete materials to building site or to processing point; also, hauls precast concrete sections to building site.

DISCUSSION TOPICS

1. What is the difference between cement and concrete?
2. When was concrete first used? Under what conditions?
3. How is portland cement manufactured? What natural materials are used to manufacture portland cement?
4. List some aggregates.
5. What do admixtures do for controls?
6. How does water affect the strength of concrete?
7. How are reinforcement bar sizes given?
8. How are lightweight aggregates formed?
9. Name the advantages of precast concrete components.
10. What is prestressed concrete?

Chapter 21

MASONRY MATERIALS

Fig. 21-1. This masonry structure uses brick which is manufactured from clay. Note low wall of natural stone. (Azrock)

Masonry construction uses building materials of stone, brick, concrete block, gypsum block, glass block and various clay products. See Fig. 21-1. Stone is a natural material quarried from the earth. All other masonry materials are manufactured usually by heating earth products to high temperatures. This heat vitrifies (melts) the materials together.

STONE

Natural materials most readily available for construction have generally determined the character of the architecture produced by any culture. Stone has been used as a structural material, as a finish material and as roofing throughout the centuries.

With the development of new materials and new structural techniques, stone is now used primarily as a decorative material. See Fig. 21-2. The range of color, texture and finish is almost inexhaustible. Some variety of stone is available in nearly all areas of the world.

CLASSIFICATION OF STONE

Geologically, stone is classified as igneous, sedimentary, and metamorphic. The characteristics of each type have a definite bearing on their durability and use.

Igneous rock is produced by heat and pressure. Such rock is produced naturally through volcanic activity and the pressure exerted by shifting of the earth's surface.

Sedimentary rock is made up of silt (bits of eroded igneous rock) or the skeletal remains of marine life that have been deposited by ancient seas. Such rock always lie in layers called strata.

Metamorphic rock is formed by the gradual change in the character and structure of igneous and sedimentary rock. Three forces produce metamorphic rock:

1. Pressure and heat.
2. Water action which dissolves and redeposits minerals.
3. Action of hot magma on old rocks. (Magma is molten material from within the earth.)

137

Fig. 21-2. This attractive building was constructed partially of "rubble" stone which refers to the method of laying the sandstone used.

All types of stone are used in building. The most important ones used in the construction industry are granite, sandstone, slate, limestone and marble.

GRANITE

Granite is an igneous rock that has been formed and cooled beneath the earth's surface. It varies from a finely granulated form to one that is crystalline (like crystals) in nature. The color of granite varies from a deep bronze to black. It is usually cut and polished at the quarry, Fig. 21-3.

Fig. 21-3. Granite is usually cut to a specific size and polished before being shipped to the construction site. (Cold Springs Granite)

Granite is used where strength or hardness is needed. Steps and paving of granite stand up well under heavy wear and abrasion. The highly durable, crystalline surface of polished granite makes a colorful, lasting finish for the outside of buildings, Fig. 21-4.

SANDSTONE

Sandstone is a sedimentary rock made up of angular or rounded grains held together by a cement-like material. This cement-like material contains a high percentage of non-oxides giving the sandstone a red or brown tone. Sandstones are soft

Fig. 21-4. High rise structure is faced with special pink granite. Quarried in Spain, it was transported to this site. (United States Fidelity & Guaranty Co.)

and easy to work but not always durable. One type, called Ohio sandstone, has been formed under extreme pressures making it harder and more durable. Ohio sandstones are usually light grey or buff colored.

SLATE

Most slates are composed of high-silica clays laid down as silt in ancient sea bottoms. This clay has formed into rock. It can be split into thin sheets with smooth, regular faces. Splits run along natural cleavage lines. The colors of slate are caused by small amounts of impurities such as iron, carbon and chloride.

Slates are used for outside and inside wall coverings, platforms, walks, roofing, flooring and counter tops, Fig. 21-5.

LIMESTONE

Limestone is a sedimentary rock that contains carbonate of lime. Limestone is the remains of shells or the skeletons of prehistoric animals. It may be white, cream, buff, grey or variegated patterns of these colors. (Variegated means marked in different colors.) Limestone is comparatively soft when taken from the earth. It hardens with exposure to the air.

MARBLE

Marble is a metamorphic, crystalline limestone. It has color, texture, grain, working qualities and finished properties. In modern construction, marble is used for exterior and interior veneer over structural frames. Marble has, in the past, been used as a structural material. Quarries in the Aegean Islands have operated continuously since 66 BC. Marble is produced from Vermont to California by more than 100 quarries. American marble varies in color from creamy white to black.

Fig. 21-6. Brick building materials are manufactured by burning certain earth minerals into small masonry blocks. When cemented together by mortar bond, bricks can produce a wide variety of structural designs.

Fig. 21-5. Thin pieces of slate make a lasting roof material. Holes must be pre-drilled for nailing.

BRICK

Building bricks are small masonry blocks of inorganic nonmetallic material hardened by heat or chemical action, Fig. 21-6. Building bricks may be solid or may have core openings to reduce the weight. Bricks are produced in a wide variety of colors, shapes and textures.

CLASSES OF BRICK

Bricks are usually classified as:
1. Adobe, made of natural, sun dried clays and a binder.
2. Kiln burned, made of clay or shale and burned to harden them.
3. Sand-lime, mixtures of sand and lime hardened under pressure and heat.
4. Concrete, made of cement and aggregates.

BRICK TYPES AND SIZES

Bricks are available in many different sizes and types. The most commonly used bricks are made of solid clay. They are classified as common building brick, face brick, special brick and custom brick.

COMMON BRICK

Common brick is the most widely used for building. It is the ordinary red brick used for walls, backing and in structures where a special color, shape or texture is not required.

MODULAR BRICK

Brick which can be laid to modular dimensions are available just about anywhere. Modular means that any dimension of a unit is a multiple of its smallest dimension. In brick the smallest dimension is 4 in. including the mortar joint. All other dimensions are multiples of 4 in. including all mortar joints. See Fig. 21-7.

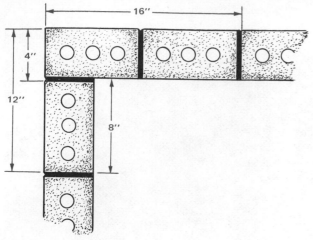

Fig. 21-7. Modular brick are so called because they can be combined in various ways to make up widths or lengths that are double, triple or four times the smallest dimension. In the illustration above, combining a length and a width equals three times the smallest dimension. Combining two lengths equals four times the smallest dimension.

FACE BRICK

Face bricks are very carefully made. Dimensions are kept very accurately to a standard size. Their structural quality is high. There is a large range of sizes, colors and textures to produce different architectural effects.

GLAZED FACE BRICK

Glazed bricks are produced to the same close tolerances as other face brick. They have a hard, impervious face with a dull, satin or glossy finish, Fig. 21-8. (Impervious means almost impossible to damage.) The brick is sprayed with a

Fig. 21-8. Glazed bricks have a hard, glass-like surface. Often produced in various colors, they provide special appearance to a finished building. Often they are used for finished walls of rest rooms and kitchens of institutional buildings.

ceramic glaze and then fired to temperatures of 2000 F (850 C) to fuse the glaze to the brick. Many colors, textures and finishes are produced.

FIREBRICK

Special clays are used where brick will be subjected to extreme heat, like a fireplace. These clays contain a high percentage of aluminum or silica, flint and feldspar (a crystalline mineral of aluminum silicate). Firebrick, containing such clays, are used to line incinerators, chimney stacks, industrial furnace fireboxes and smelting furnaces. Firebrick are usually softer than common or face brick. They are light beige or brown in color.

CLAY BUILDING TILE

Structural clay tiles are made of burned clay. Larger than brick, they have large, cored openings and thin walls. Building clay tile has many uses in the construction industry. The hollow-celled units are produced in load-bearing or nonload-bearing forms. Structural clay tile may be used to fireproof structural steel or as a furring or filling material inside curtain walls.

Clay tile may be glazed with either a transparent or colored glaze. This produces a glass-like, moisture-proof surface.

TERRA-COTTA

Terr-cotta, meaning "baked earth," is a nonload-bearing burned clay building unit. Similar to brick, it has been used for centuries as an ornamental substitute for stone. It is attractive as floor tile and as a veneer for inside and outside walls.

CONCRETE BLOCK

Hollow masonry units of portland cement, sand and fine gravel aggregates are called concrete blocks, Fig. 21-9. Concrete blocks are used for interior and exterior load-bearing and nonload-bearing walls, partitions and backing for veneer. Veneers are thin pieces or sheets of decorative masonry or wood. Weight, color and texture of concrete block depends largely on the type of aggregate used in its manufacture. Blocks made with sand and gravel weigh from 40 to 45 lb. (18 to 20 kg) for each 8 in. by 8 in. by 16 in. unit. Blocks are strong and durable. Standard units have the typical light gray color of concrete.

Fig. 21-9. Concrete building blocks are modular in size. They are often used as filler material to be faced with brick.

DETAILED BLOCKS

Detailed blocks have a patterned face. Some have vertical and horizontal grooves to simulate brick. Others have triangular, or rectangular indented areas. Such block may be laid in different directions to create interesting patterns.

SCREEN UNITS

Use of concrete masonry units to form open grilles is new. Screen-wall masonry units can be molded with an almost endless variety of pierced openings. The screen units are set in mortar in the same way as standard blocks.

GYPSUM BLOCK

Gypsum blocks are hollow units made of gypsum and a binder of vegetable fiber, asbestos or wood chips. Never used where they would be subjected to moisture, they are suitable for nonload-bearing, fire resistant interior partitions. Another use is as a fire proofing material around structural steel beams and columns. They are usually finished with a plaster coating. Easily cut and light weight, gypsum blocks are relatively inexpensive to install.

GLASS BLOCK

Glass blocks are formed by fusing together two glass shells that have been pressed to shape in forms. The shell is airtight to hold a partial vacuum inside. This partial vacuum provides good insulation. Additional advantages are low maintenance and controlled daylight. Glass block may be used as windows, interior partitions, screens, or entire walls.

SUMMARY

Masonry materials have been used as building units since ancient civilizations first dried clay in the sun to make bricks. Today, modern construction technology uses masonry units such as stone, brick, tile, glass block, gypsum block and concrete block as basic building materials.

Architectural design can be improved by the shape, size, color and texture of the blocks. The designer/builder has a large variety of natural stone or fired products to work with.

CAREERS RELATED TO CONSTRUCTION

BRICK MASON lays building materials such as brick, structural tile, concrete block and terra-cotta block to construct or repair walls, partitions, arches and other structures.

CERAMIC ENGINEER does research, designs machinery, develops processing techniques and oversees technical work concerned with manufacture of ceramic products.

STONE CUTTER cuts and shapes sandstone, marble and other soft stone. Draws designs on stone with compass, straightedge and other drawing tools. Cuts and shapes designs with hand or pneumatic stone-cutting tools.

STONE MASON builds stone structures and lays walks, curbstones or special types of masonry. Shapes and cuts stone and lays it.

STONE MECHANIC lays out work and uses hand tools and machines for cutting, polishing and sandblasting buildings and monumental stone.

DISCUSSION TOPICS

1. Describe the geological classification of rock commonly used in construction.
2. Where are screen blocks used?
3. What are the advantages of glass block?
4. Where would you use gypsum block?
5. What is terra-cotta?
6. What is a detailed concrete block?
7. What is the difference between common and face brick?
8. How is glazed brick different from common brick?
9. What is texture? How does texture differ in face brick?
10. How does adobe differ from brick?
11. What is the principal use of firebrick?

Chapter 22

WOOD AND LUMBER PRODUCTS

Fig. 22-1. Ponderosa pine and white fir forests cover thousands of acres in the Pacific Northwest. (Weyerhaeuser Co.)

Fig. 22-2. Redwood is one of the most serviceable of woods. It resists decay and ages well. (California Redwood Assoc.)

Wood has always been one of our most important construction materials. Americans have come to depend upon wood products from the great forests of this country, Fig. 22-1. Lumber production has always been one of our largest industries. Sawmills have been in production since 1610. Forest products have been used for fuel, furniture, ships and buildings.

Millions of board feet of lumber are produced annually. (One board metre is equal to 10.763 board ft.) Lumber is widely used for modern homes. Wood for pulp and paper make life more comfortable and interesting. Wood helps insulate walls. Used on outside walls it makes them more pleasing to the eye, Fig. 22-2. It is the base material for many products that make life more pleasant.

Lumber destined for construction purposes may be classified into three basic groups:

1. Lumber for primary construction purposes, Fig. 22-3.
2. Lumber used for secondary construction purposes such as building forms.
3. Lumber for architectural purposes, Fig. 22-4.

CLASSIFICATION OF WOOD

Woods are generally classified for construction purposes as either hardwood or softwood. The terms may be a little

Fig. 22-3. Lumber for primary construction uses actually becomes a part of the structure. In this house, lumber is used as a support. It is exposed to the weather.

Fig. 22-4. Lumber for architectural purposes is used as wall paneling, as ridge and side beams, and as V-grooved decking spans which form the ceiling.   (Weyerhaeuser Co.)

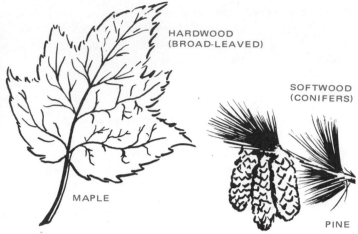

Fig. 22-5. Hardwoods are produced by broad-leaved trees. Softwoods are produced by conifers.   (Frank Paxton Lumber Co.)

Fig. 22-6. Hardwoods are used primarily for interiors. Their natural grain figuration adds warmth and beauty to the home, office or business. Flooring is of oak while furniture and cabinets are of some type of close-grained woods.

misleading. There is no direct relationship to the relative hardness of the wood. Balsa wood is classified as a hardwood, but is softer than most woods in the softwood classification. Softwoods are woods produced by evergreens or conifers (cone bearing trees). See Fig. 22-5. Hardwoods are woods produced from the broadleaf or deciduous trees. (A deciduous tree is one which loses its leaves.)

HARDWOODS

Hardwoods are generally used for wood furniture, decorative interior paneling and as interior trim. See Fig. 22-6. Some hardwoods are extremely hard and strong, but are seldom used as structural materials. The scarcity, high cost, and beauty place these woods at a premium so that they are used for interior purposes only. Veneers of hardwood bonded onto a softwood core provide the industry with an economical use of hardwood.

Many different kinds of hardwood, some native to the United States and some imported, are used in fine furniture and for interior wall paneling.

Irregularities in tree growth and defects in the wood may provide a variation within each species. Many are prized for such variation of appearance.

The important hardwoods of North America are ash, basswood, beech, birch, cottonwood, elm, gum, hickory, maple, oak, black walnut and yellow poplar.

SOFTWOODS

Softwoods are the evergreen, cone-bearing and needle-leaved trees. Wood or lumber from these trees is generally used for construction, Fig. 22-7. It provides framing for homes and forms for concrete. Softwood lumber is often laminated to serve as structural members and as plywood covering.

Important softwoods of North America are cedar, cypress, fir, hemlock, larch, pine, redwood and spruce. The proper

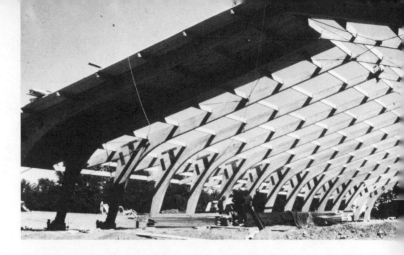

Fig. 22-7. Softwood structural beams are fabricated to span long distances. (Southern Pine Assoc.)

SPECIES	Comparative Weights	Color	Hand Tool Working	Nail Ability	Relative Density	General Strength	Resistance to Decay	Wood Finishing	Cost
HARDWOODS									
Apitong	Heavy	Reddish Brown	Hard	Poor	Medium	Good	High	Poor	Medium High
Ash, brown	Medium	Light Brown	Medium	Medium	Hard	Medium	Low	Medium	Medium
Ash, tough white	Heavy	Off-White	Hard	Poor	Hard	Good	Low	Medium	Medium
Ash, soft white	Medium	Off-White	Medium	Medium	Medium	Low	Low	Medium	Medium Low
Avodire	Medium	Golden Blond	Medium	Medium	Medium	Low	Low	Medium	High
Balsawood	Light	Cream White	Easy	Good	Soft	Low	Low	Poor	Medium
Basswood	Light	Cream White	Easy	Good	Soft	Low	Low	Medium	Medium
Beech	Heavy	Light Brown	Hard	Poor	Hard	Good	Low	Easy	Medium
Birch	Heavy	Light Brown	Hard	Poor	Hard	Good	Low	Easy	High
Butternut	Light	Light Brown	Easy	Good	Soft	Low	Medium	Medium	Medium
Cherry, black	Medium	Medium Reddish Brown	Hard	Poor	Hard	Good	Medium	Easy	High
Chestnut	Light	Light Brown	Medium	Medium	Medium	Medium	High	Poor	Medium
Cottonwood	Light	Greyish White	Medium	Good	Soft	Low	Low	Poor	Low
Elm, soft grey	Medium	Cream Tan	Hard	Good	Medium	Medium	Medium	Medium	Medium Low
Gum, red	Medium	Reddish Brown	Medium	Medium	Medium	Medium	Medium	Medium	Medium High
Hickory, true	Heavy	Reddish Tan	Hard	Poor	Hard	Good	Low	Medium	Low
Holly	Medium	White to Grey	Medium	Medium	Hard	Medium	Low	Easy	Medium
Korina	Medium	Pale Golden	Medium	Good	Medium	Medium	Low	Medium	High
Magnolia	Medium	Yellowish Brown	Medium	Medium	Medium	Medium	Low	Easy	Medium
Mahogany, Honduras	Medium	Golden Brown	Easy	Good	Medium	Medium	High	Medium	High
Mahogany, Philippine	Medium	Medium Red	Easy	Good	Medium	Medium	High	Medium	Medium High
Maple, hard	Heavy	Reddish Cream	Hard	Poor	Hard	Good	Low	Easy	Medium High
Maple, soft	Medium	Reddish Brown	Hard	Poor	Hard	Good	Low	Easy	Medium Low
Oak, red (average)	Heavy	Flesh Brown	Hard	Medium	Hard	Good	Low	Medium	Medium
Oak, white (average)	Heavy	Greyish Brown	Hard	Medium	Hard	Good	High	Medium	Medium High
Poplar, yellow	Medium	Light to Dark Yellow	Easy	Good	Soft	Low	Low	Easy	Medium
Prima Vera	Medium	Straw Tan	Medium	Medium	Medium	Medium	Medium	Medium	High
Sycamore	Medium	Flesh Brown	Hard	Good	Medium	Medium	Low	Easy	Medium Low
Walnut, black	Heavy	Dark Brown	Medium	Medium	Hard	Good	High	Medium	High
Willow, black	Light	Medium Brown	Easy	Good	Soft	Low	Low	Medium	Medium Low
SOFTWOODS									
Cedar, Tennessee Red	Medium	Red	Medium	Poor	Medium	Medium	High	Easy	Medium
Cypress	Medium	Yellow to Reddish Brown	Medium	Good	Soft	Medium	High	Poor	Medium High
Fir, Douglas	Medium	Orange-Brown	Medium	Poor	Soft	Medium	Medium	Poor	Medium
Fir, white	Light	Nearly White	Medium	Poor	Soft	Low	Low	Poor	Low
Pine, yellow longleaf	Medium	Orange to Reddish Brown	Hard	Poor	Medium	Good	Medium	Medium	Medium
Pine, northern white (Pinus Strobus)	Light	Cream to Reddish Brown	Easy	Good	Soft	Low	Medium	Medium	Medium High
Pine, ponderosa	Light	Orange to Reddish Brown	Easy	Good	Soft	Low	Low	Medium	Medium
Pine, sugar	Light	Creamy Brown	Easy	Good	Soft	Low	Medium	Poor	Medium High
Redwood	Light	Deep Reddish Brown	Easy	Good	Soft	Medium	High	Poor	Medium
Spruces (average)	Light	Nearly White	Medium	Medium	Soft	Low	Low	Medium	Medium

Fig. 22-8. Wood selection chart.

144

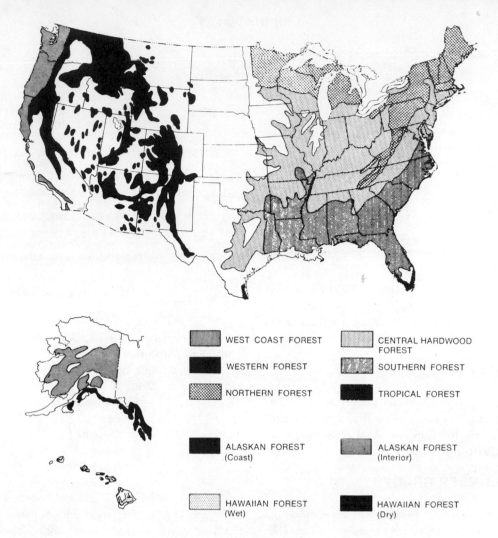

Fig. 22-9. Major forest regions of the United States.

FOREST AREAS

Lumber for the construction industry comes from all parts of the country as shown in Fig. 22-9. Trees do not grow in abundance in the grasslands and semi-desert regions of the central great plains; nor are they found in the arid (very dry) basins of the west.

WEST COAST FOREST

The west coast forest produces largely softwoods. They are known as Douglas fir, sugar pine, lodgepole pine, western hemlock, western red cedar and Sitka spruce. Common hardwoods grown in this area are red alder and big-leaf maple.

WESTERN FOREST

The western forest consists of almost 90 million acres in the Rocky Mountain regions yielding largely softwood lumber. Included are ponderosa pine, sugar pine and Idaho white pine,

Douglas fir, Engelman spruce, western larch and western red cedar. Aspen is the principal hardwood produced.

NORTHERN FOREST

The northern forest extends from Minnesota to Maine. Softwoods produced from this region are Norway pine, jack pine, white spruce, eastern hemlock, white cedar, white pine and balsam fir. Hardwoods of this region are birch, maple, oak, cherry, gum, basswood and aspen.

CENTRAL HARDWOOD FOREST

The central hardwood forest has the largest area — approximately 131 million acres. It is located in mid-America. It is primarily a hardwood forest of oaks, hickory, poplar, maple, beech, elm, ash, cottonwood, sycamore and walnut. Softwoods of this region are pine and red cedar.

SOUTHERN FOREST

The southern forest of southeast United States is noted for its softwoods of shortleaf yellow pine, longleaf yellow pine, loblolly pine, slash pine and cypress. The southern forest is equally noted for its hardwoods, oak, willow, cottonwood, ash and pecan.

TROPICAL FOREST

The tropical forest is essentially noncommercial. It lies on the tip of Florida and southern Texas. Notable are hardwoods of mangrove, eucalyptus and mahogany in this smallest of all U.S. forests.

ALASKAN FORESTS

The Alaskan forests are divided into two areas:
1. Coastal.
2. Interior.

The coastal forest consists of western hemlock, Sitka spruce, western red cedar and the Alaskan yellow cedar. The interior forest produces white and black spruce. Hardwoods of birch and aspen may also be found.

HAWAIIAN FOREST

The Hawaiin forest is divided into wet and dry areas. In the dry forest grow algaroba, koa haole, wiliwili and manbrypod. The wet forest produces furniture wood of ohia, koa, tree fern, mamani, tropical ash and eucalyptus.

LUMBER GRADES

Lumber is graded according to strength, appearance or usability. This has been the subject of extensive testing. The National Bureau of Standards, Department of Agriculture and various lumber associations have spent many years in establishing lumber grades. Grades and sizes are provided to protect the user and to establish a degree of uniformity.

When lumber is cut from the log, the pieces are very different in appearance and strength. Much depends on the number and size of imperfections. Grades are based on the number, size and location of these defects. Defects include knots, checks, splits and pitch pockets. The highest grades are practically free from these imperfections. Lower grades allow more and more imperfections.

HARDWOOD GRADES

Hardwood grades are based on the amount of usable lumber in each piece of a standard length, from 4 to 16 ft. (1.2 to 4.9 m) in length. See Fig. 22-10.

HARDWOOD GRADES	
FIRSTS	91 2/3% CLEAR FACE CUTTINGS 8 TO 16 FT. LONG, 6 IN. WIDE OR WIDER (2.5 TO 5 m LONG, 15 cm WIDE OR WIDER)
SECONDS	83 1/2% CLEAR FACE CUTTINGS 8 TO 16 FT. LONG, 6 IN. WIDE OR WIDER (2.5 TO 5 m LONG, 15 cm WIDE OR WIDER)
SELECTS	91 2/3% CLEAR FACE CUTTINGS 6 TO 16 FT. LONG, 4 IN. WIDE OR WIDER (1.7 TO 5 m LONG, 10 cm WIDE OR WIDER)
NO. 1 COMMON	66 2/3% CLEAR FACE CUTTINGS 4 TO 16 FT. LONG, 3 IN. WIDE OR WIDER (1.2 TO 5 m LONG, 7.8 cm WIDE OR WIDER)
NO. 2 COMMON	50% CLEAR FACE CUTTINGS 4 TO 16 FT. LONG, 3 IN. WIDE OR WIDER (1.2 TO 5 m LONG, 7.8 cm WIDE OR WIDER)

Fig. 22-11. Hardwood grades.

The percentage of the board free of defects determines the grade. This percentage is called the clear face cutting. Grade is determined on this base plus the length and width of the board. See Fig. 22-11.

Standard Grades
Firsts ⎫
Seconds ⎬ FAS
Selects
No. 1 Common
No. 2 Common

Firsts and seconds are grouped together and marketed as one grade (FAS). Hardwood whose unique appearance is the result of worm holes and other imperfections is called sound wormy and classed as No. 1 Common.

SOFTWOOD GRADES

The American Lumber Standards for softwood lumber divide softwood lumber into three basic groups:
1. Yard.
2. Structural.
3. Factory and shop. See Fig. 22-12.

Softwoods are graded on the condition of the whole board. A simple standard has been made to simplify the grading rules for common thickness, widths and lengths.

Yard lumber softwood grades are:
1. Select A, B, C and D.
2. Common boards, Nos. 1, 2, 3, 4 and 5.
3. Dimension lumber, Nos. 1, 2 and 3.

Select grades are used in visible construction. Common

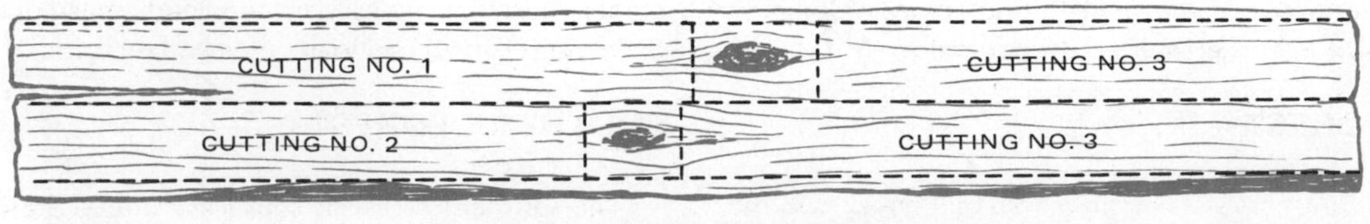

Fig. 22-10. The number of clear face cuttings in a hardwood board determines its grade. Brown lines show how knots and cracks are removed by cutting.

	SOFTWOOD GRADING		
YARD	FINISH	GRADE A SELECT (FREE OF NEARLY ALL DEFECTS) GRADE B SELECT (VERY FEW IMPERFECTIONS) GRADE C SELECT (INCREASINGLY MORE IMPERFECTIONS) GRADE D SELECT (INCREASINGLY MORE IMPERFECTIONS)	
	COMMON	NO. 1 BD. CONSTRUCTION NO. 2 BD. STANDARD NO. 3 BD. UTILITY NO. 4 BD. ECONOMY NO. 5 BD. ECONOMY	
	DIMENSION	PLANKS SCANTLING JOISTS	NO. 1 DIMENSION NO. 2 DIMENSION NO. 3 DIMENSION
STRUCTURAL	JOIST AND PLANK — 2 TO 4 IN. (5 TO 10 cm) BEAMS AND STRINGERS — 5 IN. AND OVER (13 cm AND OVER) POST AND TIMBERS (6 x 6 IN ANY LENGTH)		
FACTORY AND SHOP	FACTORY PLANK	FACTORY CLEANS SHOP LOWER	NOS. 1 AND 2 CLEAR FACTORY NO. 3 CLEAR FACTORY NOS. 1, 2 AND 3 SHOP
	SHOP LUMBER	ONE INCH ALL THICKNESS	SELECT SHOP TANK AND BOAT NO. 1 SHOP BOX NO. 2 SHOP BOX

Fig. 22-12. Softwood grading.

boards are used for sheeting, rough flooring, general utility and construction purposes.

Dimension grades are used for framing where strength and support is needed. Such lumber is usually 2 to 5 in. (5 to 13 cm) thick.

Large support timbers may be left as round poles or may be sawed to size and shape. They are graded according to strength. Sawed timbers may be either rough or planed to make them smooth. Structural lumber is used for beams, posts and joists. Round timbers are sometimes used in framing but are more often employed as poles and piling. Fig. 22-13 shows them used as foundation support for a house.

The factory and shop lumber is for special uses, such as windows, sashes and doors. The boards are usually 1 1/4 in. (32 mm) or greater in thickness.

MANUFACTURING OF LUMBER

Moving lumber from the forest to the construction site involves many different occupations. Fig. 22-14 shows a loader

Fig. 22-13. An all wood pole house. Poles are sunk into ground to provide a foundation. Other poles are used for framing members. All structural lumbers used here are pressure treated to protect wood against decay and termites. (Koppers Co., Inc.)

Fig. 22-14. Heavy lift truck loads lumber mechanically in a mill yard. (Allis-Chalmers)

Fig. 22-17. Logs collected along logging road are ready for loading and transporting to mill.

Fig. 22-15. Trees are felled and cut into logs with large chain saws. (Weyerhaeuser Co.)

Fig. 22-18. This 40-ton unloader speeds handling of large logs at the mill. (Simpson Timber Co.)

Fig. 22-16. Large caterpillar type tractor skids log out of forest.

in a mill yard placing lumber on a truck. The truck will move it to the construction site or to a distant lumber yard where it will be sold to customers.

To reach this point it has gone through a number of processing steps. We will go through these processes now.

In the forest, trees are felled, Fig. 22-15 and skidded, Fig. 22-16, to central collection sites like the one shown in Fig. 22-17. Here they are picked up and put on large trucks that move them to the mill. At the mill, Fig. 22-18, trucks are unloaded and logs are placed in a holding yard. When the mill is ready to process them they are moved up a ramp into the sawmill, Fig. 22-19.

DEBARKING

Logs have their bark removed either before entering the mill or immediately upon entering the mill building. This is done mechanically by large rough-surfaced wheels called

Fig. 22-19. Logs travel up log slip in preparation for sawing into sections of desired length. (Weyerhaeuser Co.)

Fig. 22-20. Mechanical debarkers remove the bark from log. Generally the bark is recovered and converted into a variety of other products. Paper and paper boxes are just two of the items made from bark.

Fig. 22-22. Bark is blasted from logs by strong jets of water. (Weyerhaeuser Co.)

Fig. 22-21. Ringbarker removes bark from larger logs.

debarkers. Different kinds of debarkers handle logs of different sizes. The mechanical debarker shown in Fig. 22-20 handles logs 6 to 26 in. (15 to 66 cm) in diameter. Larger logs are sent to a semiautomatic debarker. It has metal rollers and rough fingers to remove bark, Fig. 22-21.

In some mills bark is blasted from logs by jets of water under pressure of 1500 psi. This is equal to 10.3 kilopascals (kPa). Fig. 22-22 shows a water debarking operation. Sometimes the logs are debarked in the forest. In this way the bark is recycled. It is left on the forest floor to decay thereby enriching the soil.

In any case, removing bark is basically a conservation measure. It permits recovery of slabs and edgings for conversion to chips for pulp and other wood products.

Fig. 22-23. Headrig or headsaw cuts log into cants. Carriage moves past saw on track. Carriage is moved by hydraulic ram shown at lower left.

SAWING LOGS INTO LUMBER

Inside the mill shed, logs are positioned on a riderless carriage, Fig. 22-23. Air-operated, this carriage can hold logs up to 20 ft. (6 m) long. The carriage carries the log past the headrig (headsaw) or a band mill (band saw) which cuts the log into cants. A cant is a log that has been slabbed (cut) to square it up on two or four sides.

Fig. 22-24. Gang saw cuts four cants together. About 56 individual pieces of lumber are made from this single cut.

The cants are then moved to the gang saw, Fig. 22-24. The gang saw has many separate band saw blades, cutting several boards at the same time. A vertical resaw machine cuts an individual plank into separate boards, Fig. 22-25.

From the gang saws and vertical resaws, the lumber travels on chain conveyors to trimmer saws, Fig. 22-26. Usable waste is sent to chippers.

Waste utilization is a part of any modern mill. Waste wood is reduced to chips. The chips are, in turn, converted into paper, paper boxes, hardboard, particle board and many other products. Waste which cannot be used in any other product is burned, Fig. 22-27.

Fig. 22-25. Vertical resaw makes single cuts in planks.

Fig. 22-26. Trimmer saws, operated by remote control, square ends, remove defective sections and cut board to desired length.

METHODS OF CUTTING

The manner in which a piece of wood is cut from the log affects its appearance and strength. Lumber sawed in slices parallel to one side of the log is said to be plain sawed or flat sawed, Fig. 22-28. Flat-sawed lumber is produced with less waste. It is therefore cheaper. Growth ring patterns are more distinct, while the grain and decorative patterns of many hardwoods are brought out when plain sawed.

But flat or plain-sawed lumber has certain disadvantages

Fig. 22-27. Waste burner consumes wood waste without belching smoke into atmosphere. (Simpson Timber Co.)

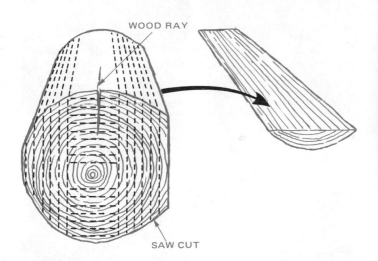

Fig. 22-28. Plain or flat-sawed lumber is cut tangent to the annual growth ring. (A tangent cut runs at right angles to a line running through the center of the log.) (Frank Paxton Lumber Co.)

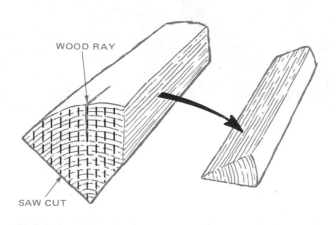

Fig. 22-29. Quarter-sawed lumber is cut more perpendicular to the annual growth rings.

too. It is more likely to warp, check, split and cup.

If the log is first quartered and boards are cut from each quarter perpendicular (at right angles) to the exterior of the log, it is said to be quarter sawed, Fig. 22-29. Quarter-sawed lumber is more expensive because of the added labor and waste in cutting. It swells and shrinks less than plain-sawed lumber. During seasoning it develops fewer cracks and checks.

SEASONING LUMBER

The process of drying lumber to the point where it is ready to be used is called seasoning. Freshly sawed (green) lumber contains as much as one hundred percent of its dry weight in moisture.

Fig. 22-30. Stacking lumber for air drying. Stacking strips allow air to circulate freely through stack. (Allis-Chalmers)

AIR DRYING

To dry properly lumber must be stacked in covered piles. Each layer is separated by one-inch strips to allow air to flow between layers. See Fig. 22-30. Drying time to bring lumber to the proper moisture content varies with the area and the weather conditions. In the United States, softwood is seldom stored longer than three or four months.

The recommended moisture content varies from seven percent in dry areas to 18 percent in damp areas. It is helpful to store lumber on the job site for a period of time. This practice allows the lumber to reach a point at which moisture content is the same as the surrounding air. This is called "equilibrium." At this point the shrinking and swelling will be at a minimum.

KILN DRYING

Lumber may be kiln dried to speed up reduction of moisture. It is placed in ovens or kilns and exposed to an elevated temperature and controlled humidity. Drying temperatures of 70 to 120 F (20 to 50 C) for periods of time ranging from 4 to 10 days are needed for softwoods. More drying is needed for hardwoods. Kiln drying reduces cracking and checking of certain types of lumber.

DEFECTS IN WOOD

Defects are imperfections in wood which detract from its appearance or strength. These include knots, decay, stains and cracks.

KNOTS

Growth of limbs or branches of a tree disturbs the symmetry of the annual rings creating knots. Knots may have deposits of pitch (gummy residue of softwood trees) which add further to the defects.

DECAY

Wood fiber may also have weakened areas caused by certain insects or fungi. This decay may be small isolated spots or large soft areas. Decay in lumber is usually called rot or dote.

STAINS AND CRACKS

Stains are discolorations that penetrate the wood. Cracks are separations running along or across the grain. They are classified according to the direction in which they run. A shake is a separation along the grain between the annual growth ring. A check is a small separation in the end grain at right angles to annular rings. It usually occurs during seasoning process as a result of shrinking. Lumber will bow, warp, cup and twist. Each of these defects may affect its strength or usability.

LUMBER SIZES

Softwood lumber is sawed in dimensions (width and thickness) of even inches. The length may vary from 8 to 24 ft. with increments of 2 ft. These are nominal dimensions of the lumber. The actual sizes, due to sawing, planing, and surfacing are 1/4 to 3/4 in. less than nominal dimensions. Standard sizes of lumber are established by the U.S. Department of Commerce, Fig. 22-31.

TREATED LUMBER

Chief causes of deterioration in wood are decay, insects and fire. Some species of wood are more durable and resist all these factors better than others. However, lumber may be treated to protect it against deterioration.

Preservatives may be brushed on the surface, or the lumber may be dipped into large tanks. These methods only partially treat the wood. The preservative must be drawn or forced into the wood for best results. See Fig. 23-32. To increase penetration of preservatives, lumber is placed in pressure tanks where the preservative is forced into the wood, Fig. 22-33.

SOFTWOOD LUMBER SIZES		
NOMINAL SIZE FOR BOARD LUMBER	ACTUAL SIZE (INCHES)	ACTUAL SIZE (CENTIMETRES)
1 x 4	3/4 x 3 1/2	1.9 x 9
1 x 6	3/4 x 5 1/2	1.9 x 14
1 x 8	3/4 x 7 1/4	1.9 x 19
1 x 10	3/4 x 9 1/4	1.9 x 22
1 x 12	3/4 x 11 1/4	1.9 x 28
FOR DIMENSION LUMBER		
2 x 4	1 1/2 x 3 1/2	3.8 x 9
2 x 6	1 1/2 x 5 1/2	3.8 x 14
2 x 8	1 1/2 x 7 1/4	3.8 x 19
2 x 10	1 1/2 x 9 1/4	3.8 x 22
2 x 12	1 1/2 x 11 1/4	3.8 x 28

Fig. 22-31. Softwood lumber sizes.

Fig. 22-32. Utility poles enter pressure vessel for treatment against insects and decay. (Koppers Co. Inc.)

Fig. 22-33. Lumber that will be exposed to soil, air and high moisture conditions is treated to increase its serviceable life.

TYPES OF PRESERVATIVES

Preservatives used to reduce damage from destructive agents of wood are usually one of three types:
1. Oil based.

Fig. 22-34. Glued-laminated timbers may be designed and built to span distances and acquire shapes not possible with solid timbers. (American Institute of Timber Construction)

2. Water soluble salts.
3. Solvent based.

Creosote is the oldest of our preservative materials. It forms an effective barrier against decay and insects. However, the objectionable odor limits its use to pilings, utility poles and waterfront piers. Creosote does not mix with water. Thus, moisture does not affect it. Creosoted wood cannot be painted successfully.

Water soluble salts, give good protection against decay and insects. Lumber can be painted after treatment.

Organic solvent-soluble materials can be painted over. Materials such as pentachlorophenol (penta) are used as a preservative for fence posts, structural lumber, sashes and doors. When the solvent evaporates, a coating of nonsoluble materials is left on the wood.

GLUED-LAMINATED LUMBER

Construction with solid timber is limited by the size and shape of timbers available. With modern methods and the development of new adhesive materials we can fabricate wood members of almost any size and shape, Fig. 22-34.

BUILDING UP BEAMS

Small pieces of wood are bonded together to form a laminated structural member that is strong and durable, Fig. 22-35. High strength, good quality material is used at stress points. Less expensive lumber can be used where strength is not important. Better grades may be selected for appearance where it can be seen. Less attractive pieces will be used when appearance is not a factor.

Laminated members are fabricated in buildings that have facilities to control temperature. Pressure must be exerted to form the bond between the glue and the wood. Complex shapes require special forms that can be firmly secured. See Fig. 22-36.

GRADES OF LAMINATED PRODUCTS

Laminated members are graded according to use and appearance. There are three appearance grades:
1. Industrial.
2. Architectural.
3. Premium.

Appearance is based on such factors as grain, fillers and surface imperfections. These characteristics do not affect the strength of the member.

Industrial grade is suitable for industrial plants and other locations where looks are not important. Architectural grades

Fig. 22-35. Worker trims large beam made up of many small pieces of wood bonded together. By using thin pieces workers can shape graceful curves while building up beams and structural members of huge size.

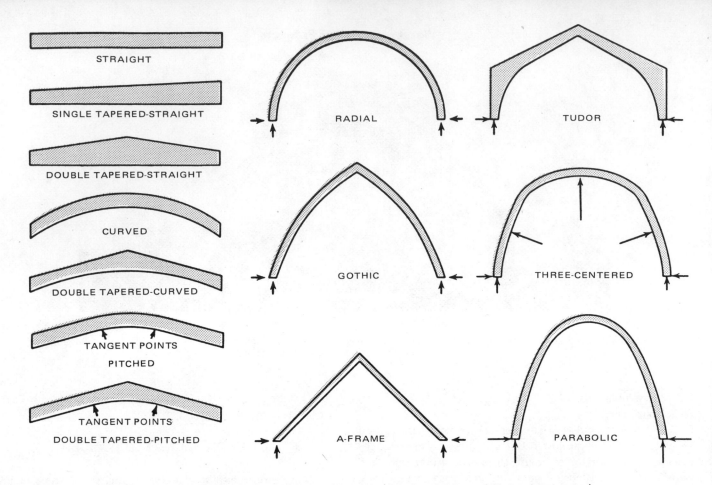

STRAIGHT

SINGLE TAPERED-STRAIGHT

DOUBLE TAPERED-STRAIGHT

CURVED

DOUBLE TAPERED-CURVED

TANGENT POINTS

PITCHED

TANGENT POINTS

DOUBLE TAPERED-PITCHED

RADIAL

TUDOR

GOTHIC

THREE-CENTERED

A-FRAME

PARABOLIC

Fig. 22-36. Various shapes of laminated arches and beams. (American Institute of Timber Construction)

are used where appearance is important. This grade allows some imperfections such as small voids, tight knots and other small defects. These defects are filled or connected on the job. Architectural grades are suitable for painting.

Premium grade is the best obtainable. It is used where natural finishes are desired.

SUMMARY

Wood was one of the earliest materials used by humans in the construction of shelter. At one time great stands of large trees covered most of the United States and Canada. It was possible for the builder to go into the forest and select a tall, straight tree for beams or structural members of almost any size. Today, this is not possible. The modern builder must adapt the wood that is still available to his requirements. Builders have learned to use wood as either a primary building material or as a secondary building material. Wood can be glued, laminated and bonded to other materials for strength and durability. Wood is still a basic material for construction.

CAREERS RELATED TO CONSTRUCTION

LOG GRADER judges quality of logs at mill or in the woods. May estimate board footage.

LOGGER fells trees and prepares them for final logging operations. Cuts felled trees into lengths. May drive tractor to skid logs through forest.

LOG INSPECTOR determines quality and estimates board footage; buys logs for mill.

GANG SAWYER controls carriage (platform on rails) to feed log to machine equipped with many saws to cut lumber into dimensions.

LOG SCALER estimates marketable lumber content of logs by measuring log and using scale stick or other measuring devices. May grade and mark logs for use as veneer.

WOOD CHEMIST experiments with wood to develop new products and improve old products made of wood.

WOOD TECHNOLOGIST conducts research in seasoning, preserving and using wood and wood by-products. Determines methods for curing lumber. Develops and improves wood treatment methods. Investigates processes for converting wood into other products.

DISCUSSION TOPICS

1. Explain the methods of sawing lumber.
2. What is the difference between softwood and hardwood?
3. Explain the advantages of kiln dried lumber.
4. What is treated lumber and where is it used?
5. What region of the United States produces large quantities of lumber for light construction?
6. What region of the United States produces large quantities

of hardwood for interior uses?

7. Name the three basic groups of softwood lumber.

8. How is structural lumber different from yard lumber?

9. What is a cant?

10. Why is wood dried?

11. What are lumber defects?

12. What is the nominal size of a 2″ x 4″?

13. What is the metric size of a 2″ x 4″?

Chapter 23
PLYWOOD AND SHEET STOCK

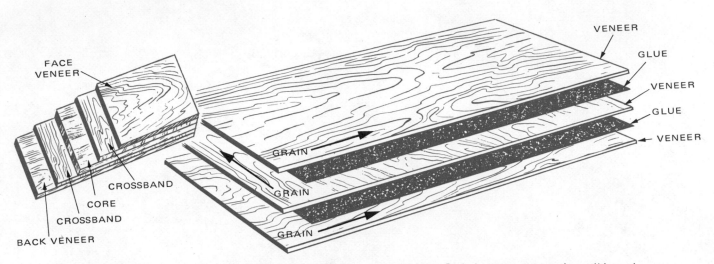

Fig. 23-1. Each plywood layer has grain running at 90 deg. to the next layer. Such sheets are stronger than solid wood. (Frank Paxton Lumber Co.)

Ancient Egyptians cut thin strips of wood and formed plywood sheets as long ago as 3000 BC. Sheets discovered in excavations have survived for centuries.

Today the use of plywood is very much dependent on the improvements of adhesives. *Plywood is construction material made of thin sheets of wood (veneers) bonded together with glue.* Each sheet of veneer is assembled so that the grain of each layer is at right angles to the next. See Fig. 23-1. Plywood is stronger than solid wood of the same thickness.

Plywood resists splitting, checking and does not swell or shrink as much as solid wood. Plywood is also less likely to warp and twist than solid wood of the same size. It is easy to work with ordinary tools and this helps speed up construction.

PLYWOOD SIZES

Plywood is generally produced in 4 ft. by 8 ft. (122 cm by 244 cm) sheets. Standard thicknesses are 1/4, 5/16, 1/2, 5/8, 3/4, 1 and 1 1/8 in. (6, 8, 12, 16, 18, 25 and 28 mm). Plywood sheets have three, five or seven plys or veneers. Having an odd number of sheets leaves the grain of both outside layers running the same direction.

METHODS OF CUTTING VENEER

Most veneers are cut by:
1. Rotary cutting.
2. Slicing. They are cut from a log or part of a log called a flitch. See Fig. 23-2.

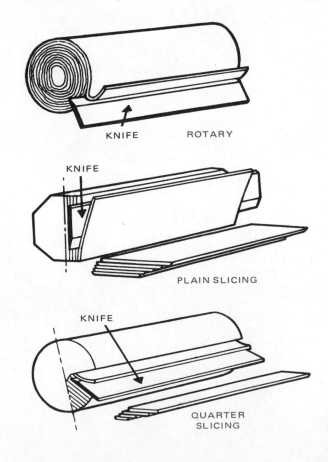

Fig. 23-2. Veneer cutting methods.

Fig. 23-3. Logs are soaked in huge steam vats to soften wood fibers before cutting veneer. (Simpson)

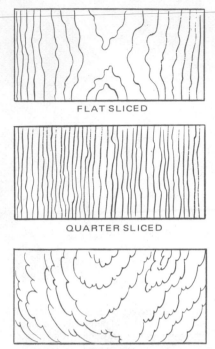

Fig. 23-5. Characteristic grain patterns using various methods of cutting.

ROTARY CUTTING

Rotary cutting is the most economical. To soften the fibers, Fig. 23-3, the flitch is first steam cooked in huge vats.

The log is then placed on a huge lathe. The lathe rotates the log slowly against a sharp knife, Fig. 23-4. This produces a thin continuous sheet of wood, somewhat like unwinding a roll of paper.

SLICING

When special grain effects are wanted, particularly for face veneers, the slicing method is used. See Fig. 23-5. Different grain patterns and variation of wood-grain structures are obtained by slicing.

Flat slicing or plain slicing cuts are taken directly through the heart of the log. Veneers so produced have a combination of straight grain and swirling figures. This method is often used for cutting the fine hardwood veneers.

Quarter slicing cuts the veneer at right angles to the annual growth rings. This method produces a striped effect, with a grain pattern of straight, parallel lines.

Fig. 23-6. Operator grades and cuts dried veneer after it comes out of the "in-line" drying machine. (Weyerhaeuser Co.)

MANUFACTURE OF PLYWOOD

Cut veneers are dried and trimmed into standard size sheets as shown in Fig. 23-6. Defects like small knots are cut out and patched with thin wooden plugs. This is done automatically by the machine shown in Fig. 23-7.

PLYWOOD CORES

The inside of the plywood sheet, called the core, is one of three types of core buildup:

Fig. 23-4. From steam vats logs are taken directly to a huge lathe for rotary cutting.

Fig. 23-7. This machine removes defects in veneers and replaces them with a thin plywood patch called a plug. (Frank Paxton Lumber Co.)

Fig. 23-9. This veneer stitching machine holds core veneers in position, closes core gaps and controls quality. (Weyerhaeuser Co.)

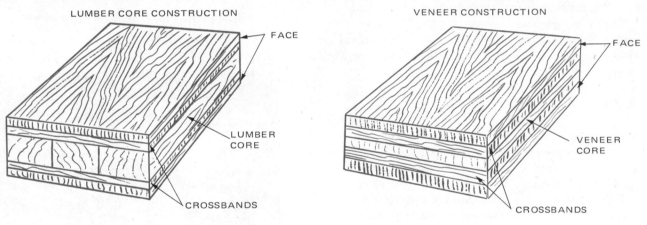

LUMBER CORE CONSTRUCTION — FACE, LUMBER CORE, CROSSBANDS

VENEER CONSTRUCTION — FACE, VENEER CORE, CROSSBANDS

Fig. 23-8. Two methods of core preparation for plywood.

1. Several layers of veneer.
2. Solid wood.
3. A composition made up of wood particles.

Fig. 23-8 illustrates two types of core. Fig. 23-9 shows veneer core being graded.

Plywood that has a core of solid wood is called lumber core plywood. It is used where low warpage is needed, such as in cabinet doors.

The composite or particle board cores are exceptionally warp free. Their dimensional stability makes them valuable for sliding and hinged cabinet doors. Wall paneling may also have particle board cores.

TYPES OF PLYWOOD

Plywood is either exterior or interior type. It depends on the type of glue used to bond the plys together.

Exterior plywood will hold its original form, shape and strength with repeated wetting and drying. It is made with a hot-pressed adhesive insoluble in water. (Insoluble means

water will not mix with it.)

Interior plywood is bonded together with glue that is not waterproof. However, such plywood will maintain strength when subjected only to occasional moisture.

SURFACE VENEER

Surface or outside veneers are made from both softwood or hardwood. Surface veneer is noted for its grain patterns. Softwood plywood is manufactured with rotary cut veneers. Hardwood plywood surface veneers are either plain or quarter sliced. Matching of hardwood veneers will yield many pleasing patterns developed by the grain, Fig. 23-10.

HARDWOOD PLYWOOD

The term hardwood plywood covers a wide range of products. Hardwood plywood is manufactured with a core of softwood and outer layers of hardwood veneer.

Many expensive and rare hardwoods have color, figure, or

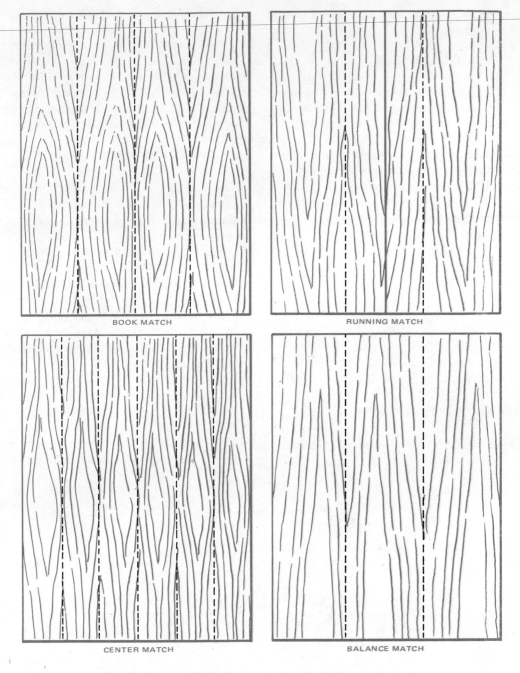

BOOK MATCH

RUNNING MATCH

CENTER MATCH

BALANCE MATCH

Fig. 23-10. Basic veneer and panel matching for hardwood plywoods.

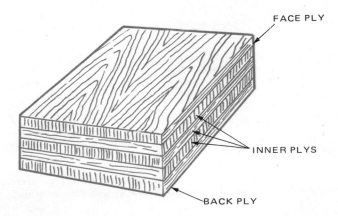

FACE PLY

INNER PLYS

BACK PLY

INTERIOR GRADES

PANEL GRADE DESIGNATION	FACE	MINIMUM BACK	INNER PLY
A—A	A	A	D
A—B	A	B	D
A—C	A	C	D
A—D	A	D	D
B—B	B	B	D
B—C	B	C	D

EXTERIOR GRADES			
A—A	A	A	C
A—B	A	B	C
A—C	A	C	C

Fig. 23-11. Softwood plywood surface grades in relation to inner ply grades.

grain characteristics that make them highly prized for paneling, cabinetwork and furniture.

PLYWOOD GRADES

Each layer of plywood veneer is graded for allowable defects, Fig. 23-11. Grades are based on the number of defects present, knots, splits, checks, stains and open sections. Repairs are permitted in some grades of veneer.

Softwood plywood is graded for appearance. Sanded plywood is graded primarily by the appearance of its front and back faces. The grade is designated by a letter.

1. N Special order cabinet grade veneer 100 percent heartwood, free of any knots.
2. A Highest standard grade. No open defects.
3. B Presents a solid face, tight knots, repairs allowable.
4. C Plugged, splits limited to 1/8 in. (3 mm) wide, holes limited to 1/4 in. by 1/2 in.
5. D Poorest grade, open space, no repairs.

The grade of the panel is indicated by two letters. An A—C panel is one having an A-grade on one face and a C-grade on the second face.

In exterior plywood, the core plies must be grade C or better. Interior plywood panels may have D grade core plies.

CONSTRUCTION PLYWOOD GRADES

Unsanded plywood panels used for sheathing, subfloors, underlayment under resilient floor coverings or carpeting and

structural applications are graded according to strength. These grades have been established to meet the needs of construction. The grade is stamped on every panel.

Most softwood is Douglas Fir although Southern Pine, Western Larch, Western Hemlock, White Fir, Cedar and other species are used, Fig. 23-12. The American Plywood Association has grouped the several species into groups according to their strength and stiffness. See Fig. 23-13.

HARDWOOD PLYWOOD GRADES

Hardwood plywoods are divided into three grades, based on the type of glue used and the resistance to moisture:

1. Constructed with a waterproof glue, able to withstand full exposure to moisture.
2. Weather resistant, will withstand occasional thorough wetting.
3. Suitable for application not subject to moisture.

Face veneers are graded by appearance, as in Fig. 23-14. The matching of individual panels of hardwood can alter the appearance of a finished installation as much as the method of cutting veneers from a log. The grading system for hardwood veneer surfaces follows:

Custom Grade	Selected or Matched Grain
Good	The highest standard quality, suitable for natural finish.
Sound	Suitable for painted surface.
Utility	Panels have open defects.
Backing	No selection for color or grain.

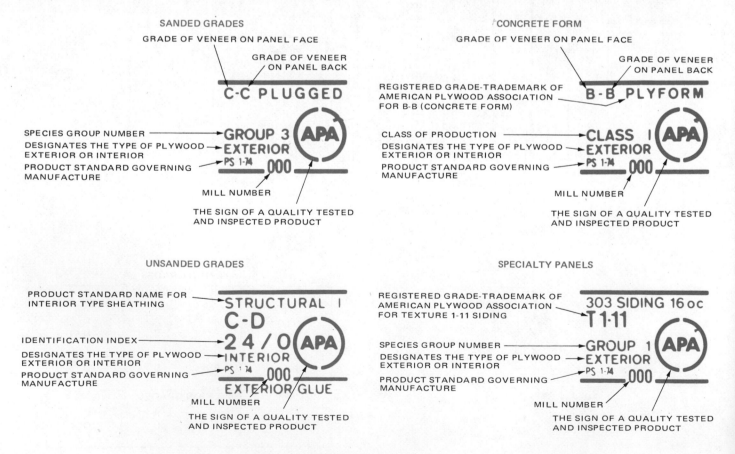

Fig. 23-12. Typical grade trademarks for softwood plywoods.   (American Plywood Assoc.)

Group 1	Group 2		Group 3	Group 4	Group 5
Birch	Cedar, Port Orford	Maple, Black	Alder, Red	Aspen	Fir, Balsam
Yellow	Douglas Fir 2	Meranti	Cedar	Bigtooth	Poplar, Balsam
Sweet	Fir	Mengkulang	Alaska	Quaking	
Douglas Fir 1	California Red	Pine	Pine	Birch, Paper	
Larch, Western	Grand	Pond	Jack	Cedar	
Maple, Sugar	Noble	Red	Lodgepole	Incense	
Pine, Caribbean	Pacific Silver	Western white	Ponderosa	Western Red	
Pine, Southern	White	Spruce, Sitka	Spruce	Fir, Subalpine	
Loblolly	Hemlock, Western	Sweet Gum	Redwood	Hemlock, Eastern	
Longleaf	Lauan	Tamarack	Spruce	Pine	
Shortleaf	Red		Black	Sugar	
Slash	Tangile		Red	Eastern White	
Tanoak	White		White	*Poplar, Western	
	Almon			Spruce, Engelmann	
	Bagtikan				

* Black Cottonwood

Fig. 23-13. Classification of softwood species by groups.

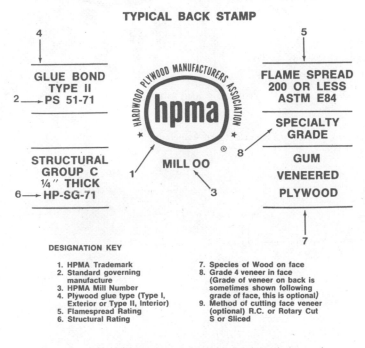

TYPICAL BACK STAMP

4 → GLUE BOND TYPE II
2 → PS 51-71

STRUCTURAL GROUP C ¼″ THICK
6 → HP-SG-71

MILL OO

1

3

5 → FLAME SPREAD 200 OR LESS ASTM E84

SPECIALTY GRADE

8

GUM VENEERED PLYWOOD

7

DESIGNATION KEY

1. HPMA Trademark
2. Standard governing manufacture
3. HPMA Mill Number
4. Plywood glue type (Type I, Exterior or Type II, Interior)
5. Flamespread Rating
6. Structural Rating
7. Species of Wood on face
8. Grade 4 veneer in face (Grade of veneer on back is sometimes shown following grade of face, this is optional)
9. Method of cutting face veneer (optional) R.C. or Rotary Cut S or Sliced

Fig. 23-14. Typical grade trademark for hardwood plywood. (Hardwood Plywood Manufacturer's Assoc.)

Several standard types of veneer matching are used for stock panels. Panels may be random matched, slip matched, or center matched. Not all of these patterns will be available from all manufacturers. Numbered sets of panels marked to indicate flitch are available on special order. Numbered sets have the same color and grain figure. Numbered sets allow the architect/designer to have a room finished with each panel alike.

PRESSED BOARDS

Pressed boards may be composed of any vegetable, mineral, or synthetic fiber mixed with a binder and pressed into a flat sheet. They are widely used in the construction industry as insulation, sheathing and finished panels.

The pressed boards may be soft-textured panels of loosely held fibers. These have little strength but excellent insulating properties. Other panels may be pressed into dense sheets which are not hurt by water or will not burn.

FIBERBOARD

Cane fiberboard is made from the stalks of sugar cane. Cane board, known as Celotex is used for insulation in walls and roofs. Given a finish, it is used for bulletin boards.

In order to increase its water resistance, the board is impregnated with asphalt. This asphalt impregnated board is also used for roof insulation or in locations where moisture is present.

MINERAL FIBERBOARD

Mineral fiberboard is manufactured of asbestos fibers or rock wool. These mineral fibers, combined with a binder, form boards that are completely fireproof. Mineral fiberboard is not recommended for location where there is moisture.

PARTICLE BOARD

Panels made of wood fibers bonded with glue and pressed into sheets under high pressure and temperatures are called particle boards. The small particles or chips are positioned in a random crisscross arrangement to form boards of outstanding uniformity in thickness and dimensional stability, Fig. 23-15. Particle board may be used for cabinetwork and underlayment. It can be engineered for strength and durability. It is ideal backing material for plastic laminates and wood veneers.

HARDBOARD

Hardboard panels are made of wood chips which have been exploded, leaving the fibers and lignin (nature's basic building block of wood). These are fused under heat and pressure into a hard, long-lasting board. Hardboard is manufactured in thicknesses of 1/10 to 5/16 in. (2.5 to 8 mm) with the faces smooth or in a screen pattern. Various additives may be used to give it special characteristics. Hardboard may be used for interior purposes. Its smooth surface is ideal for painting or veneering.

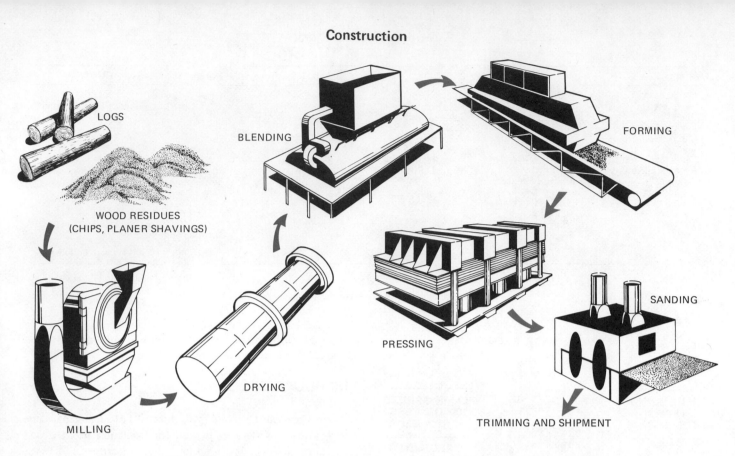

LOGS

WOOD RESIDUES
(CHIPS, PLANER SHAVINGS)

BLENDING

FORMING

MILLING

DRYING

PRESSING

SANDING

TRIMMING AND SHIPMENT

Fig. 23-15. Production steps for making particle board. (National Particle Board Association)

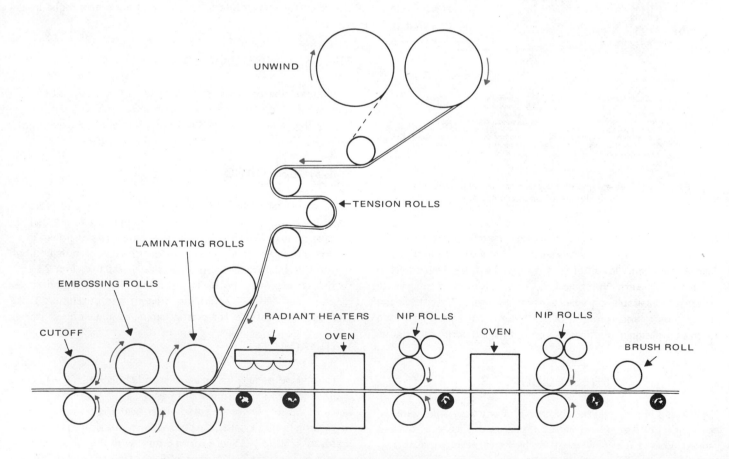

UNWIND

TENSION ROLLS

LAMINATING ROLLS

EMBOSSING ROLLS

CUTOFF

RADIANT HEATERS

OVEN

NIP ROLLS

OVEN

NIP ROLLS

BRUSH ROLL

Fig. 23-16. Typical direct rolling coat system for application of vinyl film to hardboard.
(American Hardboard Assoc.)

Embossed vinyl film is sometimes attached to a hardboard back producing a material with a wood-like appearance. A wood grain resemblance is printed on while the texture of wood pores is pressed into the surface of the film. The result is a durable surface resistant to wear and stain. Fig. 23-16 shows steps in producing the film.

Hardboard is used in exterior applications for siding. It is molded to give it the appearance of wood grain.

CORRUGATED PANELS

Corrugated paper makes up the core of the board in Fig. 23-17. These panels are made up for thicker boards. The surface skin is generally 3/16 or 1/4 in. (4.5 to 6 mm) plywood. Panels are used as passage doors for interior locations, as table tops and even for exterior siding and structural members.

Fig. 23-17. Corrugated panels provide strength for an overall light building material. (The Verticel Co.)

SUMMARY

Plywood has been used as a construction material for centuries. It is made of thin sheets of wood, rotary cut or sliced vertically from the log. These thin sheets are glued together with the grain of each layer running at right angles to the one preceding it. Thus, the finished sheet has greater strength than solid lumber. Plywood sheets are available for construction uses in various thicknesses and sizes. The common size is 4 by 8 ft. Plywood may be used for exterior or interior applications.

Grade markings are stamped on the surface of each sheet. These are determined by number and size of defects found on the surface veneer.

Pressed boards are made of waste materials of the lumber industry, including saw dust, wood chips and many other vegetable and mineral materials. Pressed boards have many uses and are manufactured for a particular use.

CAREERS RELATED TO CONSTRUCTION

DRY KILN OPERATOR operates kiln to dry lumber; directs loading of kiln; plans drying cycle; sets dials and uses instruments; keeps records of drying operations.

PLYWOOD PATCHER repairs defects in surface layers of plywood panels; examines surface to determine type of repair.

VENEER CLIPPER tends power shear that cuts veneer sheets to size; may match grain or veneer sheets before clipping.

VENEER GRADER inspects sheets of veneer for color, texture of grain and knots, worm holes and unsanded spots; marks grade on sheets.

VENEER LATHE OPERATOR sets up and operates lathe that cuts veneer from log.

VENEER SLICING MACHINE OPERATOR operates machine to slice flitches into veneer strips.

DISCUSSION TOPICS

1. From what species of wood is exterior plywood made?
2. What kind of wood is used for plywood intended for architectural purposes?
3. In what sizes are sheet stock available?
4. How does hardwood veneer paneling differ from construction plywood?
5. What is the difference between exterior and interior plywood?
6. Discuss the methods of matching the face veneer of hardwood plywood.
7. List the grades of construction plywood.
8. How is particle board made?
9. How are hardwood veneers cut?
10. How is particle board manufactured?

Chapter 24

STEEL AND RELATED PRODUCTS

Fig. 24-1. Structural frame of this 110-story building is erected completely of steel. (The Port of New York Authority)

Traditionally, metals are classified as ferrous and non-ferrous. Ferrous metals contain a large percentage of iron. Nonferrous metals contain little or no iron.

When two or more metals are combined while in the molten state, they form a metallic substance called an "alloy." Alloys, likewise, are classed as either ferrous or nonferrous, depending on the percentage of iron they contain. Ferrous metals are used as a construction material because of strength characteristics. Steel, for example, is an alloy of iron, Fig. 24-1.

Iron probably has had more influence on civilization than any other material. Iron knife blades and swords have been found in ancient tombs dating back to 3000 BC. These weapons indicate that the civilizations of Assyria, Babylon and Greece had learned to separate iron from ore and forge it into useful tools.

It is believed that iron was first forged in open limestone fire pits. Ore mixed with charcoal was added to the fire, and limestone was added to act as flux to promote separation of impurities. The ore mixed with impurities melted and flowed to the bottom of the pit, where it formed a spongy mass. This ancient iron, for the most part, was made into weapons, armor and small tools.

PROPERTIES OF IRON

Iron is one of the most abundant metals in the earth's crust. About five percent of it is iron. This tough, silver-white metal

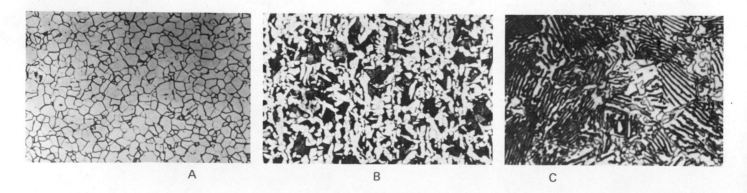

Fig. 24-2. Photomicrographs. A—Low carbon steel. B—Medium carbon steel. C—High carbon steel. (American Iron & Steel Institute)

is ductile (capable of being drawn out into very thin wires) and malleable (capable of being hammered or rolled into shape).

Iron readily oxidizes (rusts) when left exposed to air. The differences in appearance, strength, brittleness and workability of various irons depends to a great extent on the percentage of carbon alloyed with the pure iron.

Usually, the alloy is referred to as low, medium or high carbon steel, Fig. 24-2, depending on the carbon content. The electron microscope, Fig. 24-3, is an example of equipment for studying the alloying contents of steel. This powerful microscope is necessary equipment to help understand the steel that goes into the construction of our largest buildings.

Iron that contains less than 0.1 percent carbon usually is referred to as wrought iron. Iron that contains 2 to 4 percent carbon is referred to as cast iron (very brittle). Steel refers to iron alloys that contain not more than 2 percent carbon and also some manganese.

The physical properties of iron are governed by several factors:
1. Percent of carbon in the finished product.
2. Alloying of other metals during the melting process.
3. Manner in which the metal is allowed to cool (tempering).
4. Method of shaping the metal after cooling.

MATERIAL FOR IRON PRODUCTION

The production of iron requires iron ore, fuel and a flux, Fig. 24-4. The fuel used in the blast furnace is coke; the flux is

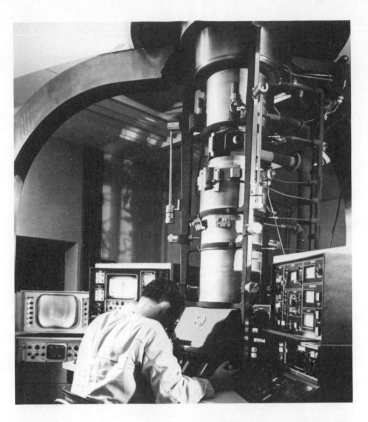

Fig. 24-3. Electron microscope and related equipment "looks" into steel. Through such studies designers can better understand this building material. (American Iron & Steel Institute)

Fig. 24-4. In an open pit mine on Minnesota's Mesabi Range, a power shovel loads iron ore into a railroad car.

limestone. Approximately 350 tons (320 metric tons) of iron ore, 200 tons (180 metric tons) of coke and 100 tons (90 metric tons) of limestone are used to produce 200 tons (180 metric tons) of pig iron (iron with some impurities).

IRON ORE

Most iron ore mined in the United States is in the form of iron oxides, Fig. 24-5. These ores usually contain phosphorus, silica, sulfur, manganese and silicon. Most of these impurities are removed during the production of iron.

COKE

The fuel used in the production of iron is "coke." Coke is a light, porous material made from coal. In the process, coal is placed in large firebrick-lined ovens, where it is heated to drive off the volatile portions as gases. The residue is coke.

By-products of the coking ovens are extremely valuable. Coking of one ton of coal produces 600 cubic feet (17 cubic metres) of gas that can be used in firing the blast furnace or sold as a by-product. Ammonium sulfide, tars and several light oils are also produced.

LIMESTONE

To remove impurities from the iron ore, crushed limestone combines with the impurities to form slag. Slag is lighter than iron. It floats on top of the molten iron and is removed.

BLAST FURNACE OPERATION

Today's blast furnace, Fig. 24-6, consists of a 20 to 30 foot (6 to 9 metre) round steel stack 100 to 200 feet (30 to 60

Fig. 24-6. Iron-making "blast" furnace takes its name from the blast of hot air and gases forced up through iron ore, limestone and coke. (American Iron & Steel Institute)

Fig. 24-5. Iron ore from mines is loaded on special ore boats up to 1000 feet long for transportation to steel-making plants.

metres) high, lined with a special refractory firebrick. The furnace is loaded at the top with iron ore, coke and limestone. Once it is put into operation, the blast furnace continues to operate 24 hours a day, seven days a week. If allowed to cool, the brick liner will crumble.

In order to reduce the ore, a blast of hot air is introduced into the lower part of the furnace at a temperature of 1100 deg. F (592 deg. C). Two holes are located at the bottom of the furnace, one above the other. The molten slag floats on top of the iron and is drawn off into ladles through the top hole. The bottom hole is tapped, and molten iron is allowed to flow into large ladles. The iron is transported, in liquid form, for manufacture into steel and other alloys, or it is cast into ingots (mass of metal cast into a convenient shape for storage). During operation, the blast furnace is tapped five or six times a day.

MANUFACTURE OF STEEL

The making of steel involves the oxidation and removal of unwanted elements from pig iron, the product of the blast furnace. This second step in steel production adds certain other elements to produce a desirable composition. The molten metal is worked until the sulfur and phosphorus contents are acceptably low. These elements cause undesirable characteristics in steel. Generally, carbon and other elements are added after the steel has been poured from the furnace into the ladle.

Most steel is made by adding scrap steel to the pig iron melt. The average charge in the United States is about 50 percent scrap, which makes steel a recyclable material.

OPEN HEARTH PROCESS

The open hearth furnace is a shallow, pan-shaped unit, lined and roofed with firebrick. Limestone and scrap metal is

Fig. 24-7. A charging machine is used for adding scrap iron and other materials to pure iron. In the background, a ladle is adding molten iron directly from the blast furnace. (United States Steel Corp.)

charged into the furnace, along with molten iron from the blast furnace, Fig. 24-7. Oxidation and removal of impurities is completed by the action of a flame that sweeps over the charge.

The temperature of the open hearth furnace reaches 2912 to 3090 deg. F (1585 to 1687 deg. C). The slag prevents direct contact of the gas flame with the molten metal. American open hearth furnaces can hold 50 to 250 tons (45 to 230 metric tons) of steel. The entire refining process takes 6 to 10 hours.

At the completion of the melt, a hole at the bottom of the furnace is opened and molten metal flows into the ladle, Fig. 24-8. The capacity of this ladle is as near as possible to furnace

Fig. 24-8. Molten steel is "tapped" from an open hearth furnace and retained in a ladle. (United States Steel Corp.)

capacity. The slag (waste) flows over the molten metal into a separate ladle. Then, the molten steel is poured into "teeming" molds, Fig. 24-9.

Other principle methods of making steel are the oxygen process, the electric furnace process and the vacuum process.

SHAPING STEEL

Regardless of the type of furnace used to process steel, the output of the furnace usually is molded in teems. Teeming ingots range in size from 1000 lb. to 25 tons (450 kilograms to 110,000 kilograms). The ingots may be square or rectangular, depending on the shape needed for later processing. These ingots are held in soaking pits until they reach a uniform temperature of about 2300 deg. F (1247 deg. C). The exact temperature depends on:

1. The size and shape of the ingot.
2. The composition of the steel.
3. The planned method of rolling, Fig. 24-10.

Fig. 24-9. Molten steel is "teemed" into ingot molds. Later, these ingots will be formed into usable shapes.

Fig. 24-10. A steel slab enters the rolls of a 160-in. plate mill. (American Iron & Steel Institute)

BLOOMING MILL

When the ingot has reached the desired temperature throughout, it is conveyed to the blooming mill. The blooming mill is a set of rollers made of steel, chilled to produce a hard exterior, Fig. 24-11.

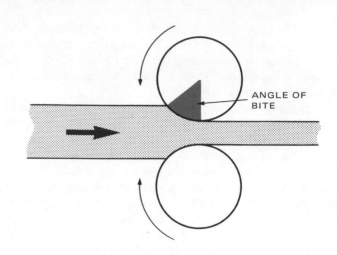

Fig. 24-11. Blooming mill rolls reduce thickness of steel slab. Maximum amount of reduction is determined by the "angle of bite." Smooth rolls can take a 30 deg. bite.

The product of the blooming mill is usually a large steel slab, ranging from 6 in. x 6 in. to 12 in. x 12 in. (15 cm x 15 cm to 30 cm x 30 cm). The initial rolling operation reduces the size of the ingot so that the second set of rolls can give it a final shape, Fig. 24-12.

Fig. 24-12. Hot strip mill operators view metal slab emerging from a pass through a stand of roughing rolls. (American Iron & Steel Institute)

STRUCTURAL MILL

Large structural members may be rolled directly from the bloom without reheating, Fig. 24-13. The steel is passed through a succession of rolls that gradually form each member into the desired shape, Fig. 24-14.

Structural shapes are formed by passing the steel back and forth through one set of rolls in the structural mill, as shown in Fig. 24-15. The space between the rolls decreases with each pass of the steel, until the member has reached the desired thickness and shape.

Fig. 24-13. Structural beams that form the framework of this future power plant were rolled directly off the bloom. (Missouri Public Service Co.)

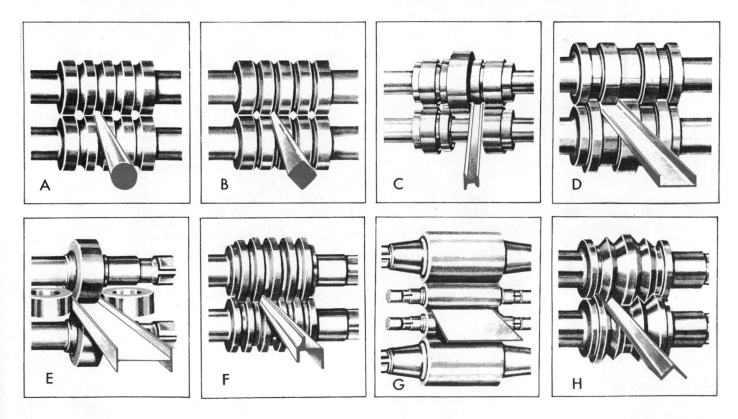

Fig. 24-14. Roll arrangement and passes cut in the rolls to produce: A—Rounds. B—Squares. C—Rails. D—Channels. E—Columns. F—Special sections. G—Sheets. H—Zee-bars.

169

Fig. 24-15. Structural mill produces wide-flange structural shapes. It uses both horizontal and vertical positioned rolls to produce desired shapes. (American Iron & Steel Institute)

PIPE

Pipe is an important structural material, for plumbing and support columns. It is classified according to the process by which it is manufactured.

The two principle forms of pipe are seamless pipe and welded pipe. Seamless pipe is produced by forcing a steel rod over the pointed nose of a piercing mandrel. The mandrel punches a hole in the center of the rod, while rolls on the outside control the diameter, Fig. 24-16. The process usually consists of two piercing operations and two or more passes through the plug mill, which gives the seamless pipe a uniform size and finish.

Welded pipe is produced by forming a flat strip into a U-shape, then closing it into a tube, Fig. 24-17. The seam is welded by passing an electric current between the edges.

Welded pipe can be produced to very close tolerances, both on the inside and outside diameters. The welded seams are closely inspected. The manufacturing process is carefully

Fig. 24-17. Large diameter pipe is made from plates that have been formed in successive shaping processes. Final step after "O-ing" (shown) is to weld a pressure tight seam.

Fig. 24-16. Technician tests a piece of seamless pipe for internal defects. (American Iron & Steel Institute)

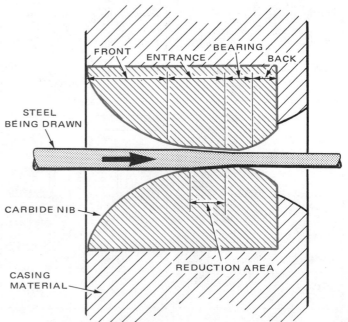

Fig. 24-18. Steel enters wire drawing die and is pulled against a carbide nib. The emerging wire is smaller than the steel fed into die.

controlled, and the welded area is not necessarily a point of weakness.

Most pipe used for structural and mechanical purposes is produced in relatively few sizes, both in seamless and welded forms.

WIRE

Wire for construction is produced by drawing a steel rod through a series of dies, Fig. 24-18, each reducing the diameter. Wire is usually thought of as round. However, it also comes in square, rectangular, polygonal and many other shapes. A variety of strengths, hardnesses and finishes are available to the construction industry.

CLASSIFICATION OF STEELS

Carbon steels are usually classified according to their carbon content. (See chart that follows.) Up to a point, the more carbon, the stronger the steel. As the carbon content increases, the steel also becomes less ductile and more brittle, Fig. 24-19.

CARBON-STEEL CHARACTERISTICS AND ITS USES

Very Mild 0.05 to 0.15 percent carbon. An easily worked steal that is tough and soft, used for sheets, wire, rivets, pipe and fastenings.

Mild Structural 0.15 to 0.25 percent carbon. A strong, ductile, machinable steel used for buildings, bridges, boats and bailers.

Medium 0.25 to 0.35 percent carbon. Stronger and harder than mild structural grade, used for machinery, shipbuilding and for general structural purposes.

Medium Hard 0.35 to 0.65 percent carbon. Used in locations subject to wear and abrasion.

Spring 0.85 to 1.05 percent carbon. Used in the manufacture of springs.

Tool 1.05 to 1.20 percent carbon. The hardest and strongest of the carbon steels, used for tools and cutting edges.

ALLOY STEELS

Much of the steel used for construction is low-to-medium carbon steel that is tough, strong and easily worked. Alloy steels are popular, too. An alloy steel is one in which alloying elements have been added to the molten mix in excess of those allowed in carbon steel.

By alloying elements such as manganese, nickel, chromium, vanadium and copper with steel, properties are provided that are unobtainable in carbon steels. For example, copper alloys have improved resistance to corrosion and are used for

Fig. 24-19. Steel for structural members and for wire reinforcement is of the low-to-medium classification.
(Missouri State Highway Department)

Fig. 24-20. Exterior of this building is "weathered steel," Magari-R, providing a protective oxide film that seals the surface against further corrosion.   (Babcock-Davis Associates)

products subject to moisture, such as wire cloth and sheet steel. Chromium increases strength and acts as a hardener.

WEATHERING STEELS

Recent developments in steel construction include "weathering steel," a group of patented alloy steels classed as high strength, low alloy steel. See Fig. 24-20. These alloy steels

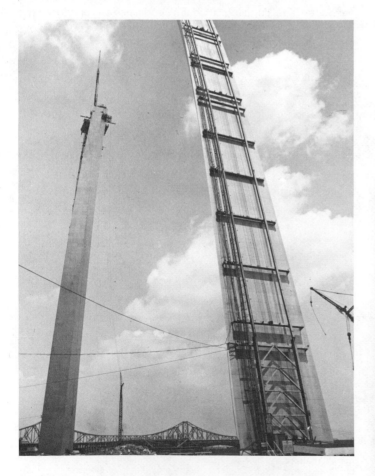

Fig. 24-21. Surface of this structure exposed to the elements is stainless steel. Stainless surface retains its beauty and resists oxidation and discoloration by atmospheric conditions.
(Jefferson National Expansion Memorial)

Fig. 24-22. Huge steel beam being removed from the hot dip zinc bath is ticketed for a bridge support.
(American Hot Dipped Galvanizer's Assoc.)

offer corrosion resistance and develop strength up to 40 percent higher than structural carbon steel. Because of increased strength, they can be used in thinner sections to reduce weight, or to support heavier loads.

The corrosion resistance of these alloys is provided by a tightly adhered, protective oxide film that seals the surface of the member against further corrosion. This protective oxide film gradually assumes a pleasing texture and darkens to a color ranging from brown to warm purple.

The atmospheric corrosion resistance eliminates the need for protection of surfaces directly exposed to the weather. The texture and color of the oxide film depends on the eroding action of wind and rain, and the drying action of sunlight. Exposure of these materials should be avoided where recurring wetting by sea water is possible.

STAINLESS STEEL

The corrosion resistance of iron is greatly improved when the quantities of chromium are increased, Fig. 24-21. At 11.5 percent, there is sufficient chromium to form an inert film of chromic oxide over the entire metal surface. This steel is considered stainless.

GALVANIZING

Steel sheets may be protected against rust and corrosion by a galvanized zinc coating. Galvanizing serves as a double protection. It forms a mechanical barrier against moisture and prevents oxidation or rusting of the steel base.

HOT DIPPING

To coat steel with zinc, cleaned steel is immersed in a bath of molten zinc, Fig. 24-22. The zinc adheres to the base metal in a smooth, even cover, usually of crystalline appearance.

Galvanized sheet may be formed, rolled, shaped or assembled with little damage to the coating. However, if edges are trimmed or welded, the area must be treated for protection.

SUMMARY

The use of iron dates back to 3000 BC. Steel is an alloy (combination of two or more metals) of iron. As methods improved for removing metals from natural earth forms, increased use of steel for building materials developed.

Steel is the product of the raw materials of limestone, coal and iron ore. Alloying iron with certain amounts of carbon yields steels of varying properties. The higher the percentage of carbon, the tougher, harder and stronger the steel becomes. However, after a given amount, steel becomes brittle.

Steel is formed by casting, rolling, drawing, bending and folding to give the various shapes used in construction.

Steel is alloyed with other elements such as manganese, nickel, chromium, vanadium and copper to provide a material with special properties unobtainable in carbon steels.

CAREERS RELATED TO CONSTRUCTION

EXTRACTIVE METALLURGIST deals with extraction of metals from ores and methods of refining and alloying them to obtain pure metal. METALLURGIST specifies, controls and tests the quality of steel during its manufacture. METALLURGICAL ENGINEER develops methods of processing and converting metals into useful products.

FURNACE OPERATOR regulates furnace temperature; adds alloying materials as specified; takes samples of molten steel for testing; taps molten steel from furnace into a ladle.

ROLLING MILL OPERATOR breaks down ingots into more easily handled semifinished shapes called blooms, slabs and billets. BLOOMING MILL OPERATOR operates rolls, or directs and coordinates the steel rolling process, by regulating opening between rolls after each pass. SHEARMAN cuts rolled steel to length.

DISCUSSION TOPICS

1. Define the following terms in relation to steel: malleable, ductile, tensile, brittleness.
2. What raw materials are necessary for steel production?
3. What is coke?
4. Explain the operation of a blast furnace.
5. List the several methods of shaping steel into construction material shapes.
6. What is a structural mill?
7. How is wire formed?
8. What are the carbon contents of the various steels?
9. What is the principle of protection in the "weathering steels?"
10. Describe the galvanizing process.

Chapter 25

WINDOWS

Fig. 25-1. Glass is an efficient and effective building material. When used with stone, brick, steel or concrete, glass gives texture, dimension and integrity to modern architecture. (Libby-Owens-Ford Co.)

Today, windows are made of a transparent material called "glass." Actually, glass was used as a clear, protective glazing material for windows as early as the First Century AD.

The first important architectural use of glass came in the Middle Ages in the form of jeweled stained glass windows installed in the Gothic cathedrals of Europe. Then centuries passed before new concepts in building design offered a vastly expanded role for glass.

Only in the Twentieth Century, with new and improved methods for making glass, has the potential for flat glass been tapped for the architecture/construction field, Fig. 25-1.

Today, glass is a many-purpose design material and functional building material used in numerous places to create better and brighter living conditions.

RAW MATERIALS OF GLASS

Natural glass is formed by volcanic eruptions. Natural glass is composed of three elements of the earth: sand, soda and lime. The same three elements in varying forms make up the basic composition of manufactured glass. About fifty other chemical elements are used in various ways in modern

Fig. 25-2. Large non-sash windows link the interior and exterior of this new building. (Libby-Owens-Ford Co.)

glassmaking, to affect the color or durability of the glass, or to impart some desired physical property of the glass.

Today, the average glass window consists of about 70 percent sand (silica), 13 percent lime (calcium oxide), 12 percent soda (sodium carbonate) and 5 percent additives:

1. Sand is converted into glass by the action of heat. It is largely responsible for the transparency of glass.
2. Soda assists in melting sand at a lower-than-normal temperature. It helps to remove impurities from the glass.
3. Lime introduced to the molten glass makes the mix more fluid. It shortens the settling time and improves the weatherability of glass.

MANUFACTURE OF FLAT GLASS

Glass has become so commomplace that its strength, durability and unique optical characteristics often are taken

for granted. To appreciate the versatility of glass, consider how flat glass is made and fabricated into many architectural products, Fig. 25-2.

The manufacturing process differs for flat glass and other glass products. Stemware, cookware, containers and lighting fixtures are blown, pressed or molded. Flat glass is manufactured in wide ribbons in a straight line. The finished glass product is ready for use in flat form, or it may undergo further manufacturing or fabricating (bending, laminating, tempering and coating) to give it additional structural properties.

THE FLOAT GLASS PROCESS

In the float forming process, molten glass leaving the melting furnace is floated on the perfectly flat surface of molten tin, Fig. 25-3. Speed of the flow, or draw, from the tank determines the thickness of the glass. The slower the

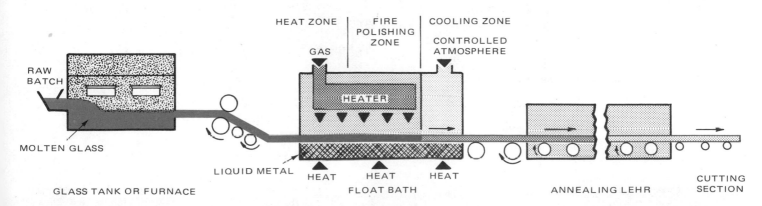

Fig. 25-3. The float glass process floats molten glass on liquid metal, then the annealing lehr cools the glass by slowly dropping the temperature to relieve stresses. (PPG Industries)

draw, the thicker the glass will be.

The tin bath is divided into three zones: heating, polishing and cooling. In the first zone, heat from the molten glass and liquid metal and from a gas heater causes the glass to float uniformly over the flat tin surface. The polishing zone gives the glass brilliant finish. The cooling zone maintains the flatness and transparency of the glass as it becomes hard enough to be conveyed by rolls without marring the surface.

The glass leaving the float enters an annealing oven called a "lehr." The lehr has a temperature range of 1200 to 4000 deg. F (650 to 2200 deg. C). The temperature of the glass must be lowered gradually, since fast cooling would cause internal stress. From the lehr, the ribbon of glass continues to the automatic cutting, inspecting and packaging area.

PLATE GLASS FORMING

Plate glass forming begins when molten glass leaves the melting tank and flows horizontally between water-cooled forming rolls. These rolls shape the glass into a semi-solid, continuous ribbon of a determined thickness. Next, the ribbon moves over steel rolls, where it cools slightly and solidifies into rough plate glass.

From the cooling rolls, the glass ribbon enters an annealing lehr (similar to the one used for float glass) that gradually reduces the heat to control the stresses. The cooled glass ribbon then enters a battery of twin grinders and high speed polishers, which give the transparency and high surface beauty associated with plate glass. Finally, the glass goes to the cutting, inspecting and packaging areas.

SHEET GLASS FORMING

Sheet glass forming has become the standard for the glass industry. The glass is drawn vertically from a pool of molten glass, Fig. 25-4. A clay block, or draw bar, submerged in the molten glass helps form the ribbon. As the glass is pulled upward, it hardens and the ribbon is given a brilliant, fire-polished finish. As with float glass, the speed of the drawing determines thickness and width.

In the sheet glass forming process, the glass is untouched (except for rolls) until it hardens. The sheet ribbon is annealed and cooled as it gradually moves upward past three inspection and observation areas to the process floor. The total height of the sheet forming machinery is about 40 feet above the melting furnace. On the process floor, the glass is cut into predetermined sizes.

SPECIAL GLASS

Special glass is available for special purposes. It is designed with a given property for a specific reason. To increase the impact resistance, for example, or to reduce heat transmission, add color or increase its reflection qualities. These are some of the many properties that can be controlled during the manufacture of glass.

TEMPERED GLASS

Tempered glass is produced by heating glass almost to its melting point, then chilling it rapidly. This process creates a glass that is three to five times as strong as ordinary glass.

INSULATING GLASS

The increased demand for weather-conditioned buildings has led to the development of double-glazed and even triple-glazed glass. Often referred to as insulating glass, these units consist of two or more sheets of glass separated by an air

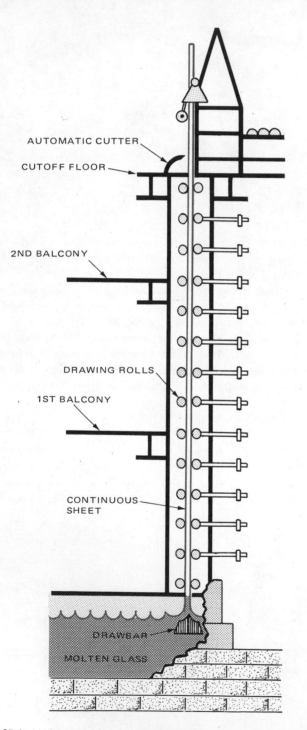

AUTOMATIC CUTTER

CUTOFF FLOOR

2ND BALCONY

DRAWING ROLLS

1ST BALCONY

CONTINUOUS SHEET

DRAWBAR

MOLTEN GLASS

Fig. 25-4. In sheet glass forming, the glass is drawn upward from a pool of molten glass.

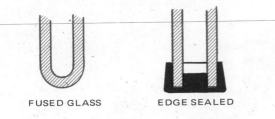

Fig. 25-5. Two methods are shown for sealing and insulating glass units.

space, Fig. 25-5. The air between the sheets of glass is free of moisture, therefore it greatly slows down the transfer of heat. The insulating air helps in overcoming the problem of moisture condensation that usually forms on the inside of glass in cold weather.

TINTED AND COLORED GLASS

The properties of glass can be controlled by adding metallic films, oxides or paint. The tints and coatings will filter light, conduct electricity and reflect heat and light. Tints may be a thin film of transparent color sandwiched between two sheets of glass laminated together, Fig. 25-6. Tints may be completed during the glassmaking process by adding certain metallic oxides to the molten glass.

The art of making and installing stain glass, sometimes called cathedral glass, dates back to the development of Gothic architecture during the early twelfth century. Windows of the great cathedrals became virtually walls of magnificient stained-glass artistry. See Fig. 8-3.

REFLECTIVE GLASS

Mirrors are types of reflective glass used to reflect your image. Most reflective glass in a structure is used to reflect heat and light. Reflective glass is opaque from the side facing the light. It is translucent from the inside during the day and

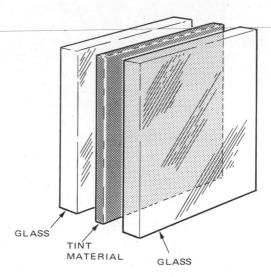

Fig. 25-6. Makeup of laminated glass with tinted film.

translucent from the outside at night.

Reflective glass also can be an important consideration in exterior design. A building clad in reflective glass will reflect outdoor sculptural elements, and it will change with the movement of passing people and vehicles, Fig. 25-7. The color of the building is a reflection of the sky overhead. On gray cloudy days, the reflected color is gray. On clear days, the blue sky and white clouds will be reflected.

WINDOW UNITS

Windows originally were used for admitting daylight and, later, for ventilation. Today, we have a great variety of window types: single hung, double hung, casement, awning, sliding, hopper, jalousie and fixed, Fig. 25-8. All except the fixed type refer to the methods by which each opens for ventilation purposes. With the increased use of air conditioning, the provision for ventilation is becoming unnecessary.

Fig. 25-7. Highlight of this commercial building is the reflective glass exterior.
(Libby-Owens-Ford Co.)

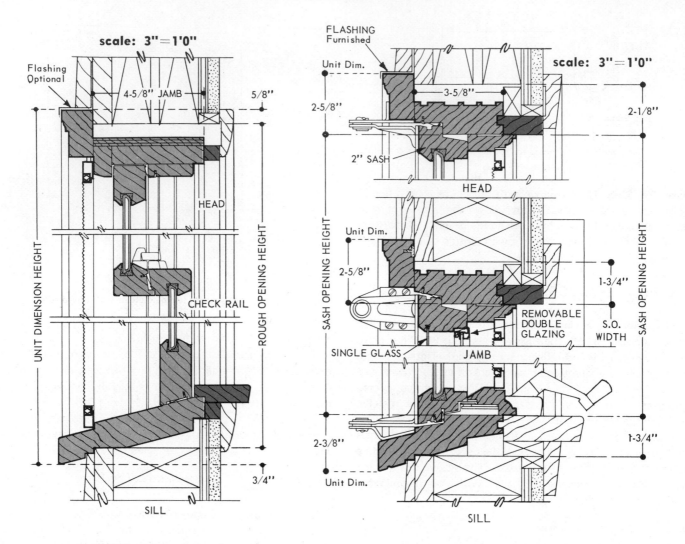

Fig. 25-8. Detail drawings of typical residential class windows. Left—Double hung window. Right—Casement window.

Generally, all windows are glass panels secured in metal or wood frames. Each frame, referred to as a unit, is standardized as to size. The units are manufactured complete with hardware, weatherstripping and operating mechanism, ready to be installed in a building. They are manufactured to receive 1/8 in. window glass, 1/4 in. thick plate and 1/4 in. special glass or insulating glass. Windows are manufactured in four classes, based on their use in building construction:

1. Residential, Fig. 25-9.
2. Commercial, Fig. 25-10.
3. Industrial, Fig. 25-11.
4. Monumental, Fig. 25-12.

These classes have established standards developed by the associations of window manufacturers and various governmental agencies.

WOOD FRAME WINDOWS

Standards have been established for the different types of wood window units. These standards cover material used, frame and sash sizes, hardware and weatherstripping. The standard also requires that all wood parts receive preservative treatment against fungi, insects and water.

Most wood units are made of pine that is kiln-dried to a moisture content of 6 to 12 percent. All exterior metal fasteners should be galvanized aluminum or stainless steel to resist corrosion. Exterior nail holes should be sealed with putty. The glass is glazed in place, then secured by either:

1. Stop bead.
2. Glazing compound.
3. Rubber gasket, Fig. 25-13.

Wood windows are used in most residential applications. The double hung type of wood window is the most widely used. Sliding, awning and casement are also used in residential construction, Fig. 25-14.

METAL WINDOWS

All types of metal window units are manufactured as a complete unit, including hardware, weatherstripping and operating mechanism. Usually, they are glazed on the construction site.

Metal frame windows generally are used in commercial, industrial and monumental applications. Some metal frame units are manufactured for residential use.

Standards for metal frame windows are based on use. For

Fig. 25-9. Residential class of window.
(Western Wood Products Assoc.)

Fig. 25-11. Industrial class of window.
(PPG Industries, Inc.)

Fig. 25-10. Commercial class of window.
(California Redwood Assoc.)

example, windows used in an industrial building do not necessarily have to meet the rigid requirements that are required for hospital use.

FINISHES FOR WINDOW UNITS

Window units are finished in a variety of ways. Some are totally pre-finished in the manufacturing plant. Some are only

Fig. 25-12. Monumental class of window.
(Cupple's Products)

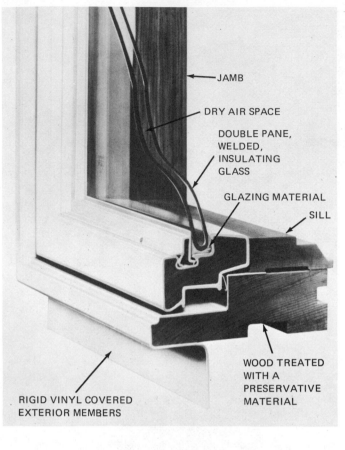

Fig. 25-13. Double pane insulating glass window parts. (Andersen Corp.)

Labels on figure:
- JAMB
- DRY AIR SPACE
- DOUBLE PANE, WELDED, INSULATING GLASS
- GLAZING MATERIAL
- SILL
- WOOD TREATED WITH A PRESERVATIVE MATERIAL
- RIGID VINYL COVERED EXTERIOR MEMBERS

Fig. 25-14. Gliding type of window. (Andersen Corp.)

Window Sizes

UNIT DIM.	2-0	2-4	2-8	3-0	3-4	3-8	4-0	4-4
RGH. OPG.	1-10	2-2	2-6	2-10	3-2	3-6	3-10	4-2
SASH OPG.	1-8	2-0	2-4	2-8	3-0	3-4	3-8	4-0
GLASS*	16½	20½	24½	28½	32½	36½	40½	44½

*Exposed glass surface —for actual glass size add ½" to width and height.

Window unit numbers shown in chart:
- 20210, 24210, 28210, 30210, 34210
- 2032, 2432, 2832, 3032, 3432
- 20310, 24310, 28310, 30310, 34310
- 1842, 2042, 2442, 2842, 3042, 3442
- 1846, 2046, 2446, 2846, 3046, 3446, 3846, 4046
- 2052, 2452, 2852, 3052, 3452, 3852, 4052
- 2456, 2856, 3056, 3456, 3856, 4056
- 3462, 3862

Fig. 25-15. Window size chart for double hung windows.

prime finished in the plant. Others receive no finish until installed at the site.

The following table gives typical finishes for common window units:

Window Unit Material	Factory Finish	Permanent Finish
Wood	Preservative against fungi, insect and water	Paint, stain, vinyl clad, aluminum clad
Steel	Skip prime coat or corrosion resistant prime coat	Paint, baked-on enamel, vinyl
Aluminum	Protective coating	Smooth, high polish finish, oxidized, colored, baked-on enamel
Stainless Steel	Protective coating	Colored, smooth high polish finish

STANDARD SIZES

For exact sizes, check the manufacturer's specifications. Wood windows use four sets of measurements:

1. Unit dimension.
2. Rough opening.
3. Sash opening.
4. Glass size, Fig. 25-15.

The unit dimension gives the exact overall dimension of the window unit, including the exterior casing. Height dimensions are taken from top of the head casing to the bottom of the edge of the sill. Width dimensions are taken from the back of left side casing to the back of right side casing.

The rough opening is the opening size in the frame to allow the installation of the unit. The height of the rough opening is approximately 3/4 in. (18 mm) larger than the exact height of the sash unit. The width is 1 1/2 in. (36 mm) larger than the exact width of the sash unit.

The sash opening is the exact measurement of the unit's height and width, less the jamb.

The glass size is the exact size of the individual glass openings.

SUMMARY

Wall openings that provide natural daylight and ventilation are classed as windows. The movable sash is a system of one or more panes of glass in a movable frame of wood or metal. It admits natural light and provides a means of ventilation. The window unit is attached to the structural framing of the surrounding wall.

CAREERS RELATED TO CONSTRUCTION

CERAMIC ENGINEER researches and experiments with ceramic products. Specialist in structural glass develops new and improved methods of glassmaking.

GLASS SETTER installs window glass, plate glass and structural glass panels on walls and partitions; mirrors; automatic glass doors; shower doors and bathtub enclosures.

DISCUSSION TOPICS

1. List the three basic raw materials used in manufacturing glass.
2. What are the two primary purposes of windows?
3. Describe the two methods of forming flat glass.
4. What is meant by flame polishing? How is flame polishing done?
5. What are the special properties of tempered glass?
6. What is stained glass?
7. Describe insulated glass. What is the advantage of windows made of insulated glass?
8. What is reflective glass? Where could it be used?
9. What materials are used to make window frames?
10. Define unit dimension, rough opening, sash opening and glass size.

Chapter 26

DOORS

Fig. 26-1. Doors often are used as elements of architectural styling.

A door is a movable section installed to close a doorway, or entrance, to a room or building. Doors usually are made of wood, metal, glass or a combination of these materials. They are designed to swing, fold, slide or roll as the mechanical means of operation. The doorway includes a doorframe, usually constructed of masonry, wood or metal, and often surrounded by architectural and sculptural elements. See Fig. 26-1.

Historically, the first doors were merely animal hides or textiles hung over an opening. Ancient Egyptian wall paintings show representations of doorways equipped with matting that could be rolled up and lowered. Doors of rigid construction appeared later. Then, with the introduction of monumental architecture, doors were richly carved and decorated with stone, metal or wood.

Today, many types of doors are available to provide access, protection, safety and privacy. See Fig. 26-2. Interior doors control the passage of sound and separate interior living spaces. They may be used to conceal storage areas, Fig. 26-3, or as movable partitions to divide interior areas. Sliding metal or wood-framed glass doors are used to connect indoor and outdoor spaces, Fig. 26-4.

WOOD DOORS

Wood doors are either solid-core or hollow-core construction, Fig. 26-5. Typically, solid-core doors are used as exterior

Fig. 26-2. Entrance doors create a lasting impression if they are attractive, different and interesting.   (Koppers Co., Inc.)

182

Fig. 26-3. A decorated bi-fold door lends character to a room.
(The Verticel Co.)

Fig. 26-4. Glass doors provide a convenient and practical connection between indoor and outdoor living areas.
(Koppers Co., Inc.)

PANEL DOORS

Panel doors consist of vertical members called stiles and horizontal members called rails. The stiles and rails enclose spaces filled by panels of solid wood, plywood particle board, louvers or glass inserts. Additional horizontal or vertical members called muntins may be used to divide the door into several spaces. Each space is, in turn, filled by a panel, making panel doors very decorative for interior or exterior uses.

FLUSH DOORS

Flush doors have uniformly smooth surfaces. The flush door is made of sheets of thin veneer over a solid core of wood or particle board, or over a hollow core of wood, paper, plastic or wood strips. The veneer faces act as stressed-skin panels that

doors or where heavy service is anticipated. Hollow-core doors are used for interior applications. Wood doors are also classified according to their construction, such as panel doors or flush doors, Fig. 26-6.

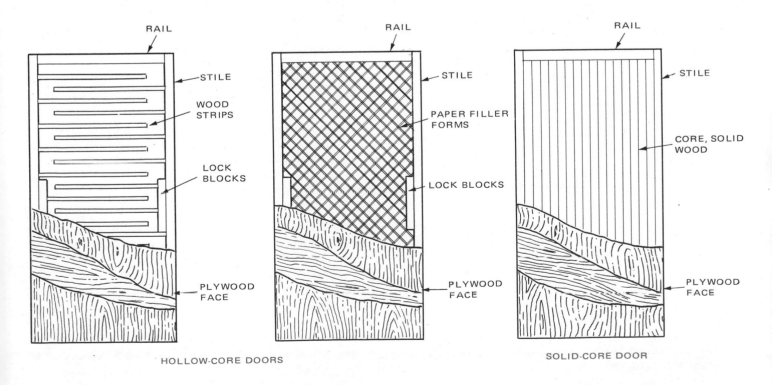

HOLLOW-CORE DOORS

SOLID-CORE DOOR

Fig. 26-5. Various methods are used to construct the interior core for slab doors.

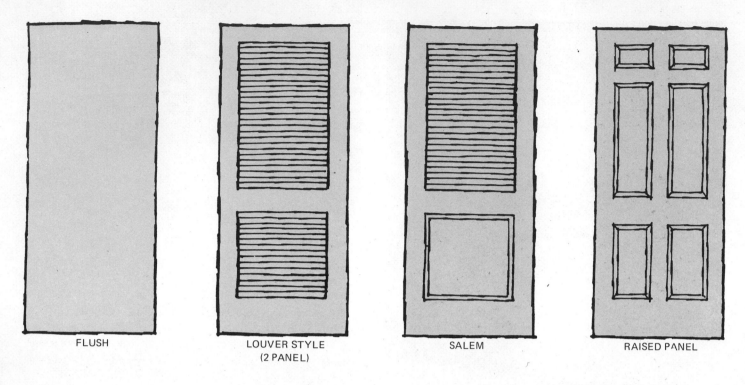

| FLUSH | LOUVER STYLE (2 PANEL) | SALEM | RAISED PANEL |

Fig. 26-6. Four popular classes of interior wood doors are pictured, each having certain advantages.

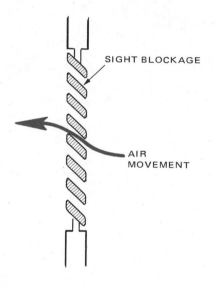

Fig. 26-7. Louver doors are attractive and provide passage for circulating air.

tend to stabilize the door against warping. The face veneers usually are hardwood suitable for a natural finish or for painting.

LOUVER DOORS

Full louver doors consist of horizontal wood slats set at an angle within a stile-rail frame. Louvers are sightproof, but they allow air to pass through, Fig. 26-7. Many building codes require louvers at the top and bottom of doors leading to closets containing gas-fired hot water heaters. Interior louver doors act as air return passages for air conditioning.

METAL DOORS

Metal doors are hollow, and usually set in metal frames. Metal doors have the advantage of being installed prefinished with baked-on enamel or covered with vinyl fabrics in a variety of patterns or wood grains.

The internal construction of metal doors consists of horizontal or vertical members welded to the inside face panels. The inside of the hollow door is filled with a sound-deadening material. Hollow metal doors are manufactured to fit a particular metal frame.

GLASS DOORS

Traditionally, glass inserts for panel doors have been used for entrance doors. Contemporary glass panel doors are used to accent architectural styling.

SLIDING GLASS DOORS

Glass panels and doors have increased in demand in modern residences and office buildings. Nearly whole walls have become glass, Figs. 26-4 and 26-8.

Sliding glass doors are large panels of glass set in wood, steel or aluminum frames. Wood frames give the owner the feeling of softness and the beauty of natural wood. Steel has the advantage of strength and rigidity. Aluminum, because of its strength and light weight, has become very popular for the frame and track of sliding glass doors. Over 50 percent of all sliding glass doors today have extruded aluminum frames.

The simplest type of sliding glass door consists of one fixed panel and a second sliding panel that bypasses the fixed panel

Fig. 26-8. Large sliding doors look good and operate smoothly by means of twin adjustable rollers on ball bearings. (Andersen Corp.)

when open, Fig. 26-9. Sliding glass doors are identified with an X to indicate movable sections and an O to indicate fixed sections. The door is identified as viewed from the outside. For example, an OX door consists of a fixed left panel and a movable right panel, while the XO door has a movable left panel and a fixed right panel.

The sliding glass door moves on an aluminum track with nylon rollers. A special nylon pile weatherstripping provides protection against weather penetration at the door jambs.

SPECIAL DOORS

Special doors, usually of the interior type, provide the answer to problem passageways, or they are used as special architectural features. Traditionally, doors move on hinges. The hinges allow the door to open and close by pivoting on one edge. Special doors move on alternate systems (sliding, folding, etc.).

POCKET SLIDING DOORS

If a single sliding door is used to close an opening, it must slide into a pocket built into the wall. Pocket doors are usually used:
1. Where passageways are to remain open or closed the majority of the time.
2. Where space is too limited for a hinged door to swing.

BYPASS SLIDING DOORS

Two or more panels may be hung on parallel tracks to serve as bypass doors. Tracks are available to accommodate doors from 3/4-in. to 1 3/8-in. (19 mm to 35 mm) thick. Bypass sliding doors usually are used on closet openings or wardrobes.

BI-FOLD DOORS

Bi-fold doors are used for closet or wardrobe openings because, in their folded position, they do not extend into the storage space. See Fig. 26-10. Also, when open, a bi-fold door extends outward only one half the width of the panel. The width of each leaf of the panel depends on the width of the opening.

FOLDING DOORS

Folding doors are hung from an overhead track with nylon rollers similar to those used with sliding doors, Fig. 26-11. Folding doors may be used for closet or space dividers. The individual folding leaves may be 3/8-in. or 1/2-in. (9 mm or 13 mm) thick and from 3 1/2-in. to 12-in. (88 mm to 305 mm) wide. Folding doors may be made of solid wood or plywood.

ACCORDION DOORS

Accordion doors consist of a framework of steel covered with a vinyl-coated fabric in a wide range of colors and

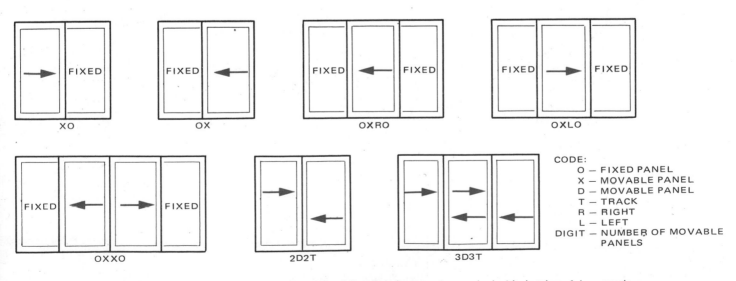

Fig. 26-9. Descriptive code for fixed and movable sliding glass doors can be matched with sketches of door panels.

BI-FOLD LOUVER TWO PANEL

Fig. 26-10. Two panel, bi-fold louver doors are ideal for closets and wardrobes.

Fig. 26-11. Folding doors provide closure for wardrobes or closets with limited space.

Fig. 26-12. Accordion doors close off closet openings, yet give full access when open.

textures. These doors operate much the same way as the folding doors, Fig. 26-12. Accordion doors are used for closet doors and space dividers.

DOOR SIZES AND SWINGS

Interior doors are usually 1 3/8-in. (35 mm) thick. Exterior doors are 1 3/4-in. (45 mm) thick, but also are available 2-in. (51 mm) and 2 1/4-in. (57 mm) thick on special order.

The standard height is 6 ft. 8 in. (203 cm), although some doors are 7 ft. 0 in. (213 cm) in height. Interior doors are available in widths of 1 ft. 8 in., 2 ft. 0 in., 2 ft. 4 in., 2 ft. 6 in. and 2 ft. 8 in. (51 cm, 61 cm, 71 cm, 76 cm and 81 cm). Interior doors through which furniture must be moved are at least 2 ft. 6 in. (76 cm).

Popular widths of major entrance doors are 3 ft. 0 in., 3 ft. 6 in. or 4 ft. 0 in. (92 cm, 107 cm or 122 cm). Secondary entrance doors may be 2 ft. 8 in. (81 cm). Doors of commercial buildings are usually 4 ft. 0 in. (122 cm) or wider and use two panels.

The most common type of door is the swinging door. Door swings are classed as either right-hand swing or left-hand swing, depending on which side is hinged, Fig. 26-13. If a person is standing outside of an exterior door (or on corridor side of an interior door or room side of a closet door), and the hinge is on the left side, the door is said to be "left-hand swing." If this particular door swings inward, it is called a "left-hand, single." If it swings outward, it is a "left-hand, reverse single."

A door that swings both ways through an opening is called a double-acting door.

SUMMARY

Doors are movable sections installed to close an opening in the wall. They give access, protection, safety and privacy to the living area. Doors are constructed of wood, metal, glass and other materials. They provide a barrier against sound, temperature and light. Doors may be hollow-core or solid-core construction. They are designed with a plain slab (flush) surface or by the stile-rail assembly with panels of solid wood, louvers, glass or other materials. Door sizes are standardized at a 6 ft. 8 in. height and come in various widths. The swing of a door may be either right hand or left hand.

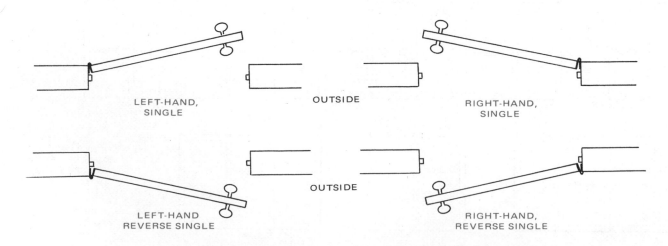

LEFT-HAND, SINGLE

OUTSIDE

RIGHT-HAND, SINGLE

LEFT-HAND REVERSE SINGLE

OUTSIDE

RIGHT-HAND, REVERSE SINGLE

Fig. 26-13. Door swings are identified by location of hinges and direction of swing.

CAREERS RELATED TO CONSTRUCTION

DOOR ASSEMBLER manufactures hollow-core or solid-core doors from cut-to-size parts according to prescribed patterns.

DOOR SETTER fits hinges and hangs wood doors in doorways, a job usually done by a carpenter. Setter also installs bi-fold, folding, sliding and accordion doors.

DISCUSSION TOPICS

1. Describe a hollow-core wood door.
2. What is a flush door?
3. List the types of doors which could be used at a closet or wardrobe opening.
4. Describe the symbol OXO.
5. Describe a right-hand swing, single door.
6. What is meant by a left-hand swing, reverse single door?
7. Sketch the stile-rail construction of doors.
8. What type of materials are used for the panels for panel doors?
9. How does a folding door differ from a bi-fold door?
10. How does a folding door differ from an accordion door?
11. What is a double-acting door?
12. What is standard door height?
13. What are popular widths of main entrance doors?

Chapter 27

HARDWARE

Hardware is a term that applies to a large number of metal items used in the construction industry. In this chapter, hardware refers to metal fastening devices, hinges, door locks, saddle straps and metal trim pieces.

ROUGH HARDWARE

In general, rough hardware consists of utility items which are not usable in the finished building. These are items such as nails, bolts, framing anchors, and many other items such as ash doors and fireplace dampers. Rough hardware goes into the construction of the basic building.

NAILS

Everyone has driven a nail into some construction material. Nails have been used as fastening devices, in construction, since the Ancient Egyptians began using hand-wrought nails. In the late nineteenth century, a wire-nail machine was developed in France. With this breakthrough, and other advancements in technology, Fig. 27-1, nails were mass produced and hundreds of different sizes and types were developed.

SIZES OF NAILS

Nails of various sizes are used in almost all construction projects. Nail size refers to the nail's length and to the

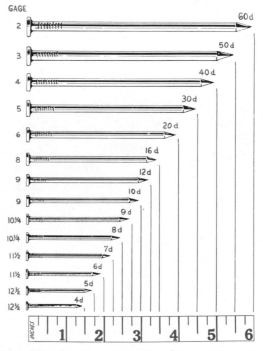

Fig. 27-2. Sizes of common wire nails. (National Forest Products)

diameter of its shank. Its length "size" is indicated by the term "penny." In any given penny size, nail length will always be the same, regardless of head or shank style. See Fig. 27-2. However, the diameter "size" of the nail shank *will* vary.

The "penny" designation is an old English term which may have come from the cost of 1000 nails of a particular size. Today, the term "penny" indicates only the length of the nail. In its abbreviated form, penny is written as "d" (8d, for example).

The shank diameter of nails varies with the length and style of each particular nail. See Fig. 27-3. The shank size is specified by a wire gauge size. Nails are made from wire cut to length, and the head is fashioned on the cut length. The nail retains the wire gauge size from which it was cut.

TYPES OF NAILS

Nails are grouped into four general classes: common, box, casing and finishing. Several subclasses of nails are to be found and used in construction. Nails have been developed for special purposes for use in special places under certain conditions.

Fig. 27-1. A heavy-duty, in-plant type of nailing machine. (Bostitch Div., Textron, Inc.)

SIZE	mm	LENGTH INCHES	COMMON NAIL GAGE NO.	BOX NAIL GAGE NO.	CASING NAIL GAGE NO.	FINISHING NAIL GAGE NO.
2d	25	1	15	15 1/2	15 1/2	16 1/2
4d	38	1 1/2	12 1/2	14	14	15
6d	50	2	11 1/2	12 1/2	12 1/2	13
8d	63	2 1/2	10 1/4	11 1/2	11 1/2	12 1/2
10d	76	3	9	10 1/2	10 1/2	11 1/2
12d	82	3 1/4	9	10 1/2	10 1/2	11 1/2
16d	88	3 1/2	8	10	10	11
20d	102	4	6	9	9	10
30d	114	4 1/2	5	9	9	– – –
40d	126	5	4	8	8	– – –

WIRE NAIL CHART

Fig. 27-3. Chart gives size and gage details on common, box, casing and finishing nails.

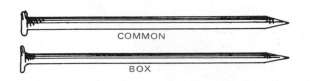

Fig. 27-4. Comparison of common and box nails.
(American Plywood Assoc.)

COMMON AND BOX NAILS

Common and box nails, Fig. 27-4, are for normal construction, particularly for wood framing. Smooth box nails will have a smaller diameter and thinner heads than common nails of the same penny size. Since their smaller diameter has less tendency to split the lumber, box nails are recommended for most uses.

Usually, 16d box nails are used in general framing. The 6d and 8d sizes are most frequently used for applications such as wall sheathing and roof sheathing. The size of the nail used depends on the thickness of the sheathing. Likewise, the length of the nail is determined by the thickness of lumber or plywood used. *Two-thirds of the length of the nail should penetrate the second piece of wood.*

Fig. 27-5. Comparison of casing and finishing nails.
(American Plywood Assoc.)

CASING AND FINISHING NAILS

Casing and finishing nails are used when you do not want large nail holes to show. Applications include interior and exterior trim work, installation of paneling and exterior siding application, Fig. 27-5. To further reduce the visibility of the nail heads, both casing and finishing nails may be tapped below the surface with a hammer and nail set. Then, the

Fig. 27-6. Scaffold nail. Note double head for ease of removal.
(American Plywood Assoc.)

remaining hole is filled with some type of filler.

Distinguishing features of the casing nail are a large diameter shank size and a larger V-shaped head. The finishing nail has a smaller shank and a smaller, barrel-shaped head.

The casing nail, generally, is used where heavier pieces of timber are involved and, most often, for exterior applications. The finishing nail is commonly used for the installation of interior window trim, door trim and paneling. The 4d, 6d and 8d sizes are most frequently used.

SCAFFOLD NAILS

Scaffold, double-headed or duplex nails can be used to save time and trouble in many operations where the fastener must be removed, Fig. 27-6. Use in the scaffolding, bracing, concrete forms and in temporary layout work makes the double-headed nail indispensable for both strength and ease of removal. The 8d and 10d sizes are most commonly used for scaffolds and bracing, also any other temporary fastening.

DEFORMED SHANK NAILS

A variety of deformed shank patterns are available, such as screw shank and ring shank nails, Fig. 27-7. All of these nails have greater holding power than smooth nails. Therefore, it is possible to use a smaller deformed shank nail than a smooth shank nail and still do the job satisfactorily.

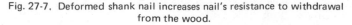

Fig. 27-7. Deformed shank nail increases nail's resistance to withdrawal from the wood.

Fig. 27-8. Construction featuring natural exposed wood, requires the use of nonstaining nails. (California Redwood Assoc.)

Ring shank nails should be used for pole type construction, and for the installation of plywood underlayment or subfloor underlayment. Ring shank nails perform very well where flexing of the individual pieces of wood occur between each other. Screw shank nails are typically used for installation of wood strip flooring.

For the fastening of 5/8 in. (16 mm) plywood subfloor to the floor joints, a 4d ring shank nail should be used. Generally, an 8d screw thread nail is used for wood flooring.

NONSTAINING NAILS

Nails, because of their iron content, will cause a discoloration of the wood when exposed to the weather. For long service and freedom from staining, nonstaining nails are used. They are necessary when exterior exposure is combined with the need for good appearance in siding, fascias, soffits, exterior trim and wood decks, Fig. 27-8. Galvanizing is the most common nail coating, and offers good protection against staining. Nails are also made of aluminum, bronze and stainless steel, but are more expensive.

DRY WALL NAILS

Dry wall nails are short nails, usually not exceeding 2 in. (5 cm) in length. Dry wall nails are similar in appearance to the box nail, having a small shank diameter and a long thin flat head, Fig. 27-9.

Fig. 27-9. Dry wall nails are designed for use in installing gypsum interior wall boards. (American Plywood Assoc.)

Fig. 27-10. Roofing nails are designed to secure and hold roofing material. Note large, flat head.

The unique feature of the dry wall nail is its color. Dry wall nails have been chemically "blued" which gives the nail a deep blue-black color. The bluing retards rusting and deterioration of the nail.

The size of dry wall nails used is determined by the thickness of the dry wall to be applied. For example, when 1/2 in. (13 mm) dry wall is applied, 4d nails should be used.

ROOFING NAILS

Roofing nails are made of noncorrosive material, such as galvanized iron or aluminum. The roofing nail has a large diameter shank and a large, thin head, Fig. 27-10. Its length is relatively short, depending upon the thickness of the roofing materials.

Fig. 27-11. Power nailer drives finishing nails with touch of trigger, countersinking nail heads in the process. This nailer is used on interior trim work and cabinet face nailing. (Duo-Fast Fastener Corp.)

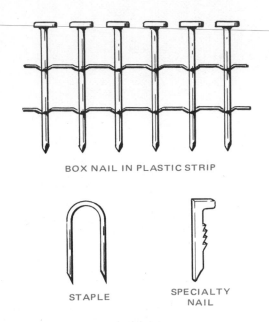

BOX NAIL IN PLASTIC STRIP

STAPLE   SPECIALTY NAIL

Fig. 27-12. Types of nails marketed for use with automatic nailers.

Roofing nails for installing sheet metal roofing often have a lead or plastic ring underneath the head which tightly seals the punched hole.

POWER NAILERS

For centuries, nails have been driven by the carpenter's hammer. Now, a large variety of power nailers are available to meet various nailing requirements, Fig. 27-11. The nailers are powered either by pneumatic force (air pressure) or by impact force (electrically powered).

Nails used with the power nailer are designed in various ways:

1. Customary box nail.
2. Staple.
3. Specialty nail, Fig. 27-12.

Box and finishing nails are held in a coil of plastic retainer and loaded into a magazine of the nailer, Fig. 27-13. The staple and specialty nail are usually lacquered into strips for easy handling.

PURCHASING NAILS

Nails are sold in 100 lb. boxes. Each size and style have a given number of nails per pound, Fig. 27-14. For the preparation of a cost analysis for a construction project, the number of pounds or boxes, can be estimated.

BOLTS AND SCREWS

Bolts are used to fasten wood to wood in heavy timber construction; also to secure wood to metal and wood to concrete. Bolts are manufactured in sizes from 1/4 in. to 1 1/4 in. (6 mm to 32 mm) in diameter and from 3/4-in. to

Fig. 27-13. Magazine type nailer   (Bostitch Div., Textron, Inc.)

NUMBER OF NAILS PER POUND

SIZE	COMMON	BOX	CASING	FINISHING
2d	876	1010	1010	1351
4d	316	473	473	584
6d	181	236	236	309
8d	106	145	145	189
10d	69	94	94	121
12d	63	88	88	113
16d	49	71	71	90
20d	31	52	52	62
30d	24	46	46	— —
40d	18	35	35	— —
50d	14	— —	— —	— —
60d	11	— —	— —	— —

Fig. 27-14. Chart gives nails per pound by size and type.

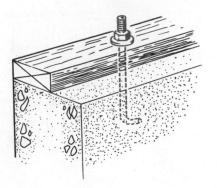

Fig. 27-15. L-shaped anchor bolt is used to secure a wood sill to a concrete foundation.

30-in. long (19 mm to 800 mm).

Bolts are used to secure wood sills to a concrete foundation, Fig. 27-15. Square and hexagonal shaped headed bolts, and square and hexagonal shaped nuts are the most common, Fig. 27-16. However, special bolts, like the anchor bolt with an L-shaped end are manufactured for specific jobs. The bent end forms a better bond to concrete and increases its holding power.

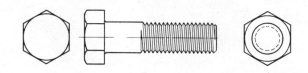

Fig. 27-16. Hexagonal head bolt and nut.

FLAT HEAD — TO ALLOW HEAD TO BE SET FLUSH OR COUNTER-SUNK BELOW SURFACE

ROUND HEAD — FOR SECURING METAL PLATES TO WOOD

OVAL HEAD — FOR FASTENING OTHER HARDWARE TO WOOD, USUALLY HEAD IS LEFT EXPOSED

PAN HEAD — USUALLY USED FOR SHEET METALWORK

Fig. 27-17. Screw head shapes shown and usage described.

Wood screws are made to secure wood to wood and metal to wood. These fasteners are available with various shapes of heads: flat, round, oval and pan. See Fig. 27-17. The screw head are either slotted or Phillips to fit the driver, Fig. 27-18.

SLOTTED   PHILLIPS

Fig. 27-18. Types of screw heads for most applications.

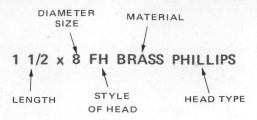

DIAMETER SIZE   MATERIAL

1 1/2 x 8 FH BRASS PHILLIPS

LENGTH   STYLE OF HEAD   HEAD TYPE

Fig. 27-19. Screw designations are given and defined.

Most screws are made of iron. However, screws also are made of aluminum, brass or steel with special coatings, depending upon their use.

The length of wood screws is measured in inches. The diameter of the shank is designated by a wire gauge size. The larger the gauge number, the larger the shank diameter. A No. 0 wood screw is 0.06 in. (2 mm) in diameter. A No. 24 wood screw is 0.372 in. (9 mm) in diameter.

When purchasing screws, the following information is needed:
1. Length.
2. Diameter size.
3. Head style and type.
4. Material, Fig. 27-19.

Lag screws or lag bolts are large wood screws with square or hexagonal shaped heads.

RINGS

To increase the strength of joints in heavy timber construction (particularly in truss construction), the split ring is used, Fig. 27-20. A split ring is a steel ring either 2 in. or 4 in.

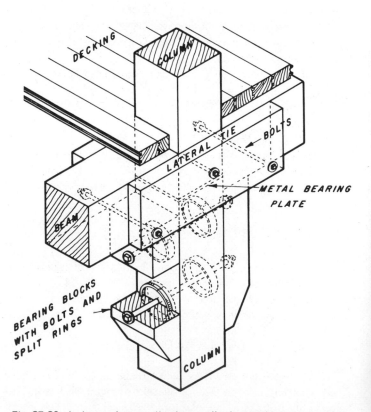

Fig. 27-20. In heavy-duty applications, split rings and bolts are used to join timbers. (National Forest Products)

(50 mm or 100 mm) in diameter, which is fitted into precut circular grooves. The ring is inserted between two pieces of wood held together by a bolt that runs through the center of the ring. The addition of the split ring greatly increases the shear resistance of the joint.

FRAMING ANCHORS

Various metal fittings have been developed for use in wood framing. They help to eliminate bulky support members otherwise used for joists butting onto a beam. These fittings are called clips, saddle straps, beam and joist hangers and framing anchors, Fig. 27-21.

The hinge connecter, for joining laminated wood beams to exterior concrete pieces is an example of an additional use of the framing anchor, Fig. 27-22. Framing anchors are made of

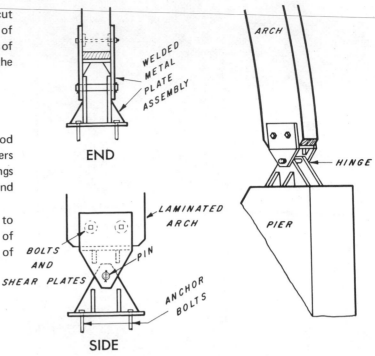

Fig. 27-22. Typical hinge connector joins laminated beam to concrete pier. (National Forest Products)

light sheet metal plates with holes punched for nails, or they may be made of welded steel plate with holes drilled for bolts, Fig. 27-23.

DECORATIVE HARDWARE

Decorative hardware is sometimes called finishing hardware. These are the items which are visible when the building is completed.

Finishing hardware serves several purposes:
1. Decoration.
2. Protection.

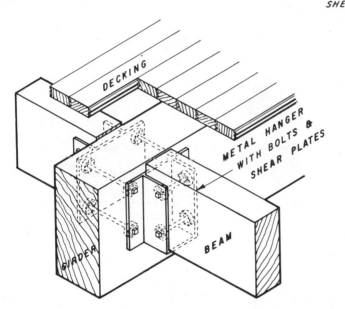

Fig. 27-21. Metal framing anchor for joining beam to girder. (National Forest Products)

Fig. 27-23. Note use of framing anchors. A—Hinge connector. B—Horizontal support bracket. C—Diagonal bracing anchor. (South Dakota Dept. of Transportation, Div. of Highways)

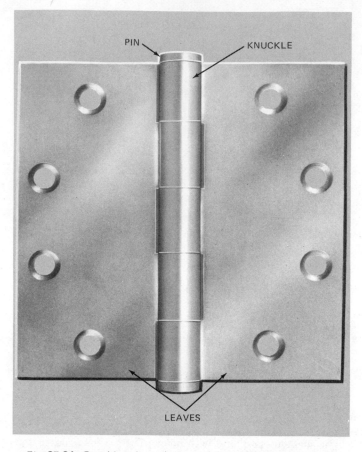

Fig. 27-24. Butt hinge is used on wood doors that swing one way.

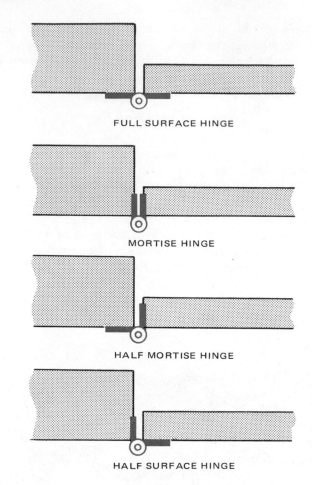

FULL SURFACE HINGE

MORTISE HINGE

HALF MORTISE HINGE

HALF SURFACE HINGE

Fig. 27-25. Types of hinges as viewed from top of door.

3. Decoration and protection.

Bath accessories such as towel bars, soap trays, paper holders and curtain tracks usually are not considered as finishing hardware. However, they do serve a decorative purpose. Finishing hardware are exposed items such as hinges, locks and door closers installed by the builder.

DOOR HINGES

A hinge is a device that allows a door or other panel to turn or swing. Usually, it is made of brass, bronze, steel or stainless steel. In some cases, chromium plated steel bushings are inserted around the pin to reduce wear.

Hinges may be exposed, semiconcealed or concealed. A hinge installed on the edge of the door is called a "butt hinge." It consists of a pin, knuckle and two leaves, or butts, Fig. 27-24. A butt hinge is mortised into the edge of the door and frame. Usually, it is classified by the type of mortise the hinge requires (full surface, half surface, mortise or half mortise). See Fig. 27-25.

On installation, the pin and knuckle of the butt hinge extend beyond the face of the door. In operation, the two leaves of the hinge turn on the pin at the center, allowing the door to swing past the door frame.

The most commonly used sizes of butt hinges are 3 1/2 in. x 3 1/2 in., 4 in. x 4 in. and 4 1/2 in. x 4 1/2 in. (9 cm x 9 cm, 10 cm x 10 cm and 11.5 cm x 11.5 cm). The measurement is taken with the leaves extended.

SLIDING DOOR HARDWARE

Sliding doors have specially designed double tracks to allow the doors to bypass each other. The doors may be supported by nylon rollers in an overhead track, or they may glide in special grooves at the top and bottom. Heavier sliding doors glide from overhead tracks of extruded aluminum, Fig. 27-26. For a single door, sometimes called a "pocket door," a single track is used, Fig. 27-27. Roller wheels usually are made of nylon, with teflon-coated steel ball bearings for quiet operation. Folding doors use similar hardware for carriers.

LOCKS AND LATCHES

Door locks and latches are used to hold doors in a closed position. This is accomplished by bolts, which are operated by a variety of ways.

The mortise lock is one of the first concealed locks set into a door. It has a pin-tumbler cylinder that is easy to remove. The pin-tumbler contains a keyed mechanism for disengaging or activating the bolt, Fig. 27-28.

A standard, five pin-tumbler lock may be set for several keys. By correlating the variation in pin length, a large number of key changes is possible.

The bored-in locksets require:

1. A horizontal hole bored into the edge of the door.

194

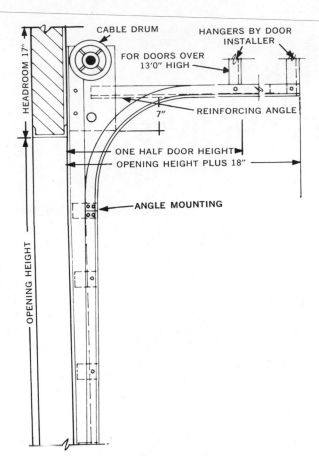

Fig. 27-26. Typical overhead door track movement. Some are counter-balanced by weights, others by a spring mechanism. (Overhead Door Co.)

Fig. 27-27. Pocket door usually glides on nylon wheels in a single track.

2. A second horizontal hole bored from one face of the door to the other, and intersecting the first horizontal hole.

Bored-in locksets have a cylindrical case that fits over a smaller latch-bolt case. The usual lockset for bored-in locks is 2 3/8 in. x 2 3/4 in. (6 cm or 7 cm) from the edge of the door. Most bored-in locksets are made for 1 3/8-in. x 1 3/4-in. (3.5 cm or 4.5 cm) thick doors.

SUMMARY

Hardware for construction may be classified as rough or finishing, depending upon the nature of its use. Rough hardware includes items such as nails, bolts, screws and

Fig. 27-28. Tubular type lock is popular in exterior doors of residences.

framing anchors. These are items which are *not* visible following the completion of the building. Finishing hardware is visible on the completed building. Since it is necessary for certain hardware items to be exposed, they usually are manufactured to be pleasing to look at. Therefore, finishing hardware serves a decorative function.

CAREERS RELATED TO CONSTRUCTION

LOCKSMITH opens, repairs and installs locks; makes original and duplicate keys; changes safe combinations; rekeys master locking systems; works on electronic burglar alarms and surveillence systems.

LOCK SETTER installs locks and lock sets in new wood doors, using various templates, drills, boring jigs and special chisels.

DISCUSSION TOPICS

1. List several items considered to be rough hardware.
2. Explain the term "penny" as related to nails.
3. What is the difference between a common nail and a box nail?
4. How is the "shank size" of a nail indicated?
5. Where are finishing nails used?
6. Describe the basic principle involved in using an L-shaped anchor bolt to secure wood sills to a concrete foundation.
7. What is meant by a nonstaining nail?
8. What are framing anchors? How are they used?
9. Describe the features and uses of split rings.
10. What are hinges?
11. What sizes of butt hinges are available?
12. What is a mortise lock?

Chapter 28

SEALANTS AND ADHESIVES

Sealants and adhesives are important materials for today's construction industry. Sealants protect against penetration of water and weather or prevent the escape of liquids. Adhesives (glues) allow two materials to be fastened together.

All materials described in this unit, with the exception of one or two, have two common characteristics:
1. Cohesiveness.
2. Adhesiveness.

Cohesiveness is the ability of particles of a material to cling tightly together. Adhesiveness is the ability of a material to afix itself to and cling to an entirely different material.

SEALANTS

Sealing compounds seal surfaces against penetration of water or other liquids. To do this, sealants must have:
1. Some adhesive qualities.
2. The ability to fill the surface pores.
3. The ability to form a continuous skin on the surface to which they are applied.

In many applications the adhesion should be permanent. In others it need only be temporary.

LIQUID SEALERS

Liquid asphalt is one common type of sealer. It has several uses. One is to coat the outer surface of concrete below ground level to keep water out of the concrete pores. See Fig. 28-1. Another similar use is to seal the inside of wood or concrete water tanks. A third use is as a sealer or primer over concrete before asphaltic tile adhesive is applied. Fig. 28-2 shows materials being applied to a floor.

An effective sealant must form a waterproof membrane. The material must be resilient or the membrane will crack and leak. This same resilience will allow sealants to expand over small cracks in the base surface without losing their effectiveness. In case of movement at the joint, the sealant must be able to bridge the joint without tearing.

Wax mixtures may be applied to form temporary seals. Such mixtures are used over concrete and terrazzo floors. Waxes prevent penetration of oil and grease into the floor surface. The wax oxidizes to form a continuous protective film. The seal is only temporary, however. As the wax continues to oxidize it becomes brittle and is worn off by traffic.

Fig. 28-1. Liquid asphalt waterproofs concrete basement wall.

Fig. 28-2. Liquid asphalt sealant materials may be applied to concrete floors by rolling, brushing or spraying. (Master Builders)

Liquid silicones seal concrete, brick and tile surfaces. Silicones prevent absorption of water which could lead to staining and efflorescence. (Efflorescence is a limelike collection on the surface of masonry walls.) The silicone sealers are particularly valuable for exposed application. Being colorless they do not affect appearance of the wall.

Epoxy resins, a plastic material, are also used as sealers over concrete, wood or terrazzo surfaces. The thin liquid adheres to and seals the surface providing a good bond for tile, paint or other permanent coverings.

PLASTIC GLAZING MATERIAL

Glazing and caulking compounds are two similar groups of sealing materials. The only difference is that when used for sealing glass into a frame, they are known as glazing materials. A caulking material has the specific quality of being able to adhere to the surface of any material it contacts. Most glazing and caulking materials are in the mastic form. This is a thick heavy-bodied consistency that must be applied with a caulking gun as shown in Fig. 28-3.

Fig. 28-3. Construction sealant is used to caulk joint between masonry wall and metal window frame. (General Electric Co.)

One-part sealants, premixed sealants with polysulfide, silicon, or urethane bases, are applied with a caulking gun. These materials undergo a chemical change at normal room temperature. They are said to be "chemically curing." Such sealants, Fig. 28-4, generally have rubberlike qualities that last for several years after they are cured.

Some require a primer coat on all glass or metal surfaces for proper adhesions. Others will adhere directly to glass or polished metal.

Polysulfides offer good resistance to most solvent materials. Silicones provide a flexible seal that continues to function over

Fig. 28-4. A one-part primeless sealant is ready to use in caulkers. This material is designed for glazing glass, plastic and metal.

a wider temperature range. Urethanes provide a seal that is resistant to abrasion and most chemicals.

Solvent release sealants do not change chemically while setting to a flexible or semirigid state. The set results from evaporation of the solvent. Many sealants of this type shrink and harden on drying. These shortcomings limit their usefulness. Their properties vary widely from compounds that remain soft and pliable to those that become hard and brittle with time.

Two-part sealants are also chemical curing. They offer the advantage of being resilient and quick curing. (Resilient means they can go back to their old shape after being stretched.) They usually consist of a polysulfide or urethane base with a catalyst, Fig. 28-5. The base material and catalyst are packaged in separate containers to be mixed when needed. Once the two have been mixed they must be applied within a short time. This time limit is known as "pot life."

Fig. 28-5. Finished sealing between precast concrete panels. The sealant is virtually unaffected by weathering agents like sunlight, rain and snow. (General Electric Co.)

197

Fig. 28-6. Windows of this 48-story building are "glazed in" with preformed glazing strips. Strips are flexible enough to absorb the structure's movements without breaking the glass or leaking. (Libby-Owens-Ford Glass Co.)

Two-part polysulfide and urethane sealants may be mixed, packed in containers and quick frozen at temperatures below —40 F. Cartons are maintained at temperatures of —20 F until ready to use. When thawed to room temperature they are ready for application.

Two-part sealants are available in a wide range of hardnesses. The softest types are used when maximum movement and a minimum of strain is expected. A medium grade is produced for applications subject to vibration and movement. In applications where a high resistance to abrasion is desired, the hardest type is used.

PREFORMED RESILIENT SEALANTS

Flexible premolded sealants are made from both natural and synthetic rubber, polyvinyl chloride and other plastics. They are produced as tapes, ribbons, beads and in many unusual shapes. Fig. 28-6 pictures a typical application. Preformed sealants may be used as a seal between two structural parts. Fig. 28-7 shows special shapes available for glazing applications. Some shapes are reinforced with fiber to help keep their shape after installation. One-ply rubber roof systems, Fig. 28-8, use preformed sealants to seal edges of elastic flashing.

ADHESIVES

The art of making and using glue has been known for a very long time. Many of the materials used in ancient times where

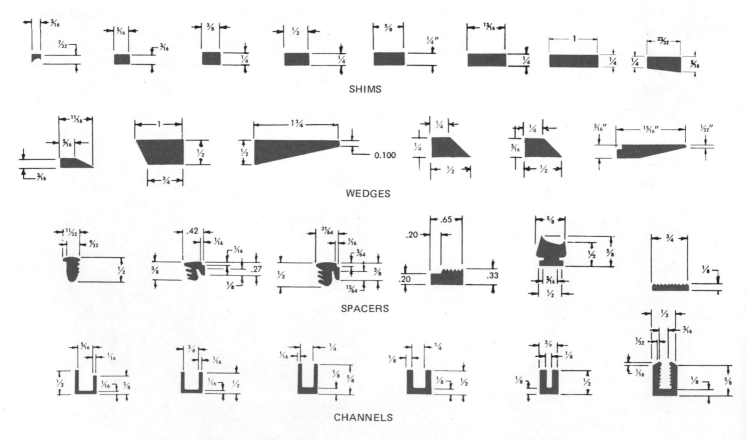

Fig. 28-7. Various preformed shapes of resilient material. (F. H. Maloney Co.)

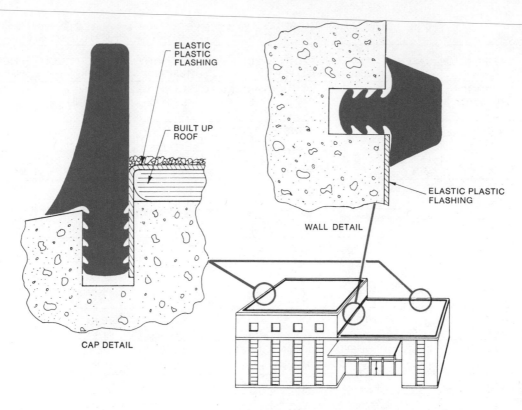

Fig. 28-8. Edges of rubber roofs are fastened down and sealed with preformed elastic material called gaskets. Gasket is pressed down into a slot in the concrete where it is held by the rough edges formed in the elastic material. (F. H. Maloney Co.)

natural products such as tree resins.

The making of adhesives was a closely-guarded secret in earlier times. It was not until the eighteenth century that glue making became widespread.

After discovery of the first synthetic resin in 1909, adhesives became more popular in construction. Growth of the plastics industry has brought many new glues made from synthetic resins, Fig. 28-9. They include resorcinol adhesive, polyvinyl acetate, polyurethane, and epoxy resin glues. Many of the old adhesives such as animal glue, casein glue and

asphalt adhesives are still used. So are a variety of vegetable and fish-origin glues.

Older glues were intended for bonding together wood, paper, leather, rubber and cloth. The new family of adhesives will bond glass, steel, concrete, ceramics and plastics. There are glues for every conceivable purpose. Fig. 28-10 gives the properties, applications and procedures for use of eight popular adhesives.

NATURAL RESIN GLUES

Animal glue is one of the oldest of glues. Its strength decreases with age and today it is used largely for paper products. Casein glue comes in powdered form. It is made from milk curd and certain chemicals and was the first water resistant glue. Other natural resin glues include fish glue, vegetable glue and blood albumin glue.

SYNTHETIC RESIN GLUE

Most of the good bonding materials are formed from a synthetic base. Synthetic glues may be classified into two groups:

1. Thermoplastic.
2. Thermosetting.

Thermoplastic adhesives may be changed repeatedly by heating. Thermosetting means that these materials will not be altered appreciably by any chemical action after they have once been heated. Thermoplastic resins include the poly-

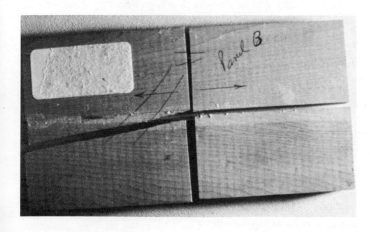

Fig. 28-9. Perhaps you have seen test panels of glue similar to this. A good adhesive, such as synthetic resin glue, will make a bond stronger than the surrounding wood materials.

COMPARATIVE ADHESIVE CHART

ADHESIVE	ALIPHATIC RESIN	CASEIN RESIN	CONTACT CEMENT	EPOXY GLUE	HIDE GLUE	POLYVINYL RESIN	RESORCINOL	UREA-FORMALDEHYDE
TYPE	THERMO-PLASTIC	THERMO-PLASTIC	RUBBER		ORGANIC AIR DRYING	THERMO-PLASTIC	THERMO-SETTING	THERMO-SETTING
MIX TIME	NONE	10 MIN.	NONE	NONE	10 MIN.	NONE	5 MIN.	5 MIN.
WORKING TIME	2–5 MIN.	15 MIN.		30 MIN.	2 HR.	1 1/2 MIN.	15 MIN.	15 MIN.
CLAMP TIME	2 HR.	8 HR.		8 HR.	30 MIN.	2 HR.	8 HR.	8 HR.
CURE TIME	10–12 HR.	24 HR.	24 HR.	8 HR.	8 HR.	10 HR.	24 HR.	24 HR.
CLAMP PRESSURE	LOW	HIGH		LOW	LOW	LOW	HIGH	HIGH
RESISTANCE TO MOISTURE	LOW	100%	100%	100%	LOW	LOW	100%	HIGH
WORKING TEMPERATURE	70 F 21 C	33 F 1 C	70 F 21 C	70 F 21 C	70 F 21 C	70 F 21 C	70 F 21 C	70 F 21 C
BENCH LIFE / STORAGE LIFE	INDEF. / 1 YR.	2 HR. / 1 YR.	INDEF. / 1 YR.	30 MIN. / INDEF.	INDEF. / 1 YR.	INDEF. / 1 YR.	2 HR. / 1 YR.	2–4 HR. / 1 YR.
ABRASIVE TO TOOLS	NO	YES	NO	MILD	NO	NO	YES	MILD
GAP FILLING	YES	NO	NO	YES	NO	YES	NO	NO
STAIN	MILD	MILD	NO	NO	MILD	NO	YES	NO
MULTI-MATERIAL	YES	NO	YES	YES	NO	YES	YES	YES

Fig. 28-10. Comparison will show which adhesive is best for the task at hand.

vinyls. The thermosetting resins include the formaldehyde group: urea, phenol, resorcinol and melamines.

FAMILY OF ADHESIVES

Many people misunderstand and misuse the term "glue." Glue is only one member of the total family of adhesives. However, it forms the largest group of materials used to adhere construction material.

Fig. 28-11 shows the four separate groups belonging to the adhesive family. Glue bonds two like materials together by creating a molecular attraction between the adhesive and the bonding material. Mastic is a cohesive material which may be used to adhere unlike materials together. For example, mastic would be used to fasten floor tile to a wood underlayment. Cement, an adhesive, undergoes a chemical change, usually by catalytic action, to create a bond. Epoxy for household use is such a material. Paste or mucilage materials are usually water-soluble adhesives. It is like the adhesive coating on postage stamps and envelopes. Each of the members of the adhesive family is used in the construction industry.

CONSTRUCTION ADHESIVES

Animal glue is sold as either solid or liquid. The solid form is heated and applied while still hot. It is slow setting and allows time for adjusting the glue joint. Animal glue has excellent bonding properties with wood, leather, paper or cloth. Its resistance to water is only moderate. It cures by air drying at room temperature.

Casein glue, made from protein material, is a dry powder. Mixed with water, it has good bonding power for wood-to-wood or paper-to-wood application. Dry-heat resistance is good and resistance to cold is moderate. It has moderate resistance to water but does not perform well when subjected to attack from molds, fungi and other wood organisms. Casein glue will set at temperatures as low as 35 F (1.7 C). Maximum

ADHESIVES

GLUE	MASTIC	CEMENT	PASTE – MUCILAGE
CASEIN	LATEX EMULSION	NEOPRENE-RUBBER	ANIMAL
POLYVINYL RESIN	CHLORINATED RUBBER	EPOXY	HIDE
UREA-FORMALDEHYDE		ASPHALT	STARCH
RESORCINOL		CELLULOSE	DEXTRIN
POLYURETHANE			
PHENOLIC			
MELAMINE			

Fig. 28-11. Family of adhesives.

strength is developed at temperatures of about 70 F (21 C) with moderate pressure.

Dextrin and starch glues are packaged in both liquid and dry form. The dry glue is mixed with water. Both forms have good bonding power with paper and leather, poor with wood to wood. They have good resistance to heat and cold but poor resistance to water or high humidity.

Asphalt cements are themosplastic materials made from asphalt emulsions or asphalt cutbacks. They bond well with paper and concrete and are used mainly on roofing and for laminating layers of wood fiberboard. These materials have poor resistance to heat but good resistance to cold and water.

Natural rubber adhesives are usually latex emulsions or dissolved crepe rubber. They produce a good bond with rubber or leather but only a fair bond with wood, ceramics and glass. Resistance to water is good but resistance to temperature changes is only fair. Room temperature is sufficient for curing.

Urea-formaldehyde resin glues are available in powder or liquid form. Urea glues are thermosetting in nature. They have excellent bonding power with wood. Heat is desirable for curing some types of resin. Other urea resins cure satisfactorily at room temperature. Rapid curing is possible with the application of high-frequency electric current applied directly to the joint. This technique is known as "wood welding."

Phenolic resin glues are made in both dry and liquid form. Phenolic resins have excellent bonding power with wood. Some set at room temperature while others require a hot press. Hot-press glues are commonly used in the manufacture of plywood. Phenol-formaldehyde, phenol-resorcinol and resorcinol-formaldehyde are all types of phenolic resin glue. All are woodworking glues used in manufacture of laminated wood structural members.

Epoxy resins are thermosetting in nature. They are manufactured in liquid form with a separate catalyst. Bonding is excellent with wood, metal, glass and masonry materials. Epoxy resins are widely used in the manufacture of laminated curtain wall panels. They are also used to repair broken concrete. Resistance to both cold and hot conditions is excellent. Some epoxy resins cure at room temperature while others require hot-press curing.

Polyvinyl resin adhesives are usually in an emulsion form. Bond is good with wood, paper and vinyl plastics, reasonable with metal. Resistance to cold is good but water resistance is low. These glues cure at room temperature.

Sodium silicate adhesives are liquids which have excellent bond with paper or glass but only fair bond to wood and metals. Resistance to cold, heat and creep is good but resistance to water is poor. Curing is at room temperature.

SUMMARY

Two common characteristics describe sealants and adhesives: cohesiveness and adhesiveness. Sealants create a coating on the surface of various construction materials to prevent the penetration or the escape of water or various liquids. To do this, the material must be cohesive enough to form a skin covering the surface. It must also have adhesive properties to afix itself to that surface. A sealant must also be elastomeric (rubber-like substance) in character, or resilient enough to be able to stretch over small cracks in the base material.

Adhesives have been put together by chemists and physicists dating back to the nineteenth century. "Glues" even developed in antiquity. The discovery of synthetic resins and the development of the plastic industry have given construction many adhesives with widely different properties. Today, adhesives can produce bonds of superior strength and durability when compared to earlier adhesives.

CAREERS RELATED TO CONSTRUCTION

GLAZIER installs glass; spreads and smooths glaze around window frames and panes.

CAULKER uses caulking gun to seal joints in and around construction components; installs preformed gaskets where used for structural seal.

CHEMIST analyzes compounds, experiments and tests chemicals for purpose of improving adhesive qualities of glues, cements and sealants.

DISCUSSION TOPICS

1. Define cohesiveness and adhesiveness.
2. What is the difference between glazing and caulking?
3. What is a preformed gasket?
4. Describe the function of sealers in construction.
5. List the family of adhesives.
6. What adhesives are used to manufacture laminated wood beams?
7. What is wood welding?
8. What is the difference between thermosetting and thermoplastic?
9. List five examples of adhesives used in the construction industry and identify the type of adhesive used.
10. Explain the difference between glue, mastic, cement and paste or mucilage.

Chapter 29

INSULATION MATERIAL

Fig. 29-1. Insulating materials can be sprayed onto ceilings to keep out heat and cold. (Port of New York Authority)

All materials resist the transfer of heat to some extent. The thicker the material the less heat transmitted. However, modern construction practices and the weight of materials limit the thickness of building material that may be used. For this reason, loss of heat from buildings in winter and the gain of heat in summer is controlled by thermal insulating materials. See Fig. 29-1. Preferred materials are spun glass, formed glass, vegetable fibers, mineral fibers and certain formed (or foamed) plastics. These materials may be made into boards, installed as loose blankets or batts. Some are blown, in granular form, into walls and ceilings.

HEAT TRANSFER

The resistance of a material to the transfer of heat depends on what kind of material it is and the material's physical characteristics. The specific resistance (R) of any material, varies with its thickness. A 2 in. thickness of material will resist twice as much heat as a 1 in. thickness. Conversely, the amount of heat that will be conducted by the material decreases with thickness. A 2 in. thickness will conduct or transfer half as much heat as 1 in. thickness of the same material.

The conductivity (K) of a material is the amount of heat it will conduct through 1 in. per unit of area. The amount of heat it will conduct through one square foot of surface area per unit of thickness is called conductance (C).

Some manufactures list the R value, Figs. 29-2 and 29-3, of their products; some list the K value; some list the C value.

C AND R VALUES FOR BUILDING MATERIALS

MATERIAL	CONDUCTANCE C	RESISTANCE R = (1/C)
AIR	1.48	0.66
AIR OUTSIDE AT 7 1/2 mph (12 km/h)	4.0	0.25
AIR OUTSIDE AT 15 mph (24 km/h)	6.0	0.17
BLANKET INSULATION 1 IN. (2.5 cm)	0.27	3.75
RIGID INSULATION BOARD	0.33	3.03
ASBESTOS SHINGLES	4.75	0.21
ASPHALT SHINGLES	2.27	0.45
WOOD SHINGLES	1.06	0.95
COMMON BRICK 4 IN. (10 cm)	1.25	0.80
DOUGLAS FIR 3/4 IN. (1.9 cm)	0.48	2.08
PLYWOOD 3/8 IN. (1.1 cm)	2.10	0.46
GYPSUM BOARD (DRY WALL)	2.44	0.41

Fig. 29-2. Chart indicates ability of certain materials to conduct heat or cold. Note that blanket insulation and rigid insulation board are very poor conductors. This makes them very good for insulating buildings.

Actual heat loss through a structure, such as a wall or a floor, is measured in British Thermal Units or "Btu." When SI metric is used, the unit is the joule. One joule is equal to 0.000 948 Btu. One Btu is the amount of heat needed to raise the temperature of 1 lb. of water 1 F.

To find the heat loss in quantity of Btu through a given structure, we must calculate the total conduction (U) of that

RESISTANCE VALUE OF WALL ELEMENTS

WALL ELEMENT	RESISTANCE VALUE (R)
OUTSIDE WALL (BRICK)	0.80
WALL SHEATHING	3.03
AIR SPACE	0.66
DRY WALL	0.41
TOTAL R	4.90

Fig. 29-3. Resistance value of building materials used in walls.

structure. The U factor of a structure is the number of Btu per hour that will flow through 1 sq. ft. of the structure when there is a one degree difference in temperature between the two sides.

The higher the U factor of a structure, the more heat will be lost through it. The lower the U factor, the less heat will pass through, that is, the better its insulation properties. Engineers and technicians use tables to determine insulation needs of a building.

THERMAL INSULATING MATERIAL

All materials used to prevent heat losses or gains are known as thermal insulation materials. There are nine basic kinds:
1. Loose fill.
2. Blankets.
3. Batts.
4. Structural insulation board.
5. Slab or block.
6. Reflective.
7. Formed-in-place.
8. Sprayed on.
9. Corrugated.

Loose fill insulation is generally bulky. There are two main types:
1. Fibrous, made from mineral wool, rock wool, glass wool, slag wool or vegetable fiber.
2. Granular, made from granulated cork or from expanded minerals such as perlite and vermiculite.

Blanket insulation is made from the same fibrous material as the loose fill. The fibers are manufactured in blankets or matts. Each blanket or matt has a variety of thicknesses from 1 in. to 6 1/2 in. (2.5 cm to 17 cm). It comes in various widths.

Batts are similar in basic manufacture to blankets, but they are made in shorter lengths. Batts are covered with paper to make installation easier. The width of each batt is 14 1/2 in. for easy placement between stud spacings. The batts usually have a paper edging along each side. It is easier to attach the batts to studs or framing members by tacking through this edge. Examples of insulating materials just described are shown in Fig. 29-4.

Structural insulation board is made from a variety of substances such as cane, wood and vegetable fibers. It is used for exterior and interior sheathing, roof insulation board and interior finishing. Fig. 29-5 shows production of rigid insulation.

Slab or block insulation is made in rigid units smaller in area than structural insulation board. This type of insulation is

Fig. 29-5. During production, large blocks of insulation are cut into thin, rigid boards by horizontal slitters. Such insulation is often used in roof construction.   (Manning, Selvage and Lee)

made from cork, shredded wood, mineral wool, vermiculite and asphalt, formed concrete, formed plastic and cellular hard rubber. A roof installation is shown in Fig. 29-6.

Reflective insulations have metallic or other special surfaces with or without some type of backing. Reflective insulations rely on their surface characteristics, thickness of air space and temperature differences for their insulating value. Batts or other types of insulation material may have a reflective surface to increase their effectiveness.

Formed-in-place (or foamed-in-place) insulation is made

Fig. 29-6. Slabs of urethane foam insulation are cemented on roof. Coated asbestos felt sheets will be applied over foam.

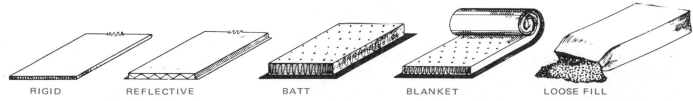

RIGID   REFLECTIVE   BATT   BLANKET   LOOSE FILL

Fig. 29-4. Popular types of building insulation.

from a liquid resin. Two ingredients are used. When mixed, they foam up and become solid. The hardened substance is filled with air pockets. These makes it a very good insulation material.

Sprayed-on insulations are produced by mixing fibrous or cellular material with an adhesive. This mixture is blown onto the surface to be insulated. Areas that are otherwise difficult to insulate, because of their shape or location, are treated this way.

Corrugated insulation is made of paper, corrugated and cemented into multiple layers. Some types are sprayed with an adhesive. When hardened the adhesive gives the product extra stiffness. Others are faced with foil to provide extra insulating value.

VAPOR INSULATION

Protecting the interior of buildings against moisture penetration from outside involves well-known procedures. Protection against condensation of water vapor produced on the inside is a different matter and is perhaps not so well understood. Protection from the outside is provided by water-repellent sealants or materials which can turn water aside. Moisture vapor, on the other hand, can permeate most ordinary building material such as wood, paper, plastic and unheated brick.

Moisture vapor comes from people cooking, laundering, unvented fuel-burning devices, humidifiers and evaporation from basement crawl spaces. Ventilation will help carry away unwanted vapor moisture. But this does not provide the total answer to the problem of condensation on the inside walls. Condensation will not take place if an effective vapor barrier (vapor insulation) is provided, Fig. 29-7.

Vapor barriers should be installed on the inside of the insulation. This barrier should be a continuous surface of asphalt, wax-coated paper, aluminum foil sheets or polyethylene film.

SUMMARY

Insulation is normally considered in relation to heat or cold transfer through a wall, ceiling or floor. Insulation may also be designed to guard against moisture vapors and other conditions that may exist such as fire and sound.

Heat loss through a wall may be calculated if the R values for each material are known.

Moving air transmits heat more rapidly than still air. However, still air has a very high insulating value. Air trapped in cellular materials serves as a very good insulator.

The fact that warm inside air will carry more moisture vapor than the cold outside air complicates the problem of insulation. When the warm moist air comes in contact with the cold outside wall, moisture will condense inside the wall structure, causing damage to the structure and the insulating material. A vapor barrier of waxed paper, polyethylene film, or aluminum foil guards against this problem.

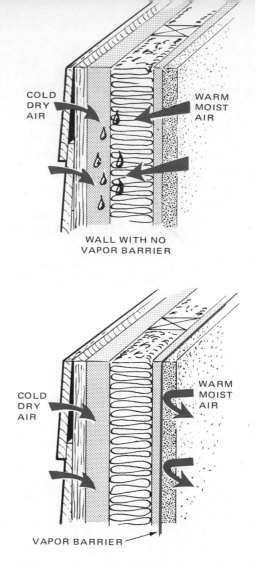

Fig. 29-7. Vapor barrier on warm side of house prevents moisture from penetrating wall and condensing on cold side.

CAREERS RELATED TO CONSTRUCTION

INSULATION INSTALLER fastens building insulation to walls, floors, ceilings and partitions. May fasten furring strips to walls, etc.

INSULATION FOREMAN supervises and coordinates work crews installing building insulation.

DISCUSSION TOPICS

1. List the reasons for insulating a wall, ceiling or floor.
2. In what forms are insulation materials available?
3. What is the source or origin of insulation materials?
4. What physical characteristics of materials are needed to make them good insulators?
5. Define a Btu.
6. What is specific resistance to heat transfer?
7. What does the U factor tell you about a structure?
8. How does reflective insulation aid in reducing heat loss?
9. What is a vapor barrier? How is a vapor barrier installed?

Chapter 30

EXTERIOR WALL COVERING MATERIALS

Fig. 30-1. Horizontal wood siding makes attractive exterior covering for this business establishment. (California Redwood Assoc.)

Fig. 30-2. Bevel siding is available in several different sizes. (Western Red Cedar Lumber Assoc.)

Outside walls of buildings serve one or more basic functions:

1. *To carry part of the load of the structure.*
2. *To protect the inside from the weather.*
3. *To present an attractive appearance.*

An exterior wall performs one, two or all three functions in varying degrees.

Walls are constructed in several different ways. The materials used to build them depend, to some extent, on the method of construction. The simplest type of wall has a wood framework of studs and plates. A second type has a framework of wood, steel or concrete. The wall material simply fills in the open spaces. The third type has walls of solid masonry or poured concrete. This wall is designed to support the weight of additional floors and/or the roof above it.

WOOD SIDING

Several species of trees are cut for the making of wood siding. Two common species are redwood and western red cedar. These species have natural properties which resist decay. Fig. 30-1 is a good example of wood siding.

Wood siding is manufactured in five basic types:

1. Tongue and groove.

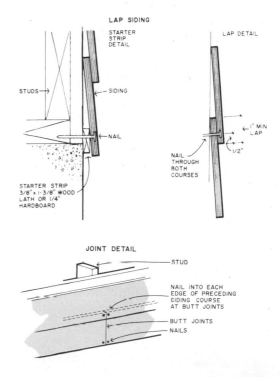

Fig. 30-3. Hardboard lap siding is used very often for residential construction. It is available in long lengths, various widths, and usually has a primer coat of paint applied. (American Hardboard Assoc.)

Construction

2. Shiplap.
3. Bevel siding.
4. Drop siding.
5. Log cabin siding.

Fig. 30-2 pictures several sizes of bevel siding. A type of manmade material is used in the hardboard siding shown in Fig. 30-3. Each type is made in several sizes and patterns. See Fig. 30-4.

In a board and batten exterior, boards are applied to a wall vertically, with narrow strips of the same material nailed over the vertical joints as in Fig. 30-5.

A variation of this system is the use of vertical plank siding. This is either 1 in. or 2 in. material. It has tongue and groove edges applied vertically, Fig. 30-6. Plank for this purpose is manufactured in both clean and knotty grades.

Red cedar shingles and shakes are commonly used as roofing materials. They can also serve as exterior wall finishing as shown in Fig. 30-7. Most shingles are made in random widths. But none are wider than 14 in. or narrower than 3 in. Lengths are standard: 16, 18 and 24.

Shingles are made by cutting a log into "shingle blocks" and then cutting the block into wedge-shaped sections. Each section is, in turn, quarter sawed into individual shingles. See Fig. 30-8.

Wood shakes are like shingles but are split rather than being

Fig. 30-5. Board and batten exterior uses cypress from the southern Gulf states. (Southern Forest Products Assoc.)

sawed from the shingle block. Splitting produces a much rougher face. Thus, shakes give the building a more rustic appearance.

Exterior plywood is used for siding, and exterior finishing. It is applied vertically in 4 by 8 ft. sheets. Joints are covered by battens.

Several types of plywood are manufactured just for exterior finishing. One of these types has shallow vertical grooves every few inches to create a vertical board appearance. See Fig. 30-9. Shallow grooves give the effect of reverse battens.

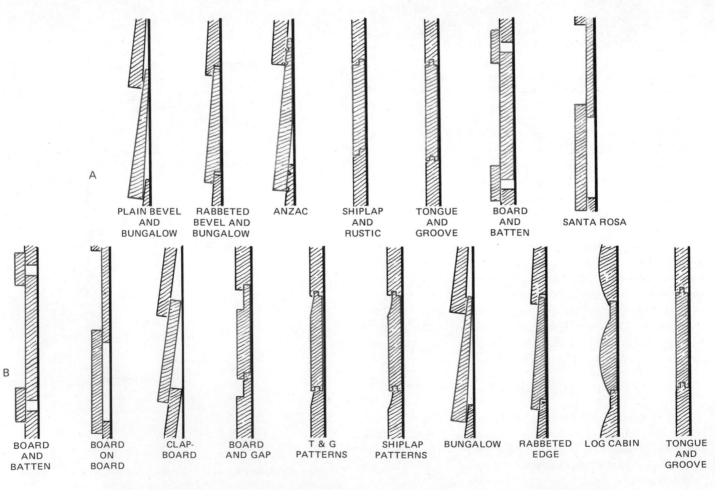

A

| PLAIN BEVEL AND BUNGALOW | RABBETED BEVEL AND BUNGALOW | ANZAC | SHIPLAP AND RUSTIC | TONGUE AND GROOVE | BOARD AND BATTEN | SANTA ROSA |

B

| BOARD AND BATTEN | BOARD ON BOARD | CLAP-BOARD | BOARD AND GAP | T & G PATTERNS | SHIPLAP PATTERNS | BUNGALOW | RABBETED EDGE | LOG CABIN | TONGUE AND GROOVE |

Fig. 30-4. Siding is available in a wide range of styles and shapes. A—Redwood siding. B—Western red cedar.

Fig. 30-6. Vertical tongue and groove siding.
(Western Wood Products Assoc.)

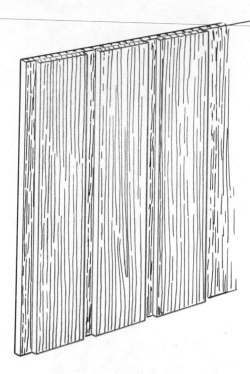

Fig. 30-9. Reverse/batten plywood siding. Grooves are usually 1/4 in.
deep and 3/8 in. wide. Panel thickness is 5/8 in.

Fig. 30-7. Red cedar singles used as a siding material.
(Koppers Co., Inc.)

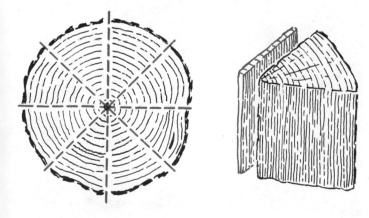

Fig. 30-8. How wood shingles are cut. Left view shows how section of
log is cut into wedges before shingles are sawed out. Right illustration
shows individual shingles sawed away from wedge.

Fig. 30-10. Building framed in reinforced concrete receives facing of
solid stone quarried in Spain and imported.
(United States Fidelity and Guaranty Co.)

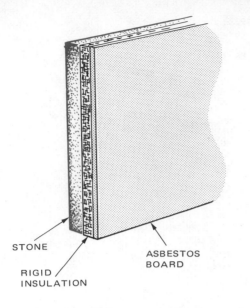

STONE

RIGID INSULATION

ASBESTOS BOARD

Fig. 30-11. A bonded stone sandwich facing panel.

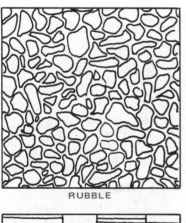

RUBBLE

ASHLAR

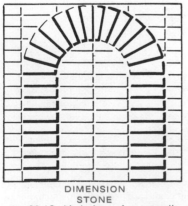

DIMENSION STONE

Fig. 30-12. Variations of stone walls.

STONE

Stone facing can be used over a builtup wall of masonry or concrete or over a framework of metal. Two types of stone facing material are available.
1. Solid stone, Fig. 30-10.
2. Slabs and insulated sandwich panels, Fig. 30-11.

Solid stone panels may be from 1 to 5 in. thick. The bonded sandwich panel shown in Fig. 30-11 may be up to 8 in. thick. The size of stone panels will depend on the specifications. In general, panels will not exceed 9 by 13 ft.

A second type of stone wall covering is known as stone masonry wall, Fig. 30-12. Stone masonry walls are classified according to shape and surface finish of the stone as:
1. Rubble.
2. Ashlar.
3. Dimension stone.

Each classification may use variations to provide interest or to bring out the characteristics of the stone. Fig. 30-13 shows a building faced in ashlar.

A rubble stone masonry wall is made up of stones as they are collected (field stone) or as the stone is broken and removed from the quarry. These stones have rounded, natural faces or angular broken faces.

Coursed rubble walls are constructed of quarried stone which is found in layers. It has a uniform thickness. When used in walls, it will result in horizontal layers giving a coursed or stiped effect.

Fig. 30-13. Exterior is cut sandstone set with an ashlar pattern.

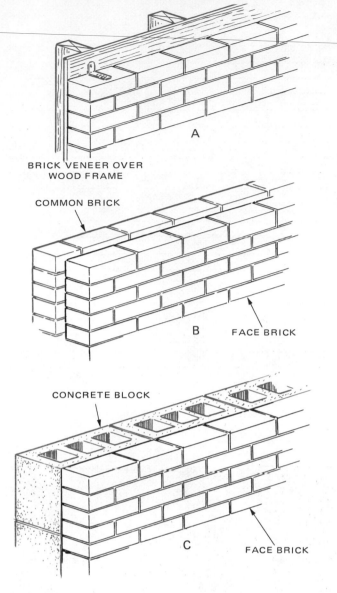

BRICK VENEER OVER
WOOD FRAME

COMMON BRICK

FACE BRICK

CONCRETE BLOCK

FACE BRICK

Fig. 30-14. Typical brick walls. A—Brick veneer over frame. B—Common brick behind face brick. C—Concrete block behind face brick.

Ashlar walls are constructed of squared stones of different sizes in random courses. The surface finish is usually a natural quarry face.

Dimension stone is cut and finished at the mill. These stones are ready to set upon arrival at the construction site. Each stone is cut, numbered and located on shop drawings and setting diagrams.

BRICK

Brick walls may be built in several different ways:
1. *As part of a steel or reinforced concrete frame building.*
2. *As a face veneer over a backup wall of common brick, tile, concrete block or wood frame.* See Fig. 30-14.

Modular brick are available along with other sizes. Modular bricks are sized so that the brick plus the mortar joint will measure 4, 8 or 12 in. The smallest dimension is the basic module. Thus, the single module brick is 2 1/2 in. by 3 3/4 in. by 7 3/4 in. Fig. 30-15 shows English and SI metric sizes of certain modular bricks.

Face brick are made under controlled conditions that produce very exact sizes and high structural qualities. Face brick are manufactured with a wide range of colors and textures for general appearance. These bricks resist weathering and are not greatly affected by temperature extremes. Produced to the same standard as face brick, glazed brick have been given a hard face with a dull, satin or glossy finish.

BRICK BONDS

The bond is the arrangement of bricks in rows, or courses. Bonds are designed for:
1. Appearance.
2. To tie the wall together.
3. To tie the outer wall to the inner wall, Fig. 30-16.
Most widely used are the common, running and English bonds.

CONCRETE

Concrete is cast in several different ways to produce materials for exterior walls. Concrete blocks are used in much the same way as structural brick or tile.

Precast concrete panels are made and assembled as curtain walls, Fig. 30-17. Such panels are made in a number of shapes. They are made plain and insulated and in a variety of surface

STANDARD BRICK SIZES AND SHAPES IN ENGLISH AND METRIC						
	HEIGHT		WIDTH		LENGTH	
	INCHES	mm	INCHES	mm	INCHES	mm
STANDARD BUILDING BRICK	2 1/2	63.5	3 7/8	98	8 1/2	216
OVERSIZE BUILDING BRICK	3 1/4	82.6	3 1/4	82.5	10	254
STANDARD FACE BRICK	2 3/16	55.6	3 1/2	89	7 1/2	190.5
NORMAN FACE BRICK	2 3/16	55.6	3 1/2	89	11 1/2	292
CONTINENTAL FACE BRICK	3 3/8	85.7	3	76	11 1/2	292
ROMAN FACE BRICK	1 1/2	38	3 1/2	89	11 1/2	292
IMPERIAL BRICK	5 3/8	136.5	3	76	15 3/8	390.5
PAVING BRICK	2 3/16	55.6	3 1/2	89	7 1/2	190.5
FIREBRICK	2 1/2	63.5	4 1/2	114	9	228.6

Fig. 30-15. Comparison chart of standard bricks.

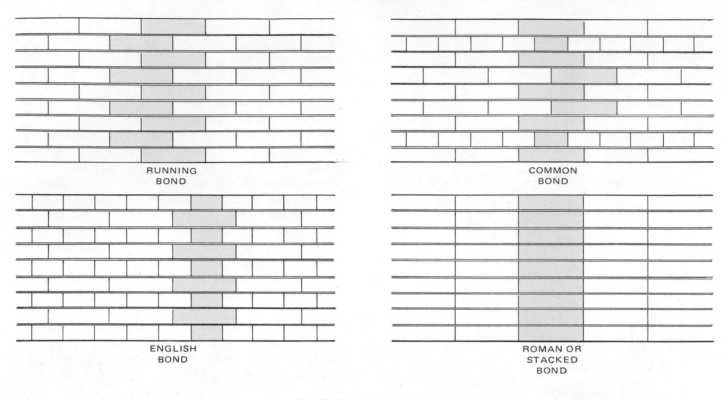

RUNNING BOND

COMMON BOND

ENGLISH BOND

ROMAN OR STACKED BOND

Fig. 30-16. Types of bonds.

Fig. 30-17. Precast concrete panels sheath exterior of this structure. Concrete panel is being hoisted into position where it will be fastened to steel framework. (Clark Equipment Co.)

Fig. 30-18. Surface texture can be created easily on concrete building panels. This wall surface was created by using a textured wood form with V-grooves. (American Plywood Assoc.)

finishes, Fig. 30-18. Surface textures can be created by the designs built into the form, Fig. 30-19. Brushing will expose the course aggregate. A layer of colored aggregate and white cement is sometimes used on the surface to produce a terrazzo-like effect. Colored mineral oxides may be used to produce colored panels.

GLASS MATERIALS

Glass is used chiefly in curtain wall construction. It comes in sheets, blocks and tiles.

Glass blocks are produced in three standard sizes, 6 by 6 in., 12 by 12 in. and 4 by 12 in. All units are 4 in. thick. Glass blocks are set in mortar like bricks or blocks.

Fig. 30-19. Poured-in-place wall with a ribbed effect.

Fig. 30-20. Transparent glass panels set in frames of wood or metal are used to form exterior wall. (California Redwood Assoc.)

Fig. 30-21. Building uses reflective coated glass for a curtain wall exterior. Reflective qualities help control heat and glare. (Libby-Owens-Ford)

corrugated metal to the Jefferson National Expansion Memorial that is sheathed in a special stainless steel. See Fig. 30-22. Different types of aluminum siding are produced from aluminum sheets. When used on homes it resembles beveled wood siding.

Tile are similar to glass block but are only 2 in. thick. They are often made up into modular units 4 or 5 ft. long by 2 ft. high. They are set in a metal frame with a rubber-like gasket.

Glass sheets are used in two ways:

1. As windows.
2. As curtain walls.

Window walls are made up of large panels of transparent glass set in wood or metal frames, Fig. 30-20. Glass sheets are manufactured to the architect's specifications. Panel surfaces may be clear and polished, patterned or aluminized. Coated with a micro thin metallic material, they produce reflective glass as shown in Fig. 30-21. Such glass helps control heat buildup and cuts down glare from the sun.

METAL SIDING

Metal is a very versatile exterior covering. It can be found on any type structure from the utility building covered with

Fig. 30-22. The surface of the Gateway Arch, St. Louis, Mo. is a special stainless steel. It will resist tarnish and oxide accumulation. (U.S. Dept. of the Interior)

Fig. 30-23. Workers are installing curtain wall panels of a natural oxidizing steel. These panels will not require any finishing and are ready to accept glass windows. (Babcock-Davis Associates, Inc.)

Fig. 30-25. Aluminum covers the 200,000-ton steel framework of World Trade Towers. (The Port of New York Authority)

Metals are often preferred for curtain wall panels, Fig. 30-23. Steel, stainless steel and aluminum are the most commonly used materials. From this group of metals, a great variety of panel styles are made. Steel skin facings, Fig. 30-24, are made by forming a flat sheet to the desired shape, normally by rolling. Stainless steel is usually rolled, while aluminum facings, Fig. 30-25, are extruded.

Metal sheet is bonded to an insulating core to make the metal sandwich panel shown in Fig. 30-26. In addition to the insulation, sandwich panels often have a stabilizing core member of hardboard, asbestos-cement board or other rigid material.

Many of the panels made of steel have exterior finishes of porcelain enamel. Aluminum facings may be plain, etched or anodized. Sizes of the sandwich panel range from 12 to 36 in. thick and up to 60 ft. long.

Fig. 30-24. Steel skins are rolled to desired shape and usually finished before leaving the manufacturing plant. (Stran-Steel Corp.)

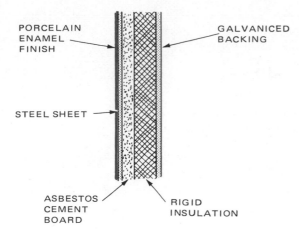

Fig. 30-26. Typical metal sandwich panel bonds insulation inside protective metal facing and backing.

MISCELLANEOUS MATERIAL

The list of materials used to produce exterior wall surfaces is a long one. For example, hardboard is sometimes applied in long, narrow strips or in 4 by 8 ft. sheets.

Fig. 30-27. Famous John Hancock Center in Chicago has facade of anodized aluminum in black and bronze tones.
(Cupples Products Corp.)

Asbestos-cement siding and siding shingles have a wood grain pattern and a wavy bottom edge. Usually only 3/16 in. thick, the shingles measure 12 by 24 in.

Plastic materials are produced in flat and corrugated sheets and in sandwich panels. Made from a fiberglass reinforced polyester plastic, they may be clear, translucent, or opaque. Color may be added during manufacture.

Often materials are used in combination. The 100-story building in Fig. 30-27 contains more than 2.5 million lb. of aluminum and 300,000 sq. ft. of glass.

SUMMARY

Many materials are used to cover the outsides of buildings. These materials are picked because of their resistance to rain, snow, direct sunlight, extreme cold, wind abrasion and, to a certain extent, air pollutants. Exterior wall coverings fall into two categories:
1. The material used for a structural wall.
2. The covering over a framework of wood, steel or concrete.

The second category is commonly referred to as a curtain wall. The curtain wall is attached to the framework and is not expected to support any of the weight of the roof or floors.

CAREERS RELATED TO CONSTRUCTION

BRICK TESTER makes brick and tile samples and tests them for strength and quality. Inspects for imperfections.

BUILDING MATERIALS SALESMAN sells building materials, equipment and supplies; knows construction and is able to read blueprints. May sell over counter or travel to sell contractors, lumberyards, etc.

QUARRY WORKER performs any of many tasks in a rock quarry. Removes mud and dirt from stone, using pick, shovel and steam hose; chips irregular shapes from stone slabs to produce rectangular shapes; loosens blasted stone; cuts notches in stone blocks, attaches hoisting cables, hooks or slings; breaks stone into pieces with sledge hammer.

BRICK YARD OPERATOR manages total brickmaking operation; directs workers who make brick; directs bookkeeping operations; organizes work schedules or directs those who organize work schedules.

DISCUSSION TOPICS

1. What are the three basic functions of exterior walls? Does a wall have to serve all three functions?
2. What species of wood are suitable for exterior siding?
3. How is natural stone obtained for construction?
4. Define "bonds" as they relate to brick work.
5. What is modular brick?
6. What is the advantage of glazed brick over standard brick?
7. What is a curtain wall?
8. List the forms of glass wall materials.
9. Name one way metal is used as an exterior wall covering.

Chapter 31

ROOF COVERINGS

Fig. 31-1. The dwellings of the colonists are reconstructed at Plimouth Plantation, near Plymouth, Massachusetts. This dwelling shows the thatched roof of reeds securely tied to roof boards. Walls are of interwoven twigs between stout pillars of wood. This lattice work is later plastered with mud called "wattle."

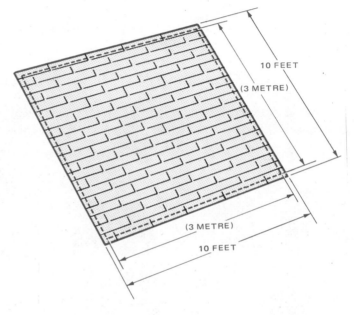

Fig. 31-2. The roofing "square" is equal to 9 m² or 100 square feet.

Early Egyptians used mats of woven reeds to roof their homes and sandstone blocks to roof their temples. Ancient Greeks fashioned clay tile for roof coverings. The Chinese and Japanese make beautiful glazed tile in a wide range of colors. Slate, a form of natural stone, roofed the Roman home and was used throughout Europe during the Middle Ages to cover the great Gothic cathedrals. Thatched roofs of grass are still used in certain areas, Fig. 31-1.

Copper was first used for roofing material about 1500 A.D. It is still used for its long-lasting qualities.

Asphalt came into use as a roofing material in the late 1800's when a chemist, William Griscom, developed an asphalt-impregnated paper in roll form.

Builders today have a large selection of roof systems to choose from. Picking the best material for the job depends on five factors:
1. Design of the structure.
2. Fire rating requirements.
3. Climate.
4. Cost.
5. Expected service life.

Another consideration in the selection of a proper roof covering is the weight of the material. Weight affects the design and cost of the supporting members.

MEASURING SHINGLES

The unit of measure used by the roofing industry is the square. A square of shingles will cover 100 sq. ft. See Fig. 31-2. In SI metric measure the square unit would be slightly smaller — 9 m² (square metres).

SLOPE AND PITCH

Proper drainage or removal of water from the roof surface is essential to good service and extended life of the roof. The word "slope" means the number of inches the roof rises in each 12 in. of horizontal run.

Pitch means the total rise of a roof divided by the total horizontal span. See Fig. 31-3. For example: If the roof spans 16 ft. and the rise at the highest point is 2 ft. the pitch is 2/16 or 1/8, a ratio of 1:8. Most architectural drawings indicate the slope of the roof rather than the pitch. A triangular symbol is placed above the profile of the roof with the rise in inches indicated for each 12 inches. This slope is written as 3:12 or 3 in. rise in a 12 in. run.

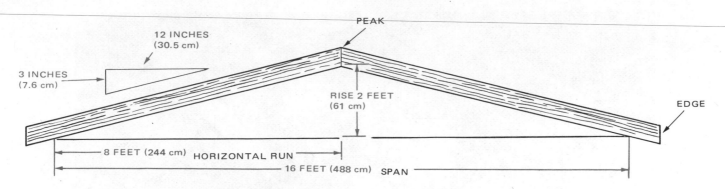

Fig. 31-3. Roof pitch means the height roof rises from its edge to its peak. Slope is inches of rise for each foot of run.

Fig. 31-4. Bituminous roof of hot melt bitumens and rolled asphalt sheets is being applied to flat roof surface.
(Manning, Selvage and Lee)

BITUMINOUS MATERIALS

Bitumens are used in several different types of roofing materials, Fig. 31-4. Bitumens come from two different sources:
1. Asphalts, a product of the distillation of petroleum.
2. Coal tar pitch, a by-product of the coking process in steel manufacturing.

Although these two materials are similar in appearance, they have different characteristics.

ASPHALTS

Roofing asphalts are graded according to their melting points. These are as low as 130 F (54 C) and as high as 200 F (94 C). The melting point of asphalt is somewhat higher than the point at which it begins to flow. This flowability provides two characteristics of asphalt roofing material:
1. The self-healing property and the lessened tendency to cracking.
2. Creeping.

Under hot summer conditions a low-melting asphalt material on a sloping roof will begin to creep.

COAL TAR PITCH

Coal tar pitch has been used for roofs in the eastern and mid-western states for many years. It is not generally used successfully in the southwest where roof surfaces reach temperatures of 140 F to 160 F under the hot desert sun. The low melting point of coal tar pitch, (140 F to 155 F) limits its usefulness.

Used on flat or low-pitched roofs in suitable climates, coal tar pitch provides a very durable roof material, Fig. 31-5. It has the property known as coal flow, or self healing. This is the ability to close up cracks or punctures through the seal. Coal tar pitch is unaffected by water.

ROOFING FELTS

Roofing felts have many uses in the construction industry:
1. Underlayment for shingles.
2. Sheathing covering.
3. Lamination in the construction of built-up roofs. See Fig. 31-6.

Roofing or building felts (felt paper) are made from a combination of materials — wood, mineral and glass fibers, saturated with asphalt or coal tar pitch. The fibers are formed into 36-in. wide rolls.

Felt papers are available in various weights or thicknesses. Seven, 15 or 30 lb. weight is generally used. These weights refer to the weight of the roofing felt required to cover 100 sq. ft. of roof surface.

ASPHALT SHINGLES

Asphalt shingles are also made from a heavy roofing felt. This is saturated with asphalt and coated with a high-melting, flexible asphalt. Mineral granules are pressed onto the asphalt coating providing a fire-resistant surface.

Shingle weights vary from 135 lb. per square for light shingles to 325 lb. for heavy shingles. One of the most common shapes of asphalt shingles is the 12 in. by 36 in. strip. The exposed edge is scored to resemble 9 in. by 12 in. tabs, Fig. 31-7.

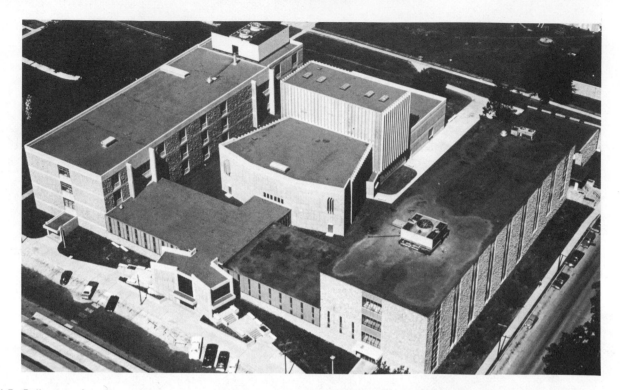

Fig. 31-5. Built-up roof on flat surface prevents water from entering structure. The built-up roof utilizes hot bitumens, rolled roofing felts and a gravel or mineral aggregate.

Fig. 31-6. Rolled roofing used in the construction of a built-up roof. Note liquid asphalt adhesive sealant and two layers of rolled roofing.

Rolled roofing is similar to asphalt shingles. The rolled roofing is available in 36 in. wide rolls and comes in weights of 90 to 120 lb. per square. Rolled roofing may be obtained with one side completely covered with mineral granules for single coverage or with only one half of the width covered when double coverage is wanted.

WOOD SHINGLES

Most wood shingles used in the United States and Canada are made from western red cedar. However, shingles are also made from cypress, redwood or other kinds of cedar. The western red cedar shingles are made from trees found primarily

Fig. 31-7. Roof covered with asphalt square tab strip shingles.

in the coastal regions of Washington and Oregon and the Canadian province of British Columbia. Western red cedar, Fig. 31-8, has the characteristics of large sizes and a fine straight grain. It has few knots and blemishes. Western red cedar does not retain moisture and is resistant to decay. Its natural color varies from yellow to brown, weathering to gray.

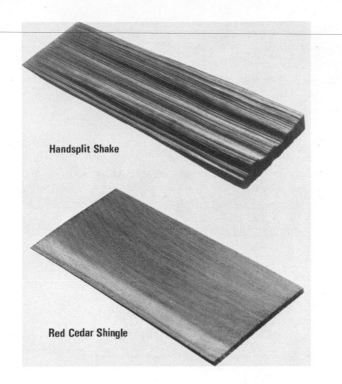

Handsplit Shake

Red Cedar Shingle

Fig. 31-8. Shingles and shakes are mostly free of knots and blemishes. They are tapered over the length with the exposed or butt end being the thickest. (Western Red Cedar Assoc.)

Fig. 31-9. A cedar shingle sawer trims edges of each shingle after it is sawed from the cedar block. (Merry, Calvo, Lane and Baker)

SHINGLES AND SHAKES

Shingles and shakes are similar to each other but the method of manufacture is different. Wood shingles are sawed from a cross section of a log called a bolt, Fig. 31-9. The shake is split from a similar bolt, Fig. 31-10.

Wood shingles are cut to standard lengths of 16, 18 and 24 in. Individual shingles are manufactured in random widths no narrower than 3 in. and no wider than 14 in. Prefabricated hip and ridge units, Fig. 31-11, are available in each length. Wood shingles are packaged in bundles. Four bundles will cover one

Fig. 31-10. "Splitterman" splits cedar sections into blanks of proper thickness. The blanks are then run diagonally through a bandsaw, producing two tapered shakes.

Fig. 31-11. Prefabricated hip and ridge units.

square (100 sq. ft.) of roof surface. Red cedar shingles are marketed under different grade labels:

1. Certigrade No. 1 blue label.
2. Certigrade No. 2 red label.
3. Certigrade No. 3 black label.

The blue label grade shingles are 100 percent heartwood and 100 percent clear, flat grain. The red label grade shingles are allowed one inch sapwood in the first 10 in. of butt width. The red label must have no less than 10 in. of clear wood above the butt on a 16-in. shingle. The black label shingle has no limitations of sapwood but must be clear wood no less than 6 in. above the butt.

Shakes are similar to shingles, except for the fact that they are split rather than sawed. Shakes have only one grade. They must be split from 100 percent heartwood bolts.

INSTALLING SHINGLES AND SHAKES

Shingles and shakes are applied almost exactly alike. Either may be installed over solid or spaced sheathing, Fig. 31-12. Horizontal strips of 1 by 3, 1 by 4, or 1 by 6 in. fir or pine are

Fig. 31-12. How to apply shingles and shakes over spaced sheathing.

spaced at the same distance as the shingle exposure.

Shingles and shakes may be laid over solid sheathing. A 15 lb. roofing felt is laid under the shingles. However, water that penetrates to the underside of the shingle becomes trapped and cannot evaporate. It will cause cupping and rotting. If a shingle or shake is properly applied, the roof has a minimum of three layers of shingles at all points over the roof surface.

METAL ROOF COVERINGS

Steel, aluminum, copper, lead and various alloys are used as roofing materials. Metal roofing has many configurations:
1. Metal shingles.
2. Corrugated sheets.
3. Sheet metal strips.

METAL SHINGLES

Metal shingles have interlocking flanges to provide a positive lock between the individual shingles. (A flange is a rim or ridge.) Metal shingles will not warp, rot, split or burn. These roofing materials are manufactured with various texture patterns. Often they resemble wood shakes. They may be finished smooth in a wide variety of baked-on colors. Beautifully colored roofs are often used on service station buildings and fast-food outlets.

CORRUGATED ROOFING

Corrugated sheets, Fig. 31-13, are widely used as roof coverings for industrial buildings, factories and farm buildings. Often the corrugated sections serve as an exposed ceiling as well as the roof. Corrugated sheets are usually galvanized steel to retard corrosion. Such roofing can be installed with a minimum of skill and is inexpensive.

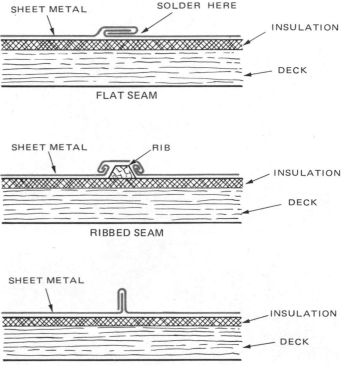

Fig. 31-14. Sheet metal roofing seams keep out moisture.

Fig. 31-13. Corrugated roof for an industrial building. Note the "valley" of preformed metal strips.
(Steel Metal & Air Conditioning Contractors National Assoc., Inc.)

SHEET METAL ROOFING

Sheet metal used for roofing includes steel, stainless steel, copper, lead, zinc and aluminum. Sheet size varies with the different types of metal, the joining technique used and the surface to be covered.

Fabrication joints in metal sheets, or "seams" can be one of three types, Fig. 31-14:
1. Flat.
2. Standing.
3. Ribbed.

Flat seams are used when the roof is flat, or nearly flat, since they can be readily soldered for additional seal. Roofs with a slope of over 4:12 (4 in. rise in every 12 in.) may have soldered standing seams. Ribbed seams are used with heavier metal for appearance or where expansion of the roof material becomes a significant factor in roof design.

CLAY TILE

Clay roofing tiles are manufactured much like bricks. Tile comes in a variety of shades from a yellow-orange to a dark red and in blends of gray and green. Clay roofing tiles are available in a number of shapes and styles. See Fig. 31-15.

Individual tiles are nailed to the sheathing through pre-punched holes, Fig. 31-16. Special shapes are available for the starter course, rakes, hips and ridges. Clay tiles are also available in the flat form and are applied like shingles.

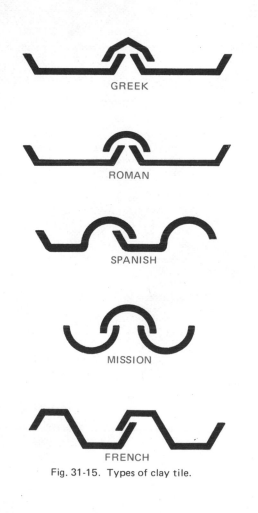

GREEK

ROMAN

SPANISH

MISSION

FRENCH

Fig. 31-15. Types of clay tile.

Fig. 31-16. Clay tile roof system is difficult to cut but lasts a long time. Usually requires a greater slope.

SUMMARY

Roofing materials vary in composition, all the way from the long grasses used for thatched roofs to the composite materials made from bitumens, mineral granules and baked clay. The ultimate roof covering material has not been discovered yet. A good roofing material must be lightweight; it must be flexible enough to expand and contract under extreme conditions; yet it must be economical.

Selection of the proper roof covering material depends upon many conditions:
1. Climate.
2. Slope or pitch of roof.
3. Intended use of structure.

CAREERS RELATED TO CONSTRUCTION

ROOFER covers roofs with roofing materials. Punches holes in hard materials such as slate, tile and terra cotta. Also installs flat roofing materials.

DISCUSSION TOPICS

1. What is the purpose of the roof?
2. What materials are used for roof coverings?
3. What is the unit of measure used for roof covering materials? Describe this unit of measure.
4. Define slope and pitch.
5. Describe bituminous materials.
6. Where is roofing felt used? What is the unit of measure for the varying thicknesses of roofing felts?
7. Describe a built-up roof.
8. Describe the difference between wood shingles and shakes.
9. List the different metals used for roof covering. What is a standing seam?
10. Name one advantage of clay tile as a roofing material.

Chapter 32

INTERIOR WALL COVERINGS

Interiors of modern structures are finished in many different ways. Many materials and techniques are used to produce color and texture. *Wall coverings include materials such as paint, paper, cloth or veneers applied over a smooth surface.* Some wall finishes quite possibly are the structural materials themselves. For instance, a brick or stone wall may be left unfinished.

GYPSUM

Gypsum is a sedimentary rock mineral which remained after the evaporation of ancient seas. In its natural state it may be gray, pink or white. The word gypsum came from the Greek word gypsas, meaning chalk.

Gypsum is found in various parts of the world. Canada, United States, France, England, Italy, China, Russia and areas of South America have large deposits.

It is obtained from open pits and underground mines, Fig. 32-1. Gypsum rock used in the construction industry is crushed and heated (calcined) in rotary kilns to temperatures of 325 F to 350 F (163 C to 176 C) until 75 percent of its combined moisture is driven off. When the powder-dry gypsum is recombined with water, it again returns to its rock-like form.

GYPSUM WALLBOARD

Gypsum wallboard is a fireproof sheathing for interior walls and ceilings. It is made of a sheet gypsum covered on each side by a heavy, specially manufactured paper, Fig. 32-2. The

Fig. 32-2. Continuous line of gypsum wallboard travels along 250-ft. long belt. "Feed" end of board machine, where the paper-gypsum core wallboard sandwich is formed, is in the background. The gypsum core is bonding to the paper and is hardening as it moves along the belt.

Fig. 32-1. Open mining of gypsum on site of ancient sea bed.   (Gypsum Assoc.)

Fig. 32-3. Gypsum panels emerging from dryer. Next, these panels will be bundled face to face, butt ends will be trimmed and bundling tapes will be applied. Bundled gypsum wallboard is ready for shipment.

paper on the exposed surface is ivory colored, while the back is gray. The board is made in a continuous sheet 4 ft. wide. The strip is cut to lengths of 8 to 12 ft. Gypsum is produced in thicknesses 3/8, 1/2, 5/8 and 1 in. Edges are recessed and reinforced with three layers of kraft paper, Fig. 32-3.

GYPSUM LATH

Gypsum lath is the same basic material as wallboard, Fig. 32-4. In gypsum lath, the same paper is used for both back and front. Laths measuring 16 by 48 by 3/8 in. are usually packaged six to a bundle.

Gypsum lath is a base for plaster, providing backing and adhesion for gypsum plaster. In order to provide better insulation, gypsum lath and gypsum wallboard are sometimes made with a sheet of aluminum foil attached to the back. This sheet reduces radiant heat losses in winter and keeps out radiant heat in the summer.

GYPSUM PLASTERS

Wall plaster contains fibers for added strength. It is a hard plaster widely used to form the first and second coats on plastered walls and ceilings, Fig. 32-5. Water and aggregate is added on the job to obtain the correct body and consistency.

Fig. 32-5. Plastering over gypsum wall board, provides smooth surface. Usually two or more coats are needed.

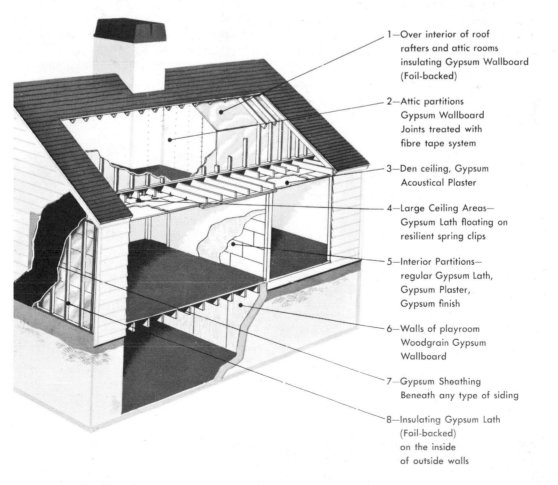

1—Over interior of roof rafters and attic rooms insulating Gypsum Wallboard (Foil-backed)

2—Attic partitions Gypsum Wallboard Joints treated with fibre tape system

3—Den ceiling, Gypsum Acoustical Plaster

4—Large Ceiling Areas— Gypsum Lath floating on resilient spring clips

5—Interior Partitions— regular Gypsum Lath, Gypsum Plaster, Gypsum finish

6—Walls of playroom Woodgrain Gypsum Wallboard

7—Gypsum Sheathing Beneath any type of siding

8—Insulating Gypsum Lath (Foil-backed) on the inside of outside walls

Fig. 32-4. Diagram shows many uses of gypsum material in residential construction.

The aggregate may be sand or a lightweight material such as vermiculite or perlite. A lightweight base coat plaster is also produced which has the gypsum and aggregate mixed in it. Water is added on the job.

Finish plaster is made especially to produce the finish coat for plastered surfaces. This plaster is mixed with hydrated lime and water. (Hydrated lime is an inert compound formed by the chemical combination of water and calcium oxide.)

JOINT FILLER

Joint filler is a gypsum material designed to fill nail holes and to fill the joints between sheets of gypsum wallboard. It is also used as an adhesive when laminating two sheets of wallboard together.

When used as a filler, the flour-like material is mixed with lukewarm water in the proportion of approximately 15 pints of water to 25 lb. of filler. The mix, which should stand for 30 minutes, is applied to the wall by hand or machine.

DRY WALL SYSTEMS

Dry wall is a term given to any of many kinds of sheets that are attached to interior walls. They take the place of plaster in modern construction.

Dry wall panels are constructed of a wide variety of materials, but the most common is gypsum. Gypsum dry wall is made of paper or vinyl-covered gypsum wallboard 1/4 to 1 in. thick and 24 or 48 in. wide. Lengths are from 7 to 16 ft. long. Edges may be squared, tapered or beveled. Many boards have factory finishes that require no taping or decorating. Finish coverings range from heavy vinyls to inexpensive papers. Sheets come in several shapes, including planks as well as panels. Vinyls give rugged protection to areas subject to heavy wear. Papers that look like wood grains are used to produce a low-cost finish, Fig. 32-6.

Dry wall systems with concealed joints used boards with tapered edges. In wood construction the boards are nailed to ceiling joists and studs with special colored nails. Wallboard must be attached with great care. Not only must the nail be driven solidly into the stud without denting the wallboard, but it must be slightly countersunk without breaking the paper

Fig. 32-6. Predecorated gypsum wallboard looks like wood paneling.

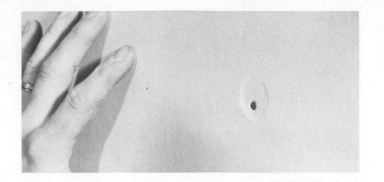

Fig. 32-7. Proper nailing of gypsum wallboard to the studs requires that the nail be driven in until hammer slightly depresses dry wall surface. (Gypsum Assoc.)

skin. See Fig. 32-7. Later it will be concealed by a nail-spotting compound.

TAPING AND FINISHING

Dry wall construction became common with the advent of concealed joints. Such joints are made possible by the development of joint treatment tapes, taping compounds and topping-finishing compounds, Fig. 32-8.

Joint treatment tapes are generally made of high quality papers. The taping compound is applied along shallow channels formed by the tapered edges where boards are butted and along both sides of inside corners. The joint treatment tapes are pressed smoothly into the wet taping compound.

Topping-finishing compounds are then applied over the tape and nailheads to make the surface smooth and flat. After the first application is dry, it is sanded where necessary, and a second, or fill coat, of topping compound is applied. The edges are feathered, and the surface is sanded to conceal all imperfections.

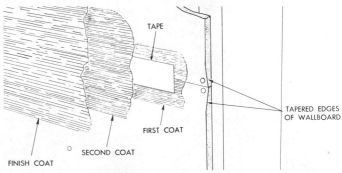

Fig. 32-8. Reinforcing joints with tape prevents cracks from appearing at filled wallboard joints. The joint fill and first coat may be joint compound or all-purpose compound. The second and third coats should be finishing compound or all-purpose compound.

LAMINATED DRY WALL

Laminated wallboard, alone or in combination with other materials, provides fire and sound protection, Fig. 32-9. On

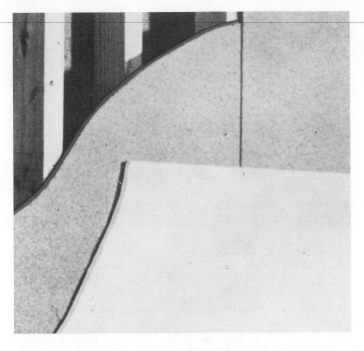

Fig. 32-9. Laminated wall made of two layers of wallboard, is stronger and more rigid. It also gives better sound control.

Fig. 32-10. Gypsum wallboard is cut by scoring the paper cover with the aid of a T-square and knife. Sheet is fractured along cut and backing paper cut with knife.

wood studs the most common lamination system is two layers of 3/8 in. gypsum wallboard. The base layer is nailed to the studs. The second layer is applied with adhesive, Fig. 32-10. Joints in the face layer must be staggered so as not to fall at the same spot as those of the base layer.

The base layer may be backing board, manufactured especially for this purpose. It may be power stapled instead of nailed. The face layer may be shop or factory finished with a wood-grained paper or vinyl.

Type X wallboard is engineered for fire protection. Dry wall on wood studs can be laminated with sound-deadening boards for soundproofing. Insulation blankets placed between studs will further deaden sound.

METAL LATH

Metal lath is formed from copper-bearing steel. It may be galvanized or coated with a paint that slows down rusting. Metal lathe may be used with any type of plaster.

Metal is second only to gypsum lath in popularity. Metal lath, with its diamond-shaped openings, provides a mechanical bond for plaster. Wet plaster is forced through the metal lath where it droops and forms "keys" on the reverse side, Fig. 32-11. The steel mesh becomes imbedded in the plaster producing a reinforced plaster slab.

Metal lath is produced by punching a series of parallel but staggered rows of diamond-shaped openings in a steel sheet. Metal lath is available in standard sheet sizes of 24 in. and 27 in. widths and 96 in. lengths.

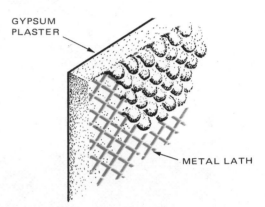

Fig. 32-11. Metal lath forms good backing for gypsum plaster.

FINAL INTERIOR FINISH MATERIAL

Wall coverings for interior walls are supplied prefinished or unfinished in either rolls or sheets. Sheet materials include metal, wood and plastic laminated panels. Roll material includes paper, cloth or vinyl plastic which may be bonded to a paper or cloth backing.

WOOD

Interior wood finishing materials include two basic groups:
1. Those used to cover walls and ceilings.
2. Those used to trim around doors and window openings and along the floor.

Wall and ceiling coverings may also be divided into two groups:
1. Solid wood coverings.
2. Plywood covering materials.

Trim is generally made from solid wood or a second material such as plastic.

Boards of various widths and thicknesses are produced for wall coverings. See Figs. 32-12 and 32-13. They are sawed from many species of wood including pine, fir, redwood, cedar, mahogany, beech, ash, walnut, maple and many others. Such boards usually are manufactured in thicknesses of 1/4, 3/8, 1/2 and 1 in. Widths will vary from 2 to 12 in. There is

Fig. 32-12. Wall covered with horizontal random length boards. (Western Wood Products Assoc.)

Fig. 32-14. Beautiful hardwood planks V-grooved and end matched, provide a warm background for comfortable furniture. (Potlatch Forest, Inc.)

Fig. 32-13. Wall covering of boards set vertically. Note the combination of material used for a pleasing atmosphere in this room. (Southern Forest Products Assoc.)

Fig. 32-15. Attractive paneling is created by combining plywood and solid wood. Panel molding was mitered, then fitted inside the squares. (Western Wood Moulding and Millwork Producers)

usually a special edge design, like tongue and groove, to allow fitting together.

Special shapes may be milled into the surface or edge. For example, a bead, channel, V-groove or bevel are produced on boards for special effects, Fig. 32-14. This type of paneling material is normally plain sawed to yield more of the natural grain figure, as in Fig. 32-15.

Plywood sheets are made from many different softwoods and hardwoods. As in structural grade plywood, the core and back sheets are made of lower grade material. The face veneer is a more desirable wood such as walnut, birch, ash, oak, mahogany, teak and rosewood.

Fir plywood paneling is produced in embossed, wire-brushed and etched surface textures. Hardwood plywood surfaces are usually plain, V-grooved or pegged.

Plywood panels are made in seven thicknesses, 3/16, 1/4, 3/8, 7/16, 1/2, 5/8, and 3/4 in. Width is uniformly 48 in. while length may be 7, 8, 9, or 10 ft. Panels may be unfinished, factory sealed, prepainted, two-toned, or completely finished.

Fig. 32-16. Modern molding factory. Finished product is inspected, tallied and bundled or packaged in cartons. (Western Wood Moulding and Millwork Producers)

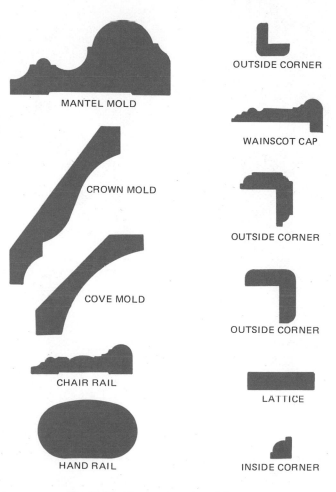

MANTEL MOLD

CROWN MOLD

COVE MOLD

CHAIR RAIL

HAND RAIL

OUTSIDE CORNER

WAINSCOT CAP

OUTSIDE CORNER

OUTSIDE CORNER

LATTICE

INSIDE CORNER

Fig. 32-17. Shapes of stock moldings.

WOOD TRIM

Wood trim is generally made from either fir, pine or mahogany. However, any species of wood is suitable. Fig. 32-16 pictures a modern molding operation. Many stock moldings are made from fir and pine. A smaller number are made from mahogany, Fig. 32-17. Lumber manufacturers and retailers have a complete list of stock moldings.

HARDBOARD WALL PANELS

Tempered hardboard (Chapter 23) can be given special facings to produce interior panels. One treatment consists of printing a wood grained pattern on the face and cutting irregular spaced V-grooves along the board to make it look like random planking, Fig. 32-18. Panels are also produced with a leather grain pattern pressed on the face.

Fig. 32-18. By printing and embossing the face of a hardboard panel the manufacturer can produce the appearance of weathered pine boards. (E. L. Bruce Co.)

Fig. 32-19. Melamine coated hardboard panels are waterproof and mildew resistant. (Formica Corp.)

Fig. 32-21. Rolled products for wall covering give the decorator a wide variety of colors and designs to choose from.

In another type of treatment hardboard is given a plastic film printed in a wide variety of wood, stone, and fabric patterns. A coating of baked or melamine plastic, Fig. 32-19, is then applied to protect the pattern. Such panels are made with tongue and groove edges. Special clips, like the design shown in Fig. 32-20, fasten the melamine sheets to wall studs without nailing through the panels.

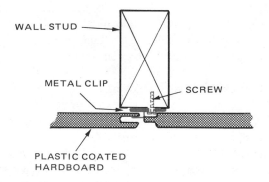

Fig. 32-20. Method of fastening paneling to stud without driving nails through it.

ROLLED COVERINGS

Wallpaper is a general term which includes any covering in roll form that is applied to a wall surface, Fig. 32-21. Today, a roll of wallpaper may be made of paper, cloth, plastic or foil.

In addition to the conventional geometric and flowered designs, rolled wall coverings are produced in a wide range of wood grains, fabric, stone, brick and mural designs. Some are already glued. After a few seconds in water, they are ready to hang. Liquid paste must be spread over the back of other papers before applying to the wall. See Fig. 32-22.

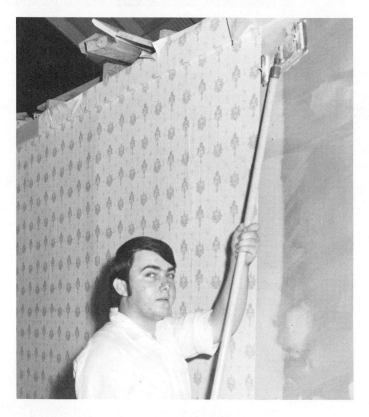

Fig. 32-22. Student practicing the skill of paper hanging. (Dallas Independent School District)

STONE AND BRICK

Stone and brick interior finishes can be produced by using solid stone or brick walls. Any type of building stone or brick may be used for this purpose.

A second type of stone or brick interior is produced with precast stone slabs. These pieces are usually from 1 to 2 in.

thick. Fiberglass sheets molded and finished to look like stone and brick are available also.

CONCRETE FINISHES

Concrete is used in various ways as exposed interior wall surfaces. Plain concrete walls may receive no special treatment in certain industrial and commercial buildings. However, concrete surfaces may be given a special finish. This consists of rubbing the surface with an abrasive stone before filling the pores with a cement grout. Textured and patterned surfaces may be produced by casting the wall in special forms. Precast concrete panels often have one face finished for interior exposure. The face may be textured, patterned, or colored, or it may have exposed aggregate.

Fig. 32-23. Ceramic wall tile provides a hard, brightly colored, stain proof, and water resistance surface, especially desirable for bathroom, restoom, and shower room walls. (American Olean Tile Co.)

CERAMIC

Ceramic products of all kinds can be used for interior wall coverings. These consist of the glazed clay products:
1. Glazed brick.
2. Glazed tile.
3. Ceramic tile.
4. Ceramic mosaic.
Ceramic tile, Figs. 32-23 and 32-24, is available in various sizes, colors and shapes. Usually it is 3/8 in. thick. Ceramic mosaic is made of plain or colored tile, generally about 1 by 1 in. square mounted on paper or cloth backing.

GLASS

Several forms of glass are offered for inside wall coverings. Glass blocks are used for decoration as well as for letting in light. Structural glass in opaque designs is used for interior partitions, room dividers and screens. Plate glass mirrors are widely used for interior decoration and to produce special effects. Stained glass windows become a means of achieving special effects as wall coverings, especially when light is diffused through them.

Fig. 32-24. Ceramic tile is used here to line a swimming pool, wall covering, and floor covering. (American Olean Tile Co.)

STEEL

Steel wall tiles are made from a thin-gauge sheet steel carefully formed to make them rigid. They are coated with porcelain enamel in a wide range of colors.

Stainless steel and aluminum are also used to produce tile or panels. Aluminum is colored by anodizing (coating by electrolytic action).

SUMMARY

This chapter describes some of the materials used to cover the interior walls of buildings and residences.

Gypsum, a very abundant material, taken from the earth provides the base over which many of the wall coverings are applied. Gypsum wallboard and joint plaster, commonly called a dry wall system, are very common materials used in surface preparation. Lath and plaster systems also rely heavily on gypsum as a basic ingredient.

Rolled wall covering products, tiles and panels are secured to the gypsum walls. Some panels are secured directly to wall studs.

CAREERS RELATED TO CONSTRUCTION

LATHER installs backing supports of metal or gypsum for plaster. He cuts and fits the lath to leave openings for plumbing and electrical fittings. He installs supports for tile ceilings and hangs suspended ceiling or sets ceiling tile.

PAPER HANGER covers walls and ceilings with decorative wallpaper or fabric. Measures walls to compute materials, mixes paste, installs scaffolding, applies sizing. May remove old materials such as paint and paper.

PANEL HANGER attaches prefinished decorative panels to interior walls.

PLASTERER applies plaster in two or three coats to inside walls and ceilings of buildings; mixes water and plaster to right consistency; sometimes produces ornamental plasterwork such as cornices and coves.

TAPER seals joints between plasterboard or other wallboard to prepare surface for painting or papering. May use sanders to smooth areas after sealing cement has dried.

TILE SETTER applies tile to walls, floors, ceilings and roof decks. Measures and cuts metal lath and attaches to walls and ceilings.

DISCUSSION TOPICS

1. What is gypsum? Where does it come from?
2. What are the sizes of plywood paneling?
3. List the two basic groups of wood paneling.
4. What is meant by stock trim?
5. What is metal lath?
6. When would you find a V-groove? Describe a V-groove.
7. How is concrete prepared to present a finished surface?
8. What are the most likely places for glazed ceramic tile?
9. How would stained glass be used for an interior wall covering?

Chapter 33

ELECTRICAL EQUIPMENT

Electrical construction equipment consists of the hardware and fixtures necessary to distribute electrical power to various locations throughout the building. Actually, electrical energy is a *planned* part of any structure, regardless of the intended use of the building. Lighting, power for mechanical devices, and energy to operate heating and cooling equipment are standard equipment for most buildings.

ELECTRICAL ENERGY

Electrical energy can be generated by forcing electrons to move in a path or circuit. This flow of electrons through a circuit is called electrical current. It is measured in amperes (amp.).

Electrical current is governed by two factors:
1. Amount of force (voltage).
2. Resistance of circuit to current flowing through it.

The voltage depends on the input from the generating device, Fig. 33-1. Resistance, on the other hand, depends on certain factors and elements of the circuit itself.

One contributing factor to total resistance is the size of the conducting wire. The smaller the diameter and the greater the length, the greater the resistance. Another key resistance factor is the material used to make the conductor. Each material has a different resistance to the flow of electricity. Aluminum, for example, has a greater resistance than copper. The resistance may either increase or decrease as the temperature changes, depending on the material.

The power-consuming device (load) is still another element of resistance. It forms part of the electrical circuit and contributes its own resistance. The final element of circuit resistance is the switch or circuit controlling device.

OHM'S LAW

There is a direct relationship between electromotive force (volts), electric current (amperes) and resistance (ohms). This relationship is expressed by Ohm's law as follows:

$$E = IR \text{ or } \frac{E}{I} = R \text{ or } \frac{E}{R} = I$$

E = electromotive force, measured in volts.
R = circuit resistance, measured in ohms.
I = electric current, measured in amperes.

E = IR means that when the electromotive force stays as is, an increase or decrease in the amount of resistance will

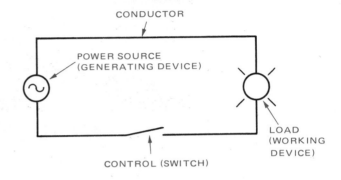

POWER SOURCE (GENERATING DEVICE)

LOAD (WORKING DEVICE)

CONTROL (SWITCH)

Fig. 33-1. Basic electrical elements are joined to form a simple circuit. Electricity will flow from generating device when switch is closed.

change the amount of current. For example: if the resistance is increased, the current decreases; if the resistance is decreased, the current increases.

The mathematical product of this relationship (I x R) is E, or the amount of electromotive force.

E = IR, then, means that in the case of a given electromotive force (volts), the current is inversely proportional to the resistance.

ELECTRIC POWER

The rate at which electricity does work is called electric power. It is measured in watts. However, one thousand watts add up to one kilowatt and this is a more convenient unit of reference for power consumed. For example, if one thousand watts of energy is expended during a period of one hour, one kilowatt-hour of power has been used.

Electric power, in watts, is the product of electromotive force (volts) and current (amperes). The following formula can be used to calculate power:

$$P = EI \text{ or } P = I^2R \text{ or } P = \frac{E^2}{R}$$
$$P = \text{watts}$$

One watt of power is consumed when one volt of electromotive force is applied to move one ampere of current through a conductor.

For example: An electric lamp is operating on 120 volts. It is drawing 2 amperes of current and has a resistance of 50 ohms. This lamp will consume 240 watts of power.

$$P = EI \qquad P = 120 \times 2 = 240 \text{ watts}$$

Fig. 33-2. Giant steel towers carry thousands of volts of electricity to to substations and then to our homes to operate the many electrical devices necessary for our comfort and well-being. (A. B. Chance)

GENERATION OF ELECTRICITY

Most electric power is produced by mechanical generators. Basically, the electromotive force is generated when wire windings are rotated through a magnetic field within the generator. The generator may be rotated by various means, from an engine-driven belt in an automobile to a giant turbine at a hydroelectric installation in a dam. Coal-powered steam turbines generate a large amount of the electrical power consumed by this nation. Nuclear reactors promise to be power sources of the future.

POWER DISTRIBUTION

Electrical power distribution has two general areas of meaning:

1. Distribution of electrical power by transformation from the energy source.
2. Distribution within the building or residence.

Electrical power may have to be transported over rough terrain and hundreds of miles from the point of generation to the user, Fig. 33-2. For easier transporation, a transformer, Fig. 33-3, may be used to increase voltage and lower the amperage. Power output as high as 600,000 volts may be transported over high tension wires.

Near populated areas where the power is to be used, the high voltage lines from the power plant are fed into a substation, Fig. 33-4. The substation is an assembly of transformers designed to lower the voltage to 13,000 volts and convert it to the desired form for local distribution.

Overhead lines or underground cables carry the electrical power throughout the city to transformers or power poles, Fig. 33-5. Overhead transformers further lower the voltage to 240/480 volts or 120/240 volts. Most residences are supplied with 120/240 volts.

Distribution of electrical power throughout the residence starts with its delivery through a service cable. The cable usually consists of three conductors twisted into a heavily insulated assembly. The service cable may be either buried or installed overhead.

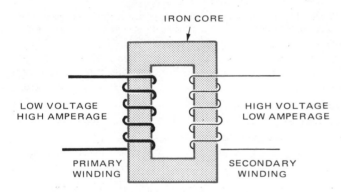
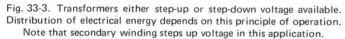

Fig. 33-3. Transformers either step-up or step-down voltage available. Distribution of electrical energy depends on this principle of operation. Note that secondary winding steps up voltage in this application.

Fig. 33-4. A substation steps down high voltages for local distribution. (Missouri Public Service Co.)

Fig. 33-5. Canister on utility pole houses a transformer that reduces voltage to 120/240 volts for use in the home. Worker is closing a master switch for community circuit. (A. B. Chance)

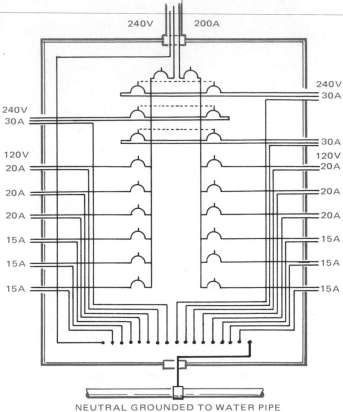

NEUTRAL GROUNDED TO WATER PIPE

Fig. 33-7. Diagram gives details of a typical residential service panel for a 200-amp. electrical system.

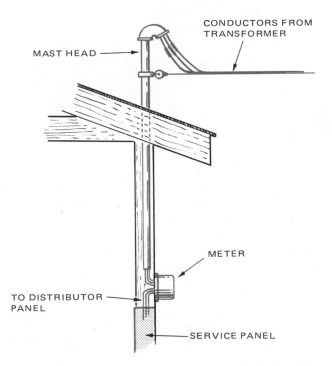

MAST HEAD

CONDUCTORS FROM TRANSFORMER

METER

TO DISTRIBUTOR PANEL

SERVICE PANEL

Fig. 33-6. Residential service entrance conductors bring electricity to home owner's service panel.

SERVICE EQUIPMENT

The service entrance conductors enter the residence through a masthead, Fig. 33-6. A minimum of a 100-amp., three-wire service must be provided for all individual residences, according to the "National Electric Code," published by the National Fire Protection Association. For larger, all electric residences, a 200-amp. three-wire distribution system is necessary, Fig. 33-7. The sizes of wire allowed for a given amperage requirement can be determined from tables published in the electrical code.

The kilowatt-hour meter measures the power consumed. The meter is installed outside the building between the service entrance conductors and the main power distribution panel. The kilowatt-hour meter is the property of the power company, and it ends the company's responsibility for maintaining the distribution system. The meter is located where it is accessible to the owner and the power company.

The service panel, distribution box or main switchboard is the heart of the electrical system of a building. The service panel is a large metal box mounted in an accessible yet out-of-the-way place in the residence.

The equipment placed in the service panel includes the main disconnect switch and a set of circuit breakers or fuses. Each hot wire of a circuit is protected by a fuse or breaker, sized for the capacity of that circuit. The neutral or ground wire is never fused, Fig. 33-8.

Branch circuits go out from the service panel to the various outlets, switches and receptacles in the house. Service panel size depends on total demand of the branch circuits. For dwellings, the National Electrical Code recommends a power supply of 3 watts per square foot of area (32 watts per square metre) for general lighting.

Three types of circuits are used for total power distribution:

1. General purpose circuit.

CURRENT-CARRYING CAPACITY FOR COPPER AND ALUMINUM WIRE

AMPERES	COPPER AWG	ALUMINUM AWG
20	12	10
30	10	8
40	8	6
50	6	4
60	4	4
70	4	3
80	3	2
90	2	1
100	1	0
150	2/0	4/0
200	4/0	300 MCM*

*MCM = 1000 CIRCULAR MILS

Fig. 33-8. Table compares the current-carrying capacity of copper wire and aluminum wire.

Fig. 33-10. A conductor is fitted with a connector designed to meet the latest Occupational Health and Safety Act standards. (Leviton Mfg. Co.)

2. Small appliance circuits.

3. Fixed appliance circuits.

The general purpose circuit includes all lighting convenience outlets except those located in the kitchen. The power requirements for a 3000 sq. ft. (280 m^2) residence is:

3 watts per sq. ft. x 3000 sq. ft. = 9000 watts

For small appliance (vacuum cleaner, toaster, mixer, etc.) power requirements, at least three additional 20-amp. circuits are necessary. These circuits are in addition to general purpose power requirements.

20 amp. x 120 volts x 3 circuits = 7200 watts

Fixed appliance requirements are the sum of power ratings of appliances such as garbage disposals, dishwashers, ranges and ovens, clothes washers and dryers, and heating and cooling units. The load is based on the rating of each appliance. These ratings are stated by the manufacturer. See Fig. 33-9.

POWER RATINGS OF RESIDENTIAL APPLIANCES

APPLIANCE	RATING		WATTS
AIR CONDITIONER	230V	20A	4600
COOKING UNIT	230V	30A	6900
WATER HEATER	230V	11A	2500
DRYER	230V	20A	4600
EXHAUST FAN	115V	4.2A	480
GARBAGE DISPOSAL	115V	8A	920
TOTAL = 20,000 WATTS			

Fig. 33-9. Typical power ratings are given and totaled for a set of residential electrical appliances.

The fixed appliance load is 20,000 watts. However, since it is unlikely that all major appliances will be operating at the same time, the load may be reduced by 25 percent. The reduced fixed appliance load, then is 15,000 watts.

The total load requirement of the residence under consideration would be:

General purpose circuits	9000 watts
Small appliance circuits	7200 watts
Fixed appliance circuits	15000 watts
	31200 watts

Therefore, based on the formula P = EI, we have:

$$\frac{P}{E} = I \qquad \frac{31,200}{230} = 135 \text{ amp.}$$

The total power requirement is 135 amp. With this amount of amperage draw, it would be wise to provide a service of at least 150 amp. This would allow for an added circuit for future expansion or the addition of another appliance at a later date.

CONDUCTORS

Copper wire generally is used for conducting electrical current. However, aluminum wire is being used in increasing amounts. Wire sizes are designated by a gauge number in American wire gauge sizes. The numbers are based on the diameter of the wire.

The current carrying capacity of wire of a given size depends on the circular mils of cross sectional area, the type of metal and the temperature of the conductor. See Fig. 33-8.

Insulation is applied to the surface of the copper or aluminum to protect the conductor from short circuits, Fig. 33-10. Rubber and various thermoplastic coatings provide insulation qualities. The type and thickness of the insulating coating depends on:

1. Maximum voltage and the nature of exposure.
2. Whether the conductor will be installed in a dry or wet location.
3. Whether it is to be buried in the ground.
4. Whether is is encased in concrete.
5. Whether it is subjected to corrosive atmospheres.

CONDUIT

Conduit is a metal tube similar to water pipe, but somewhat softer, Fig. 33-11. With insulated conductors placed inside,

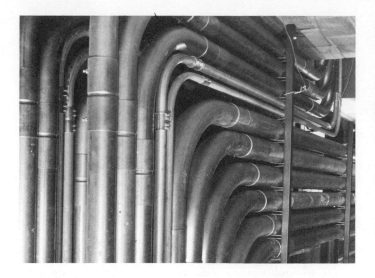

Fig. 33-11. Complex system of metal conduit houses many individual conductors carrying high voltages. (Republic Steel Corp.)

conduit provides additional protection for exposed locations. It offers the greatest possible protection against physical abuse.

It is standard practice to use steel conduit, Fig. 33-12, to conceal wiring in concrete or masonry. The inside dimensions of standard conduit sizes are 1/2, 3/4 and 1 in. (12, 19 and 25 mm). Standard unit length of conduit is 10 ft. (3 m), and a complete series of couplings, bends and fittings are available for each conduit size, Fig. 33-13.

Nonmetallic material suitable for use as conduit includes asbestos cement, polyvinyl chloride (PVC) and polyethylene. Nonmetallic conduit may be buried directly in the ground, or embedded in concrete or masonry.

Flexible metal conduit is a strong, bendable tubing for use in dry areas. Flexible conduit cannot be used as a means of grounding; a separate grounding wire must be provided inside the conduit.

RACEWAYS

Conductors also may be run through raceways, formed of metal or plastic with snap-on covers. Raceways should be

Fig. 33-12. Steel conduit for running electrical conductors through a concrete floor is being installed. Note reinforcement steel under conduit.

installed only in dry locations where voltage is less than 330 volts. Insulation should not be subjected to severe physical damage or corrosive vapors.

Raceways may be mounted on the surface of walls of an existing building or when remodeling. New structures have the raceways installed in the floor, Fig. 33-14, or in hollow cells in the steel decking forming the ceiling, Fig. 33-15. Raceways shown have the capacity for handling 184 No. 12 wires for electrical branch circuits and dozens of telephone cables.

Fig. 33-13. Conduit can be routed by making a variety of bends to permit its installation with other forms of mechanical equipment. (Republic Steel Corp.)

Construction

junction boxes usually are the first part of the electrical system to be installed. In concrete construction, the junction box and connecting conduit is placed before the concrete is poured.

Junction boxes are made square, rectangular, octagonal, or in some cases, round. They usually are 1 1/2-in. (38 mm) deep. Boxes designed for more than one device are called gang boxes.

RECEPTACLES

A convenience receptacle, slotted to receive the prongs of plugs, offers a safe means of connecting portable electrical units and appliances into the circuit. Receptacles for general purpose circuits may be rated at either 15 or 20 amp. They are designed for flush mounting. Special receptacles are available for 220V appliances, Fig. 33-16. These receptacles are slotted so that the standard 120V appliance plug will not enter.

The National Electrical Code now specifies a three-prong grounded receptacle for all 120V convenience outlets. Molded duplex receptacles having two sets of slots are commonly found throughout older residences.

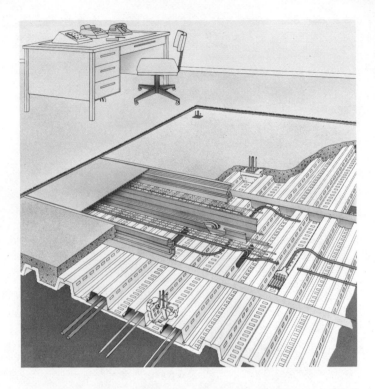

Fig. 33-14. Raceways provide a practical means of carrying electrical power and telephone lines to desks in offices or showcases in stores. (H. H. Robertson)

120V   220V   120V GROUNDED

Fig. 33-16. Slotted receptacles are designed for used with appliances or equipment having voltages indicated.

JUNCTION BOXES

A junction box or fixture is required at every point where conductors are brought out for splicing or for the connection of switches, outlets and fixtures. In wood frame construction,

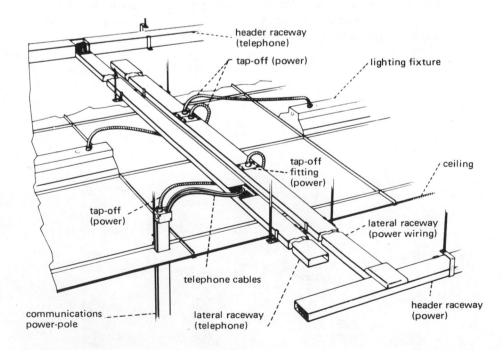

Fig. 33-15. Over-ceiling raceway has top-off fitting for power, telephone, communication and lighting fixtures. Top-off fittings terminate the flexible metal conduit at power raceway and at power source. (National Electrical Contractors Assoc.)

Fig. 33-17. Modern styling in homes and offices affects the design of wall switches and receptacles. (Leviton Mfg. Co.)

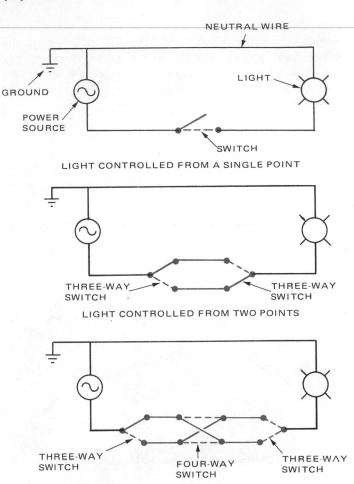

Fig. 33-18. Diagrams illustrate how circuits are controlled from one, two or three points, using correct arrangement of switches.

SWITCHES

A switch is the mechanism used to open and close an electric circuit, Fig. 33-17. Switches may be single-throw or double-throw. A single-throw switch opens or closes a single wire in a circuit. A double-throw switch can be placed in any one of three positions:

1. To open a circuit.
2. To connect a single wire to either of two other wires in the box.
3. To connect the single wire to the other wire in the box.

Switches controlling general lighting or appliance circuits are connected in the hot line of a circuit; the ground circuit is never opened. The switch may be single-throw single-pole type if only one switch is to control one or more fixtures or outlets, Fig. 33-18. If one or more fixtures are to be controlled from two point, two three-way switches are needed. If a fixture is to be controlled from more than two points, one four-way and two three-way switches are used.

Dimmer switches vary the voltage supplied to a fixture. The voltage may be lowered by a variable resistor, transformer or a transistor, Fig. 33-19. Most dimmer switches used for either fluorescent or incandescent lamps are of the transistor variety. Usually, they include a switch to complete the circuit.

LIGHTING

Lighting systems are designed for the purpose of adequately lighting the area. Different lighting tasks require different amounts of illumination for ease and efficiency of performance.

Lamps are available to meet all types of lighting requirements. However, proper lighting entails more than just providing the correct amount of light in given areas. The

Fig. 33-19. Dimmer switch is an energy-saving, light-conditioning device that is popular in dining areas. (Leviton Mfg. Co.)

human eye does not adjust readily to extreme contrast of light and darkness. Therefore, general lighting of the area surrounding the work surface must also be considered. Glare from *any* source in the visual field can cause discomfort. Reflected light may be produced by horizontal surfaces, walls or unshaded windows.

THE INCANDESCENT LAMP

The incandescent lamp consists of a filament of tungsten wire, a base and a glass globe, Fig. 33-20. In operation, the

Fig. 33-20. A tungsten lamp is equipped with a bulb life saver, which extends bulb life by cutting wattage output.

tungsten wire becomes white hot and gives off light. Incandescent lamps are rated by the number of watts of power they consume. For general lighting, they consume from 15 to 300 watts.

The amount of light given off by a given light source is measured in lumens. Manufacturers of lighting devices list the lumen output of each unit they produce.

Incandescent lamps are made in five different socket bases:
1. Miniature base, the smallest base made for lamps under 25V. Some Christmas tree lamps fall in this classification.
2. Intermediate base, for up to 125V, outdoor decorative lighting.
3. Candelabra base, for 125V, outdoor lighting and for some chandeliers using flicker type candle bulbs.
4. Medium base, the standard household bulb for general lighting.
5. Mogul base, the large-base lamp of 300 to 1500 watts used for outdoor lighting.

ELECTRIC DISCHARGE LAMPS

The electric discharge lamp operates on a different principle than the incandescent lamp. Instead of passing current through a tungsten wire, an electric discharge lamp passes the current through a gas until the gas glows. Neon, sodium vapor or mercury vapor are used.

A high voltage (1000 volts) is needed to force the current to jump from one electrode to the other through the gas. Less voltage is needed, however, if the electrode is preheated. Therefore, two types of vapor lamps, hot cathode and cold cathode, are available. These electric vapor lights are used for street lighting and outdoor lighting of factories, commercial buildings and athletic fields.

FLUORESCENT LAMPS

A fluorescent lamp is a hot cathode type of vapor lamp. The inside of the glass tube is coated with a phosphorus material that absorbs radiant energy of short wavelengths and reradiates longer wavelengths.

Fluorescent lamps come in different lengths, and the length of tubing indicates the power consumed (wattage). A nine-inch (23 cm) lamp uses 6 watts; a 15-inch (38 cm) lamp uses 14 watts; a 24-inch (60 cm) lamp uses 20 watts; a 48-inch (122 cm) lamp uses 40 watts.

SUMMARY

The electrical system of buildings and structures is a Twentieth Century innovation. It is one of the few total building concepts that is unique to our period of history. Electricity provides a clean, well organized source of energy. The mechanical equipment needed to utilize electrical energy within a structure requires very little space. Conductors can deliver electricity to any location throughout the structure, within the floor, ceiling or wall.

Manufacturers have designed and built thousands of electrical devices that utilize electrical energy. Switches, receptacles, conductors and fixtures make it possible to conveniently install the electrical system at the building site.

CAREERS RELATED TO CONSTRUCTION

ELECTRICAL PRODUCTS MANUFACTURER produces and supplies parts and assemblies needed by contractors or contractors' clients.

ELECTRICIAN installs wiring, fixtures, apparatus and control equipment; repairs and maintains electrical equipment.

ELECTRICAL ENGINEER prepares electrical system layout; plans construction and oversees electrical system installation. POWER ENGINEER supervises operation of power stations, transmission lines and distribution systems. DISTRIBUTION ENGINEER coordinates operation of facilties for transmitting power from distribution point to consumers.

ELECTRICAL TECHNICIAN diagnoses the cause of electrical or mechanical faults or failures of operational equipment; performs preventive and corrective maintenance.

ELECTRICAL GOODS SALESPERSON prepares financial and operational estimates based on blueprints, plans or other records furnished by contractors and their clients; sells electrical and electronic equipment and supplies.

DISCUSSION TOPICS

1. What determines how much resistance there is to the flow of electricity?
2. What is Ohm's law? How is it used?
3. What causes electricity to flow? What is a circuit?
4. How many amperes of current must a circuit carry to supply a 2000-watt appliance at 120V?
5. If electricity cost $.09 per kilowatt-hour, how much would it cost to run a 200-watt lamp for 24 hours?
6. What is a circuit breaker? How does it work?
7. What is a fuse?
8. What is the purpose of conduit? List kinds of conduit.
9. How is a junction box used?
10. What are conductors, switches and receptacles?
11. What is a service entrance?
12. What is incandescent lighting?
13. What is a lumen?

ELECTRICAL SYMBOLS

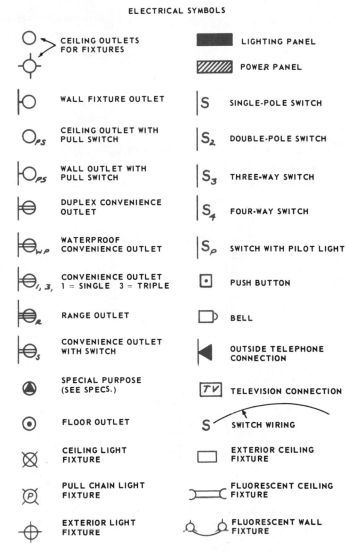

⟋○	CEILING OUTLETS FOR FIXTURES	▬	LIGHTING PANEL
		▨	POWER PANEL
⊢○	WALL FIXTURE OUTLET	S	SINGLE-POLE SWITCH
○ₚₛ	CEILING OUTLET WITH PULL SWITCH	S₂	DOUBLE-POLE SWITCH
⊢○ₚₛ	WALL OUTLET WITH PULL SWITCH	S₃	THREE-WAY SWITCH
⊖	DUPLEX CONVENIENCE OUTLET	S₄	FOUR-WAY SWITCH
⊖wₚ	WATERPROOF CONVENIENCE OUTLET	Sₚ	SWITCH WITH PILOT LIGHT
⊖₁,₃	CONVENIENCE OUTLET 1 = SINGLE 3 = TRIPLE	▣	PUSH BUTTON
⊖ᵣ	RANGE OUTLET	▭	BELL
⊖ₛ	CONVENIENCE OUTLET WITH SWITCH	◀	OUTSIDE TELEPHONE CONNECTION
▲	SPECIAL PURPOSE (SEE SPECS.)	TV	TELEVISION CONNECTION
⊙	FLOOR OUTLET	S‿	SWITCH WIRING
⊗	CEILING LIGHT FIXTURE	▢	EXTERIOR CEILING FIXTURE
⊗ₚ	PULL CHAIN LIGHT FIXTURE	⊐○⊏	FLUORESCENT CEILING FIXTURE
⊕	EXTERIOR LIGHT FIXTURE		FLUORESCENT WALL FIXTURE

Electrical symbols used in construction.

Chapter 34

MECHANICAL EQUIPMENT

The mechanical section of a set of building specifications is concerned with the plumbing, heating and cooling systems. The systems are designed by a mechanical engineer and installed under the direction of a plumbing contractor, or a heating and cooling contractor.

PLUMBING SYSTEM

A plumbing system includes all water supply and distribution pipes, soil and waste disposal piping, vent piping, gas or fuel-oil piping, water heating devices and storage tanks. Plumbing fittings and fixtures are installed by the plumbing contractor. He is also responsible for putting in oxygen, compressed air or vacuum lines.

Plumbing systems have to be installed according to a set of rules or codes such as the "Uniform Plumbing Code." This is published by the International Association of Plumbing and Mechanical Officials.

Installation of heating, ventilating, comfort conditioning and refrigeration systems is controlled by a mechanical code

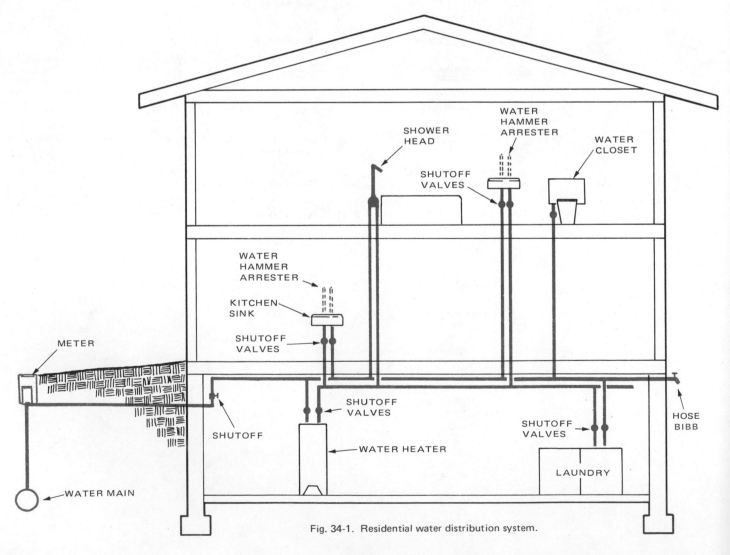

Fig. 34-1. Residential water distribution system.

such as the "Uniform Mechanical Code." It is published jointly by the International Association of Plumbing and Mechanical Officials and the International Conference of Building Officials.

WATER DEMAND

Water must be piped to each fixture in a clean and sanitary condition. Care must be taken to prevent pollutants or contaminated water from entering the water system.

The size of the water meter and of each pipe carrying water from the meter to the fixture is based on the "water demand" of that fixture. The total water demand of a house is determined by adding together the water demand of each fixture. The size of pipe required by each type of fixture is governed by local building codes.

To determine the proper size for a water supply system, the number of fixture units is determined from the blueprint and building specifications. The length of the supply lines to the most distant fixtures and the difference in elevation between the meter and highest point fixture outlet are calculated, Fig. 34-1. A half pound of pressure is subtracted for each foot of difference in elevation between the source and the highest fixture.

WATER SUPPLY MATERIAL

Water supply pipe and fittings must be made of long-lasting materials that will resist the corrosive action of water. Several metals meet these requirements:

1. Brass.
2. Copper.
3. Cast iron.
4. Malleable iron. (Malleable means being capable of being shaped by beating with a hammer or pressed by rollers.)
5. Galvanized iron.
6. Lead.
7. Plastic.

Whatever material is used, all fittings, except valves and similar devices, must be of like materials. It is not good to mix them in the same system.

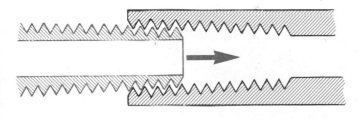

Fig. 34-2. American Standard taper pipe thread. Wedging action of the taper produces a tight seal.

IRON

Cast iron fittings smaller than 2 in. in diameter and all malleable iron fittings must be galvanized. Valves 2 in. or smaller in diameter must be brass and valves over 2 in. must

GRADES OF IRON WATER SUPPLY PIPE

STANDARD WEIGHT PIPE	USED IN DOMESTIC WATER SYSTEMS, AVAILABLE IN SIZES FROM 1/8 IN. TO 6 IN. FURNISHED WITH BOTH ENDS THREADED AND A SINGLE COUPLING.
EXTRA STRONG PIPE	USED WHEN HIGHER PRESSURES ARE USED. MAY BE FURNISHED WITH THREADED OR PLAIN ENDS. AVAILABLE IN SIZES FROM 1/8 IN. TO 8 IN.
DOUBLE EXTRA STRONG PIPE	FOR EXTREMELY HIGH WORKING PRESSURES. FURNISHED WITH PLAIN ENDS ONLY. AVAILABLE FROM 1/8 IN. TO 8 IN.

Fig. 34-3. Iron water pipe is supplied in three grades.

have a cast brass body.

Galvanized iron pipe has great strength, does not expand and contract very much and lasts a long time. Because it is very strong, iron is safe and economical to use. The galvanized coating, inside and outside, protects the pipe from corrosion and damage. Galvanized pipe used in water distribution systems is joined with a threaded connection fitting which is available in a wide variety of types, Fig. 34-2. Iron or steel pipe is produced in three grades. See Fig. 34-3.

COPPER

Copper tubing is widely used in water distribution systems. It is so popular because it does not rust and is highly resistant to corrosion. Its smooth interior surface means less water pressure is lost by friction. This allows the use of smaller pipe sizes. Since copper can be easily bent, the need for fittings and joints even at direction change is eliminated.

Copper pipe is produced in three types, in order to handle all conditions and types of application. The three types have separate color markings. See Fig. 34-4.

Copper tubing is available in both hard and soft tempers. The soft-tempered types are easily bent by hand or with a tube bender. Soft-tempered tubing comes coiled. The more rigid, hard-tempered copper tubing is sold in straight lengths. It is used for exposed piping, for horizontal runs and where a neat appearance is important.

COPPER PIPE CLASSIFICATION

HEAVY WALL K GREEN CODE	USED PRIMARILY FOR UNDERGROUND SERVICE, PLUMBING, HEATING AND COOLING AND FOR STEAM, OIL, AIR, OXYGEN AND HYDRAULIC LINES.
MEDIUM WALL L BLUE CODE	INTERIOR PLUMBING SYSTEMS.
LIGHT WALL M RED CODE	USED FOR INTERIOR LOW-PRESSURE PLUMBING SYSTEMS.

Fig. 34-4. Copper pipe is color coded for easy identification.

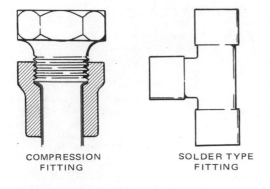

COMPRESSION
FITTING

SOLDER TYPE
FITTING

Fig. 34-5. Two methods of coupling copper pipe.

A variety of fittings are made for joining copper tubing. One type is the solder fitting which is sealed by molten solder alloy. Another design is the flared compression joint as shown in Fig. 34-5. Copper tube is not recommended where water pipe is to be embedded in concrete slabs or masonry work.

PLASTIC

Plastic pipe is light weight, has outstanding chemical resistance, high impact and pressure strength, and is easy to connect. It is well suited for water transmission systems, sprinkling and drainage systems. The most common type of plastic pipes are ABS (acrylonitrile butadiene styrene), PVC (polyvinyl chloride) and polyethylene. But not all local codes allow its use.

Heavier types of plastic pipe have threaded fittings similar to those used on steel pipe. But the most common fittings use a solvent-welded joint. The solvent softens the surface of each mating part and then rehardens to fuse the parts together. This creates a water-tight seal. Special adapters are produced for plastic-to-metal joints.

WATER HAMMER ARRESTORS

"Water hammer" refers to pounding noises and vibration which develop in a piping system. This condition occurs when water flowing through a pipe line under pressure is stopped abruptly by a closing valve. The momentary pressure suddenly created in the line may be as great as four times the operating pressure.

When water hammer occurs, a high-pressure shock wave travels back and forth through the pipe until the energy trapped is absorbed. This action accounts for the noise and vibration in pipes. Excessive surge pressure, such as this, may be prevented by arresting devices in the system.

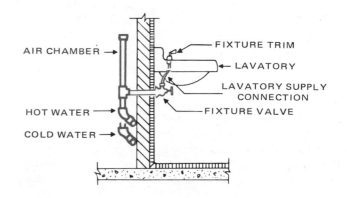

Fig. 34-6. Air chamber stops air hammer in pipes.
(Plumbing and Drainage Institute)

Capped chambers have been used as water hammer arrestors, Fig. 34-6. When properly sized they protect the system from the destructive hammering. Trapped air acts as a cushion to absorb the pressure waves.

DRAINAGE SYSTEMS

Every sink, tub and water closet in a system has a soil or waste stack, a vent system and a trap as shown in Fig. 34-7. The waste stack is made up of large pipes that carry both solid and liquid wastes out of the building. See Fig. 34-8. It connects to another large sewer line which carries the wastes to a sewage disposal plant or to a private septic system.

Fig. 34-7. Classic styled bath fixture has exposed water line and drain.   (Kohler Co.)

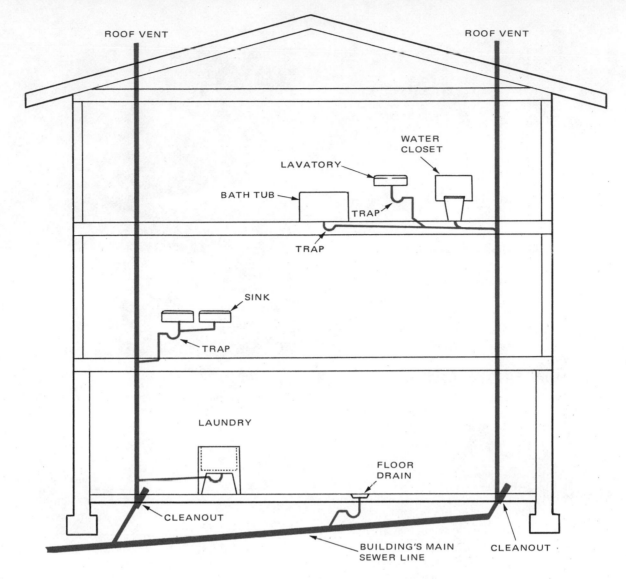

Fig. 34-8. A residential drainage system.

VENTS

The vertical vent is another pipe which provides air circulation throughout the drainage system. It picks up any sewer gases that might be present and releases them into the air above the roof. The vent also is important for equalizing pressures inside the drainage system.

TRAPS

A trap is a curve in a waste pipe that seals out gases that might come up through the pipe. If you look under the sink in your bathroom, it is the section of pipe that looks like the letter "S" lying on its side. When water drains out of a fixture some of it stays in this trap to provide the seal needed against gas. Fig. 34-16 shows a trap installed on a kitchen sink.

CLEANOUTS

Sections of horizontal drainage pipe sometimes clog up with solids and have to be cleaned out. Therefore, cleanouts must be provided at the upper end of the line. These are open sections of pipe capped with a threaded plug. One is needed wherever a horizontal run changes direction. When a pipe becomes clogged it provides the opening through which the auger or clean out tool enters the drainage line.

CAST IRON SOIL PIPE

Drainage pipe is usually called soil pipe. It is made of several different materials. Cast iron soil pipe is made of gray iron containing large flakes of graphite. A bituminous (tar-like) coating is applied to the cast iron pipe, inside and out. This protects it against acids in the waste it carries.

Iron soil pipe comes in 5 or 10-ft. (1.5 or 3 m) sections with a hub, socket or bell at one end. The opposite end is plain with a slight bead on the outside.

The bead end is inserted into the bell section and oakum, a rope-like material, is packed in the joint, Fig. 34-9. After the oakum is packed to within 1 in. of the top, molten lead is poured into the joint. The lead-oakum seal makes a tight joint. More plumbers, however, are using neoprene gaskets because of the difficulty of making a lead joint on horizontal runs.

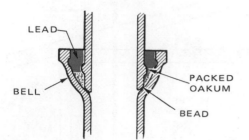

Fig. 34-9. The lead-oakum joint for cast iron soil pipe.

COPPER DRAINAGE MATERIALS

Copper drainage tube has steadily increased in popularity because it resists corrosion, is lightweight and needs no lead joints. However, copper drains are costly because copper is expensive. Rigid copper piping has thinner walls than the L or M grades of tubing. Fittings are soldered in.

PLASTIC DRAINAGE MATERIALS

Plastic pipes are faster to install than most other material. Horizontal runs must be continuously supported to prevent sagging. Joints are chemically welded assuring a leakproof connection.

CLAY SEWER PIPE

Clay pipe is used for sewer lines outside the building. It is not recommended for inside installation.

The clay used for sewer pipe is much like the clay used in brick. It must be dense enough to withstand most acids and keep out roots. Most clay pipe is produced in several diameters

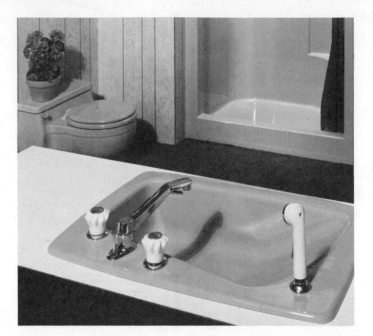

Fig. 34-11. Bathroom fixtures have porcelain coating which gives them a glass-like covering. Stains and acids do not affect such finishes. (Kohler Co.)

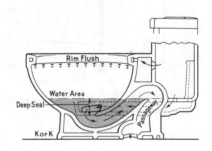

Fig. 34-12. Cross section of water closet showing water trap.

from 4 in. to 36 in. in 2-inch increments. Lengths are 24, 30, 36, 48 and 54-in.

FIXTURES

Plumbing fixtures are manufactured of dense, durable, nonabsorbent materials with smooth, glass-like surfaces. Fig. 34-10 lists six different materials used for fixtures.

WATER CLOSETS

Several types of water closets are manufactured for residential and commercial installations, Fig. 34-11. Residential types use a flushing tank. It is either attached to the bowl or cast in one piece with the bowl, Fig. 34-12.

Water closets are classified according to the method used to flush the water out of the bowl:

1. The siphon vortex.
2. The siphon jet.
3. The reverse-trap.
4. The blowout type.

MATERIALS FOR PLUMBING FIXTURES

MATERIAL	CHARACTERISTICS
VITREOUS CHINA	HIGH GLOSS, EASILY SHAPED MATERIAL. WILL NOT STAND SHOCK, TENDS TO WARP DURING MANUFACTURING.
VITREOUS GLAZED EARTHENWARE	BETTER SHOCK RESISTANCE THAN CHINA, NOT AFFECTED BY ACIDS, HEAT OR COLD. WIDELY USED IN HOSPITALS AND INSTITUTIONS.
PORCELAIN ENAMELED CAST IRON	1/8 IN. THICK GLAZED SURFACE OVER RIGID CAST IRON BODY. WILL WITHSTAND MORE SHOCK, ACID RESISTANT.
PORCELAIN ENAMELED STEEL	LIGHTWEIGHT, INEXPENSIVE MATERIAL, FLEXIBILITY OF STEEL. SHELL TENDS TO CRACK PORCELAIN COATING.
STAINLESS STEEL	HIGH-COST MATERIAL, GOOD WEAR, RESISTANT TO SOME ACIDS.
PLASTIC (FIBERGLASS)	INEXPENSIVE, LIGHTWEIGHT, SOFT SURFACE MATERIAL.

Fig. 34-10. Six materials are used for plumbing fixtures.

In the siphon vortex the rim creates a swirling action in the water to wash down the bowl. In the siphon jet type, water from the rim washes down the sides and discharges by action of a jet in the outlet. The reverse trap type is similar to the siphon jet type but has a smaller bowl. The blowout type, used only with a flushmeter valve, is efficient in water use. It is noisy and is used, therefore, mostly for institutional installations.

LAVATORIES

Lavatories come in many sizes, shapes and colors. They are manufactured from vitreous china, porcelain enameled cast iron or steel, and stainless steel, Fig. 34-13. Lavatories are held in counter tops by a stainless steel rim, are self-rimmed, cast with the counter top or they may be hung from the wall.

Fig. 34-13. Self-rimmed lavatory.   (Kohler Co.)

Fig. 34-14. A right-hand tub.   (Formica Corp.)

BATHTUBS

Porcelain enameled cast iron or steel and fiberglass are the materials most used in bathtub manufacture. There are right hand tubs and left hand tubs, Fig. 34-14. (A left hand tub is

STANDARD TUB

CORNER TUB

Fig. 34-15. Bath tubs come in many different shapes.

one which has the outlet on the left when viewed from the room.) Tubs are designed as free standing, corner, recessed and Roman style units. Fig. 34-15 shows two styles.

SHOWERS

Showers may be installed over the bath tub or as a separate unit (stall). Shower stalls may be built at the site or purchased prefabricated of steel or plastic.

SINKS

Kitchen sinks for residential use are of porcelain enameled cast iron, stainless steel or steel. They may have one, two or three compartments, Fig. 33-16. Most are rectangular but special shapes are produced to fit across a corner. Fig. 34-17

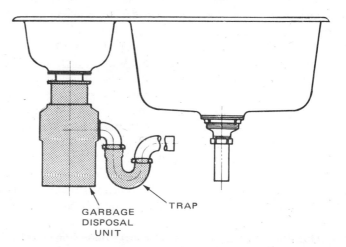

TRAP

GARBAGE
DISPOSAL
UNIT

Fig. 34-16. Two-compartment sink has special second compartment for installation of garbage disposal.

Fig. 34-17. Three-compartment sink has garbage disposal in shallow center section.

shows a three-compartment sink. One is intended for a garbage disposal. Commercial or institutional sinks are usually made of stainless steel.

HEATING SYSTEMS

Heating systems are designed to automatically heat air and deliver it to various rooms or zones. In determining the size and type of heating equipment, the first requirement is that it maintain a constant temperature of 72 F (23 C). As heat escapes through walls, floors and ceilings it must be replaced by the furnace. The amount of heat loss is measured in British Thermal Units (Btu). Heat needs depend upon the type of construction and how the building is insulated.

The capacity of a heating system is rated in two ways:
1. As input rating in Btu per hour.
2. As bonnet capacity in Btu per hour.

The input rating is used in determining the fuel consumption. Bonnet capacity is a measure of the heat output of the furnace. The difference between the two figures indicates the efficiency of the unit.

WARM AIR FURNACE

A warm-air furnace is any device that warms air and circulates it to the location needed. The original warm air furnace was called the gravity furnace. As dense cool air sinks, because it is heavier, the less dense hot air rises. Thus, the air circulates through large ducts to the areas to be heated. This type of furnace required large ducts. The furnace had to be located beneath the area it was to warm.

The gravity furnace has largely been replaced by forced air units. See Fig. 34-18. The forced air furnace has a fan to force air through a series of rather small ducts. Smaller furnaces, Fig. 34-19, used in residences and small commercial buildings consist mainly of a burner, fan cabinet, air flow filter and automatic controls. Air is moved through the small, compact duct by a "squirrel cage" fan. Such a fan can be seen in Fig. 34-18.

Operation of the forced-air furnace is quite simple. Oil, gas or electricity heat the metal fins of the heat exchanger. Air drawn through ducts by the fan is heated as it passes through the heat exchanger. Then it is expelled again through supply

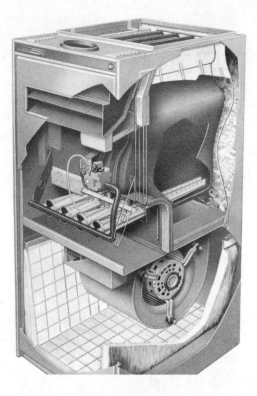

Fig. 34-18. Gas fired forced air furnace, a direct drive upflow type. Note squirrel cage fan at bottom. (Lennox Industries, Inc.)

Fig. 34-19. A direct-drive horizontal flow forced air furnace. This one is mounted near ceiling level. Box at right is called "plenum."

ducts for distribution, Fig. 34-20. Fumes from combustion are vented to the outside through a chimney.

Most forced air furnaces have removable filters in the return line. These are located just ahead of the circulation fan to catch any solid particles in the air before it is heated. The furnace may have humidifiers on the output side of the furnace. These add moisture to the heated air.

Two thermostats control the furnace. One provides automatic control of temperature within the building. The second protects the furnace from overheating.

DUCTS

Air is moved from one area to another through a system of ducts. *A plenum is the air compartment or chamber where one or more ducts are connected. Warm or cooled air collects there*

Fig. 34-20. Air distribution system for large commercial installation. (Sheet Metal and Air Conditioning Contractors National Assoc.)

Fig. 34-21. Plenum chambers allow branching of ducts and air distribution over a large area.

to be moved through the ducts. See Fig. 34-19.

A plenum may be a large duct or it may be an enclosed space above the ceiling or in a wall. Fig. 34-21 shows a plenum in a large, commercial heating unit.

Ducts may be round, square or rectangular and made of tin-plated steel, galvanized steel or flexible plastic. Galvanized steel or plastic is used if cooled air is also going to be passed through the duct. Fig. 34-22 is a view of a shop where duct work is made.

Ducts carrying both heated and cooled air should be insulated. Fiberglass or rock-wool blankets, the most common types of insulation, are wrapped around the ducts. Thicknesses of 1/2 in. to 2 in. are used.

HOT WATER SYSTEMS

Hot water (hydronic) systems have a boiler, water circulating piping, water circulation pump and radiator. The boiler is either oil or gas fired. Piping is generally copper with soldered fittings. The pump is probably electrically operated and the thermostat electrically controlled. Older systems may have large cast iron radiators. Newer units may use small copper coils embedded in concrete slabs or small baseboard radiators

Fig. 34-22. Metal duct work is cut in sections and assembled on the site.

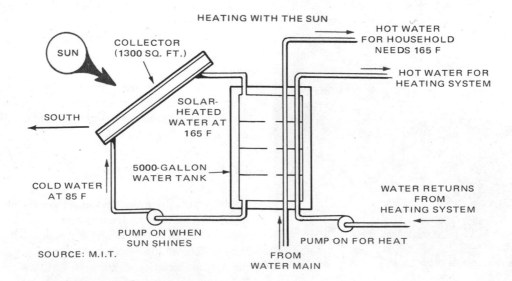

Fig. 34-24. Newly designed heat transfer module is a compact burner mounted inside a core of thousands of small steel balls fused together with oxygen-free copper.

around the walls of the area being heated.

Steam heating systems are similar to the hot water system except the water is converted to steam and moves to the radiator as a vapor.

RADIANT HEAT

A radiant heating system creates warm surfaces that cut down the loss of body heat. Radiant heat is provided by electricity flowing through a resistance wire, or by hot water flowing through pipes embedded in a concrete floor.

SOLAR HEAT

Much attention is being given to newer, cleaner sources of heat. Solar energy is such a source. All energy developed on earth has its origin from the sun. We are just now beginning to harness its daily shower of energy. Fig. 34-23 is a simple

diagram of a solar heating system.

The heat collector is located on the roof of a dwelling. Its design permits sunlight to enter and prevents heat from escaping. Heated water circulates through the house with the help of a small pump. A heavily insulated water tank can retain heat for from one to five days. Some areas receive more sunlight per day than others, making some parts of the country or world better than others for direct solar conversion into heat.

HEAT TRANSFER MODULE

Something new in furnaces is the heat transfer module. Actually, it is a miniature heat exchanger, Fig. 34-24. This small unit has the capacity to heat an average home. Steel tubing is imbedded in a matrix of steel balls, and a solution of ethylene glycol and water is pumped through to carry away heat, Fig. 34-25. A heat exchanger transfers the heat to a forced-air system for distribution.

HEATING WITH THE SUN

SUN

COLLECTOR
(1300 SQ. FT.)

SOUTH

SOLAR-
HEATED
WATER AT
165 F

COLD WATER
AT 85 F

5000-GALLON
WATER TANK

HOT WATER
FOR HOUSEHOLD
NEEDS 165 F

HOT WATER FOR
HEATING SYSTEM

WATER RETURNS
FROM
HEATING SYSTEM

PUMP ON WHEN
SUN SHINES

SOURCE: M.I.T.

FROM
WATER MAIN

PUMP ON FOR HEAT

Fig. 34-23. Solar heating system. Collector absorbs sun's rays to heat water.

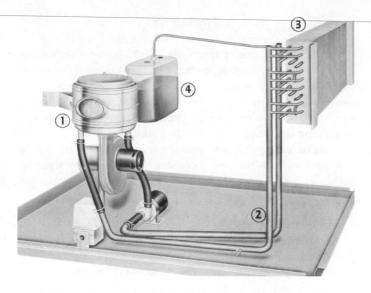

Fig. 34-25. Heat transfer module. 1—Burner. 2—Tubing to carry away superheated liquid. 3—Heat exchanger where heated air is carried to areas of home. 4—Expansion tank catches liquid which expands after heating. (Amana Refrigeration, Inc.)

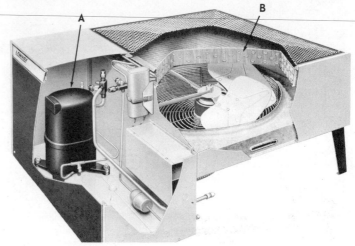

Fig. 34-27. Cutaway shows compressor and condensing coil of central air conditioning unit. A—Hermetic (sealed) compressor. B—Condenser. (Lennox Industries Inc.)

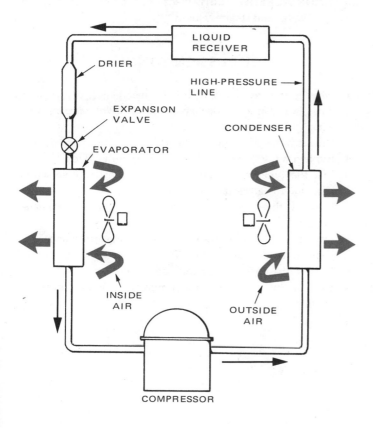

Fig. 34-26. Schematic diagram of compression cooling system.

COOLING SYSTEMS

There are several methods of cooling air. Two most commonly used are the compression system and the absorption system.

The compressor is the heart of the compression system. Fig. 34-26 is a simple example of how the system works. The compressor takes low-pressure refrigerant gas and forces it through a high-pressure line. The gas goes through a series of coiled tubes called a condenser, Fig. 34-27. A fan passes air over the coil and the refrigerant loses heat and condenses to a liquid. The liquid is then forced to a liquid receiver or reservoir. From the liquid receiver it passes through an expansion valve which regulates the flow to the evaporator. (The evaporator, like the condenser, is a series of coils.) In the evaporator the liquid changes back to a gas at a low temperature. Air or water is chilled and distributed to the space to be cooled. Most central air conditioning systems used in homes deliver chilled water to the furnace. See Fig. 34-28.

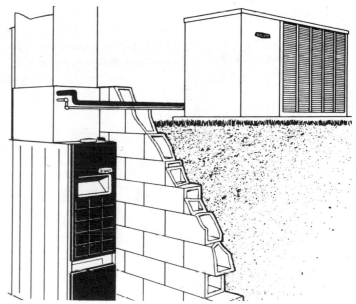

Fig. 34-28. The most common residential cooling system has its evaporating coil located inside plenum chamber of furnace. Furnace fan circulates air over evaporating coil cooling air before it is transported through ducts to various rooms of the home. (Bryant Air Conditioning Co.).

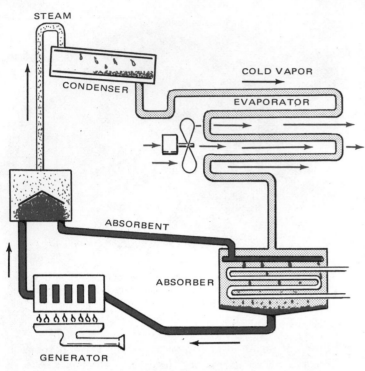

STEAM

CONDENSER

COLD VAPOR

EVAPORATOR

ABSORBENT

ABSORBER

GENERATOR

Fig. 34-29. Schematic diagram of absorption cooling system.

The absorption system's cycle begins in the generator. Here a solution of water and lithium bromide (similar to ordinary table salt) is heated to the boiling point by a gas flame, Fig. 34-29.

The solution is carried to the separator which separates the lithium bromide from the water vapor and releases it to the absorber. The steam continues on to the condenser, where it condenses back to water and flows down to the evaporator. The absorbing action of the lithium bromide causes a high vacuum in the evaporator. This allows the water to expand and become a cold water vapor. The cold water vapor is then pumped to coils at the location to be cooled and continues on to the absorber where it is reunited with the lithium bromide. The solution is returned to the generator and the cycle is repeated.

SUMMARY

No building, home or factory is built today without various electrical and mechanical systems. Water is needed for drinking, washing and operating the drainage system. Fuel gas is needed by many buildings for heating and cooking. Hospitals have oxygen delivered to each patient's room while manufacturers pipe acetylene gas for welding areas.

Various piping systems provide transportation of liquids and gases through the structure. Sheet metal duct work carries heated or cooled air for comfort conditioning.

CAREERS RELATED TO CONSTRUCTION

PLUMBING CONTRACTOR supervises crew of plumbers, estimates plumbing jobs, plans work schedule for crews.

PLUMBER assembles, installs and repairs pipes, fittings and fixtures of heating, water and drainage systems.

MECHANICAL ENGINEER plans and designs centralized heat, gas, water and steam systems.

DISCUSSION TOPICS

1. What is the responsibility of the mechanical contractor?
2. What is a "Uniform Mechanical Code?" Who prepares this code?
3. How is water demand calculated?
4. Explain water hammer. How can it be eliminated?
5. List the types of material used for piping, where each is used and the advantage of each.
6. What is the purpose of a roof vent for the drainage system?
7. What gives bathroom fixtures their shining appearance?
8. Explain the purpose of a drain trap.
9. Explain the operation of a forced-air furnace.
10. Explain the operation of a compressor and an absorbent cooling system.

Chapter 35

FLOOR COVERINGS

The choice of material for a floor depends on a number of factors:

1. The type of building involved; residential, industrial, or commercial.
2. Type of floor frame or subfloor.
3. Use to which the floor will be put.
4. Special requirements such as sound absorbing qualities.
5. Color.
6. Smoothness and resistance to abrasion.
7. Ease of maintenance.

Flooring is made of wood, concrete, clay, asphalt, terrazzo, plastic, rubber and cork.

WOOD FLOORS

Woods used for finished flooring must have hard, dense surfaces to withstand heavy wear and abrasion. Both hardwoods and softwoods have these qualities. Consideration must be given to color, texture, freedom from defects and grain pattern.

All wood flooring is carefully kiln dried. When unfinished, it must be stored in locations where the moisture content can be controlled. Wood flooring requires a protective coating of varnish, lacquer, shellac or wax to seal and protect it.

SOFTWOOD FLOORING

Several species of softwood are used to produce wood flooring. Douglas fir and southern pine are commonly used because of their superior toughness and strength. Grading rules vary somewhat from species to species but apply generally over the whole range of softwood flooring. As a rule wood is divided into vertical grain and flat grain. Vertical grain is subdivided into "B and Better," "C," and "D." Specifications for each grade are outlined in grading rules published by lumber manufacturers and lumbermens' associations. Flat-grain flooring also is divided into three divisions, "C and Better," "D," and "E." Again, grading rules specify the defects allowed.

HARDWOOD FLOORING

Many hardwood species are used as flooring, but the more common ones are maple, birch, beech, red oak, white oak and walnut. Woods are kiln dried and manufactured with tongue

Fig. 35-1. Oak is an excellent wood for flooring. Here it is put down in a herringbone pattern. (Peace Flooring)

and groove edges and ends. See Fig. 35-1.

Red oak, white oak and northern hard maple are the most-used flooring materials among the hardwoods. Oak flooring may be obtained either flat-sawed or quarter-sawed. Quarter sawing brings out the best grain characteristics and produces the best quality of flooring. Flat sawing gives a more highly figured and varied grain pattern.

Northern hard maple is a smooth, durable wood. It is widely used where wear is particularly important. The grain is not as interesting as is the oak. Maple is especially suitable where a highly polished surface is desired.

STRIP FLOORING

Strip flooring is made of individual strips in random lengths. Widths are uniform. Each piece is applied individually.

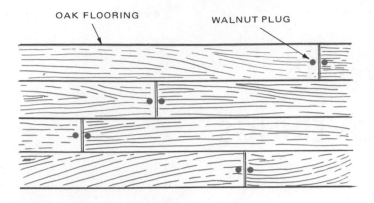

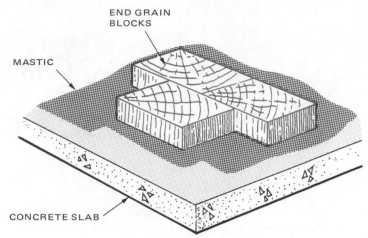

Fig. 35-2. Design of colonial plank flooring. Plugs are usually of a darker colored wood.

Fig. 35-4. End grain block flooring.

Strip flooring is also available in what is known as colonial plank, Fig. 35-2. Strips are of various widths and have round inserts of some wood of contrasting color. Usual thickness is 25/32 in. Widths range from 3 to 8 in.

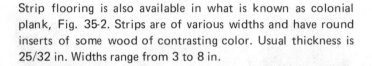

Fig. 35-3. Patterns of parquet flooring. (Arco Chemical Co.)

Fig. 35-5. Wood plastic composite material for this floor combines the warmth and charm of wood and the abrasion resistance of plastic.

WOOD/PLASTIC COMPOSITE FLOORING

Technology has combined the best of two materials to create a new flooring with the beauty of wood and the durability of plastic, Fig. 35-5.

By impregnating wood with a liquid plastic (meth-methracylate monomer) and allowing it to polymerize within the wood, a composite material is created. It is very durable, resists abrasion and is easy to maintain. This floor is desirable for heavy wear areas like commercial buildings, air terminals, and entry areas of residences.

PARQUET FLOORING

Parquet flooring is made up of blocks of hardwood of various sizes which can be laid in a number of patterns, as in Fig. 35-3. Blocks are designed both for nailing and for fastening to the subfloor with mastic.

BLOCK FLOORING

Floor blocks are individual pieces of wood with grain on end as shown in Fig. 35-4. It is made up in squares or rectangles which must be laid in mastic. Such floors are practical where heavy equipment is to be used.

CONCRETE FLOORING

Concrete is used for flooring in a wide variety of situations ranging from basement, play and recreation areas to industrial floors. As a result, they are subjected to every kind of abuse.

A particularly hard surface is possible by introducing metallic aggregates into the topping. These aggregates are normally applied to the freshly poured surface and are worked into the top by floating and troweling.

Colored concrete floors are produced by adding an inorganic coloring agent to the topping mix and floating it in.

CLAY TILE FLOORING

Clay tiles are made by a process similar to that used in making brick. See Fig. 35-6. These tiles have many different shapes, being square, rectangular, hexagonal, and octagonal. Color range is wide. They are made in thicknesses from 1/2 in. to 3/4 in.

Fig. 35-8. Where flooring will receive heavy use from traffic, tile is a good covering. Tiles are laid in an adhesive mastic which is waterproof. This floor will have a high resistance to abrasion.
(Azrock Floor Products)

Fig. 35-6. Glazed clay tile set with a cement mortar produces a very hard and durable floor. Unglazed tile is also made. (American Olean Tile Co.)

liquid.) This mixture is spread over the concrete or wood floors, screeded, compacted, and floated to a depth of 1/2 in.

Asphalt tiles are made of asbestos fibers bound together by a blend of asphaltic binders. Pigments are added for color and, in some cases, polystyrene plastic is added to produce a stronger tile, Fig. 35-8.

Fig. 35-7. Rough leveling of an asphalt mastic floor with a "come along" followed by screeding with a wooden straightedge.
(Master Builders)

Fig. 35-9. Terrazzo floors are widely used in public buildings and make a durable and attractive wearing surface.

ASPHALT FLOORING

Asphalt is the basic ingredient in two types of flooring, asphalt mastic and asphalt tile. Asphalt mastic flooring is made by mixing an asphalt emulsion with portland cement, sand and gravel to form a plastic mixture, Fig. 35-7. (An emulsion is the mixture resulting from a solid suspended in some type of

TERRAZZO FLOORING

Terrazzo is made of concrete which is ground and polished. The finish is the result of the aggregate appearing as a mottled surface, Fig. 35-9. A base slab is poured first, reinforced with

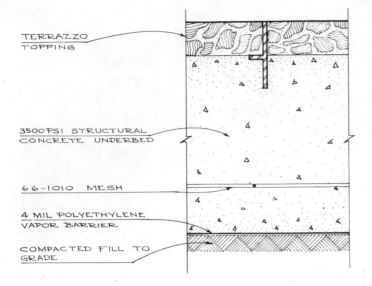

TERRAZZO TOPPING

3500 PSI STRUCTURAL CONCRETE UNDERBED

66-1010 MESH

4 MIL POLYETHYLENE VAPOR BARRIER

COMPACTED FILL TO GRADE

Fig. 35-10. Section of terrazzo floor.
(National Terrazzo and Mosaic Assoc.)

wire mesh to reduce shrinkage. Secondly, a cement and sand bed is spread over the base. Then the terrazzo topping is poured on. See Fig. 35-10.

A gridwork of thin strips of brass, bronze, aluminum, or plastic is set over the mortar bed. The gridwork is about 1 in. higher than the mortar bed. The topping is poured into this grid.

The topping mix is made up of cement, sand and marble chips. This topping is spread over the floor and compacted (packed down) until it is level with the top of the metal grid. By using white cement, colored pigments and carefully chosen marble chips of one or more colors, a great variety of effects can be produced.

CORK FLOORING

Cork is the basic ingredient in two types of flooring, linoleum and cork tile. Linseed oil is used as a binder in the manufacture of linoleum. It is mixed with powdered cork, resin gum, wood flour and color. This mixture is spread over burlap backing and rolled into sheets. Linoleum is made into three grades:

1. Grade A, 3/32 in. thick.
2. Grade AA, 1/8 in. thick.
3. Grade AAA, 1/4 in. thick.

SUMMARY

Floor coverings are classified into two groups, those forming the frame or support structure and those which are applied to the frame by a mastic. The concrete floor, itself, properly troweled, may become the wearing surface. Tiles are adhered to a subfloor by some mastic material providing a thin, durable and colorful floor covering. Floors may be plain, unattractive surfaces for industrial use, or colorful, geometrically patterned surfaces of commercial, office, or residential floors. Materials may be a single natural material such as wood or combinations of materials such as linoleum products.

CAREERS RELATED TO CONSTRUCTION

CONCRETE FINISHER smooths and finishes surfaces of poured concrete floors, walls, sidewalks or curbs. May operate machine to trowel concrete surfaces or may mix cement using hoe or concrete-mixing machine. May supervise grading and leveling and form setting.

FLOORING MATERIALS ESTIMATOR measures flooring areas and determines quantity of materials needed to cover area. May also sell flooring materials and advise on quality or suggest best material for the application.

TILE LAYER (also called TILE SETTER) applies tiles to floors, walls, ceilings and decks. Reads blueprints, measures areas and lays out work.

DISCUSSION TOPICS

1. List the things taken into consideration when deciding what kind of material is to be used for a floor.
2. What type of wood is normally used in the home?
3. What is parquet flooring?
4. What is a wood/plastic composite?
5. Figure the cost of flooring material needed to cover a room 35 by 29 ft. (10.7 m by 5.8 m) when the material is sold at the rate of $15.25 per sq. yd. ($18.25 per m^2).
6. What are the three grades of linoleum?
7. Where are terrazzo type floors used most extensively?
8. Where would clay tile be used?
9. What type of floor is normally used for a basketball court?
10. How is cork utilized as a flooring material?

Chapter 36

PAINTS AND FINISHES

Fig. 36-1. All coating materials are intended to give protection and decoration to the surface. Brushing is an age-old method of applying this coating.

Fig. 36-2. Latex paint applied by roller will hide the drab color of this block masonry wall making it clean, sanitary and colorful. (National Paint and Coatings Assoc.)

People have been coating their buildings for centuries to protect them against the elements and to make them look attractive. *Surfaces are coated with a variety of materials, not only for protection and decoration, but also for sanitation, improved heating and lighting, safety and economy.* See Fig. 36-1.

Paints consist of white or colored solids suspended in a liquid. When the paint is applied to a suitable surface, the solids form a continuous film. The finely divided solid particles used in paint are known as pigments. The liquid portion of paint is called the vehicle. Driers speed the formation of the film. Thinners evaporate easily at low temperatures and leave a solid film.

PAINT

A paint system consists of separate coats of paint applied at suitable intervals to build up a protective film. The thickness of a coating is measured in "mils," or thousandths of an inch. (A coating of 6 mils has a thickness of 6/1000 or 0.006 in.)

Six mils, in the SI metric system, would be 1500 μ m (micrometres).

The body of a paint is the solid, finely ground material which gives a paint the power to hide, as well as color, a surface as in Fig. 36-2. In white paint, this body is also the pigment. Products most widely used for paint body are zinc oxide, lithopone, and titanium dioxide.

OIL BASE PAINT

Linseed oil is extracted from (squeezed out of) flaxseed. It is produced both as raw and boiled oil. Boiled linseed oil has been heat treated and has certain metallic driers added. Linseed oil is slow drying. Like other drying oils, it oxidizes (combines with oxygen) when exposed to air, forming a hard film. Other oils used in paint are tung (pressed from seeds of the Tung tree), soybean oil, castor oil, perilla oil and fish oil.

Pigments give the paint its color. The white body pigments were mentioned before. Colored pigments are taken from vegetable dyes and metallic oxide.

Thinners are the volatile (easily vaporized) solvents which cause the paint to flow better. They evaporate when the paint is applied.

One of the most common thinners for oil-based paints is turpentine. It is distilled from the gum of pine trees. Some petroleum solvents, such as naphtha and benzene, are also used as thinners.

Driers are organic salts of various metals, such as iron, zinc, cobalt, lead, manganese, and calcium which are added to paint. They speed up the oxidation and hardening of the oil vehicle.

ALKYD PAINT

Alkyd paints are so called because of the synthetic resin used in the paint formation. (Synthetic resins are made from chemicals.) Alkyd resin is obtained by the combining of an alcohol and an acid. Alkyd paints are produced by combining a drying oil, such as linseed oil or dehydrated castor oil, with the alkyd resin.

Alkyd paints, in general, have mild alkali resistance but excellent water resistance. They also have the ability to produce lighter colors and retain color better than other paint. Because of their excellent weathering ability, alkyd paints are often used for porches and decks. Altogether, some 50 types of alkyds are used in paint manufacture. See Fig. 36-3.

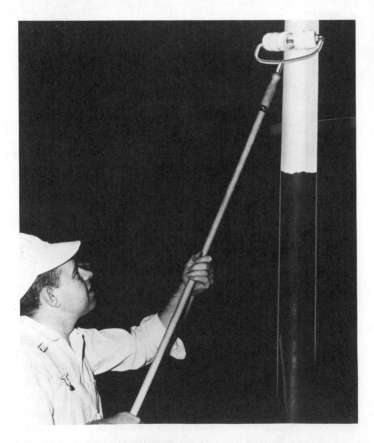

Fig. 36-3. Paints are specially mixed for many jobs to withstand heat, cold, moisture and many acids.

In latex paints the vehicle is a synthetic resin emulsion. These emulsion-based paints tend to foam, so a defoaming agent is added. Latex emulsions can be used as outside (exterior) as well as interior paint. One of its important uses is in exterior finishes for masonry and stucco.

METALLIC PAINT

Metallic paint consists of a metallic pigment and a vehicle. Very fine flakes of aluminum, copper, bronze, zinc or tin make up the pigment. They are suspended in a vehicle which may be a natural or synthetic resin.

Spraying is the best method of applying metallic paints, Fig. 36-4. Spraying permits spreading a uniform film and encourages even depositing of the metallic flakes. Metallic paints are used for many decorative purposes. Aluminum paint is an excellent primer for other exterior paints.

Fig. 36-4. Metallic paints are excellent coatings for the structural steel of this bridge. Spraying saves time.   (National Paint and Coatings Assoc.)

SILICONE

Silicone acts as a colorless water-repellent coating for new brick and concrete masonry units. It is also used on porous concrete to prevent staining and is said to be able to protect the surface up to 10 years. Silicone is sometimes referred to as "waterproofing" material. However, it is more properly termed a water repellent. It sets up a surface tension which causes water to bead on the surface and roll off. The material will bridge microscopic openings in masonry walls, Fig. 36-5.

VARNISHES

Varnishes make up a group of transparent liquids used to provide a protective coating in much the same way as paint

Fig. 36-5. Silicone treatments prevent moisture from being absorbed by brick and other masonry. (Bil-Jax, Inc.)

Fig. 36-6. Painter uses modern equipment for professional application of varnish to kitchen cabinets. Thinned varnish is delivered under high pressure to the small orifice (opening) of the spray tip.

without hiding the surface. *All varnishes have basically the same components as paint: body, vehicle, thinner and a drier.* Varnishes are used for woodwork, cabinetry, paneling, and passage doors of the residence. See Fig. 36-6.

LACQUERS

Lacquer is a newer product made from synthetic material. It is often used in place of varnish for clear finishes.

Most modern lacquer is based on nitrocellulose used with a mixture of volatile solvents. The solvents evaporate rapidly, leaving a protective film.

Lacquers dry in 5 to 15 minutes, forming a fine film in about 30 minutes. They are mixed for either brushing or spraying. Spraying grade lacquers dry too rapidly for brushing.

Clear lacquers are used where thin, colorless, tough films are desired. They are not generally considered as durable as high grade varnish on exposure to sunlight and moisture. Lacquer with a high solid content will give a lasting smooth finish to stained or natural wood.

STAINS

Stains give color to wood surfaces. They do this without concealing the natural grain texture and beauty of wood.

Stains are not intended to provide a protective coating. Thus, staining must be followed with a protective material such as varnish or lacquer.

There are a number of kinds of wood stains available. The difference is in the kind of solvent used to dissolve the coloring material. There are water soluble stains, spirit soluble stains, penetrating oil stains, and pigmented wiping stains.

WATER STAINS

Water stains are synthetic dyes in powder form, usually made from coal tar. They are dissolved in water before use.

Water stain is easy to apply by brush, sponge or spray. It is non-fading and non-bleeding giving deep, even penetration of color. However, it has the tendency to raise the grain, thus roughening the surface. Water stain dries in about 12 hours.

SPIRIT STAINS

Spirit stains are made from dyes which are soluble in alcohol. They are sold in powder form or in ready-mixed liquid form.

This stain produces the brightest and strongest colors, but is likely to fade. Because these stains dry rapidly, they should be sprayed. Drying time is usually about 15 minutes.

PENETRATING OIL STAINS

Penetrating oil stain is made by dissolving oil soluble dyes in coal tar solvents such as toluol, benzol or xylol. Although they are available in powder form, oil stains are usually produced as a ready-mixed liquid.

Stain is easy to apply by brush, sponge, dip method or spraying. Excess stain should be wiped from the surface after application. Oil stains tend to bleed into finish coats and are not color fast. Drying time varies from 1 to 24 hours.

PIGMENTED WIPING STAINS

Pigmented wiping stains are made from mineral pigments ground into a drying oil. They can be brushed, sprayed, or wiped on.

They have good light resistance, good color uniformity, and little or no tendency to raise the grain. However, they lack the staining capacity of many other stains. These stains are not as transparent as some others and tend to hide the fine grain structure of wood. This hiding tendency is useful in blending different woods and in masking undesirable grain patterns.

FILLERS

Fillers are finishing materials used on wood surfaces to fill the pores and provide a perfectly smooth, uniform surface. Filler is generally used on those woods which have large open pores. Filler is also used to impart color to the wood pores and thereby emphasizing the grain structure.

There are two general types of fillers:
1. Paste filler, used on open grained woods.
2. Liquid fillers, used for closed grain woods.
Paste wood fillers consist of a base or body pigment, vehicle and thinner.

Filler is brushed, sprayed or dipped. It must be thinned to the proper consistency. It is wiped off, across the grain, before it hardens on the surface.

SUMMARY

Paints and finishes are liquid materials mixed to coat surfaces and provide protection against decay and weathering elements. A second reason for finishing is decoration. Decoration provides color for the enjoyment of the owner. Many different materials have been mixed to answer all of our coating needs.

CAREERS RELATED TO CONSTRUCTION

CHEMICAL ENGINEER in the paint and varnish industry, designs equipment and sets up processes for manufacturing paints, varnishes, stains, fillers, solvents and removers.

CHEMIST researches and tests new compounds in effort to find new products and improve quality of old products; evaluates changes in paint color, texture, strength and durability; maintains records and prepares reports on results.

PAINTER applies paint, varnish or enamel to inside and outside of houses, plants and other structures. Prepares surface by removing loose paint using scrapers, wire brushes and torches. Applies paint using brushes, rollers or sprayers. Sometimes mixes colors to order or matches colors. May specialize in certain types of painting.

PAINT SALESPERSON may sell all kinds of hardware as well. Displays, mixes and matches paint and sells it to customers. Advises customer on paint quality and special needs. Advises on painting techniques. Estimates amounts of paint required.

DISCUSSION TOPICS

1. List the reasons for applying a coating to a surface.
2. How is thickness of paint measured?
3. What are the functions of pigment in paint?
4. Where does linseed oil come from?
5. What is an alkyd resin?
6. List the methods of application for paints.
7. What are stains?
8. How do lacquers and varnishes fit into the family of finishes?
9. What are the benefits of silicone coatings?
10. What is paste wood filler?

Paint adds the finishing touch to most construction projects. Paints are both decorative and protective. (American Plywood Assoc.)

SECTION 4

THE TECHNOLOGY OF CONSTRUCTION PROCESSES AND PRODUCTION

Construction processes and production relate to the many steps and procedures required in building any structure. This section gives basic information on preparation of the construction site, poured-in-place concrete, waterproofing and dampproofing. It covers masonry and wood framing of floors, walls, ceilings, roofs and special framing techniques, including metal framing. Also covered are plumbing, cabinetry, landscaping and safe work practices.

Beams and girders form a strong skeleton for this massive structural steel building under construction. Note crane at the top, used to raise and position steel members.   (Missouri Public Service Co.)

After the construction site is prepared and the foundation is poured, wood framing begins, usually with the help of specialized equipment like this high-lift forklift machine. (Allis Chalmers)

Chapter 37

SITE PREPARATION

Fig. 37-1. One of the first activities in construction area is clearing of trees, underbrush and old buildings on the site. (Allis Chalmers)

Fig. 37-3. Semitrailer vans are used for site storage of valuable fittings and other small items.

Site preparation is the activity involved in clearing debris from a building location. All unwanted objects, trees, underbrush, old buildings, rock and high ground must be removed, Fig. 37-1.

While the construction site is being cleared, temporary facilities are set up and temporary utilities are installed. Immediately after the site is cleared, surveyors stake out the location of the structure, safety installations are made and earth moving begins.

TEMPORARY FACILITIES

The size and complexity of a construction project determines the number and kind of facilities needed. Shelters on the site may include:

1. Office space, Fig. 37-2.
2. Storage for materials, Fig. 37-3.
3. Equipment storage.
4. Sanitary facilities.
5. Laboratories.
6. Maintenance shops.

Weather is a serious problem in any construction project. Protection is needed for new materials that are susceptible to damage from rain. Temporary protection is needed for workers to guard against rain and cold. Specialized tools and equipment also must be protected from thievery and from damage by the weather. Tents, canvas tarpaulins or plastic covers are often used to provide shelter for construction activities.

In planning for construction, the contractor must plan for shelter. This becomes a part of the total cost of construction. In construction projects that last for several months, or even years, the shelters take on a more permanent look. These units are built in sections to allow them to be taken down and moved at the completion of the project.

Mobile trailers are a popular type of shelter for the office facility. The contractor will use the trailer to house the

Fig. 37-2. Small mobile homes or trailers are utilized for construction site offices.

superintendent, foremen, timekeeper, storekeeper and, sometimes, the resident architect.

TEMPORARY UTILITIES

Today's construction jobs depend on the availability of certain utilities, including electricity, communication and water.

Electricity is needed to operate lights and, especially, pumps for water removal. It also is used to operate machines and equipment such as electric arc welders, saws and many small portable power tools. Electricity is needed in the site's maintenance shop.

Communication systems, telephone and radio are musts for any construction job, large or small. Telephones are the connecting link between the contractor and materials suppliers. Radios are used within the site for rapid relaying of information from the superintendent to the foreman. Even workers directing the operation of heavy equipment can benefit by radio communication.

Water is important to any construction site for drinking, sanitation and mixing material. Water is needed for cleaning, cooling equipment, fire protection and tempering concrete. If water mains are near the site, temporary lines are connected. If this source of supply is not convenient, then water may be hauled in by tank truck.

When working in some remote area (road building, for example), water may be pumped from nearby lakes or streams. In other cases, it might be necessary to drill a well close to the site to obtain an adequate supply of water. Drinking water from alternate sources must be purified by special filters.

CLEARING THE SITE

Site clearing practices include demolition, salvaging and growth removal.

Demolishing is the act of safely bringing down any old structures that may be on the site. Salvaging is the removal of any valuable object from the site. This is done by tearing down and taking apart the various old fixtures and materials or the easy removal of natural material of value from the site. Generally, growth removal is accomplished by hauling away, Fig. 37-4.

SURVEYING

Before construction begins, location of the project has been shown on a drawing. Once the site is cleared, workers must establish the exact location and size of the structure to be built. A surveyor is trained to use the equipment necessary to accurately locate the position of the building, Fig. 37-5. Using a transit and measuring tapes to measure horizontal distances, the surveyor shows construction workers the exact location and size of the structure they are to build.

Fig. 37-5. With the help of instruments, surveyor can take measurements from architectural plans and mark exact location of structure on the site. (David White Instruments)

SAFETY

Safety on the construction job is important. Personnel safety is backed by the Occupational Safety and Health Act (OSHA), and contractors take this seriously. OSHA provides legislation and requirements necessary to guarantee the highest possible degree of safe working conditions.

Site safety is the contractor's responsibility to protect the general public as well as the workers. In heavily populated areas, temporary fencing is used to control trespassing. Covered walkways provide pedestrians with protection against falling material.

EARTH MOVING

Earth moving is a means of preparing the site, Fig. 37-6. Many kinds of earth moving equipment are used. This equipment ranges from small tractors to huge power shovels, Fig. 37-7, some capable of removing enough material in one "bite" to fill a large truck.

Earth removal is done to:

Fig. 37-4. Debris is hauled from construction area by trucks as second step in site preparation. (Allis Chalmers)

Fig. 37-6. In large construction areas, such as road building, a scraper is used to remove and transport earth.   (Allis Chalmers)

Fig. 37-7. A power shovel loads trucks and clears rubble from site of future road bed.   (Northwest Engineering)

Fig. 37-8. Earth and rock have been removed in site preparation for setting a dam for future flood control.
(Corps of Engineers, Kansas City District)

Fig. 37-9. A motor grader is used in site leveling operation. (Allis Chalmers)

Fig. 37-10. Replacement of earth, called back fill, is carefully packed around the structure to provide support and prevent damage. (Missouri State Highway Dept.)

1. Reach a good base for a foundation.
2. Make cuts for highways.
3. Dig for a key to bedrock, while building a dam, Fig. 37-8.
4. Level uneven ground, Fig. 37-9.

Tough, dense material (rock) must be loosened before it can be removed. Common techniques of rock removal include ripping, breaking and blasting. When blasting is done (mostly in solid rock), explosives are placed in special locations to blast apart layers of rock to break it into pieces.

Often, earth is relocated around certain structures to lend support. The relocated earth is called "back fill." The back-filling operation typically takes place around completed foundation. See Fig. 37-10.

SUMMARY

Site preparation, in general, is clearing the construction area of unwanted debris. Disposal of this material by hauling away is common practice. The equipment and procedures used in earth removal are similar, regardless of the kind of construction. In many cases, the only difference is the size of the equipment used and man hours required to complete the job.

CAREERS RELATED TO CONSTRUCTION

CONSTRUCTION SUPERINTENDENT confers regularly with contractor, engineers and local building inspector; directs foremen in their duties and oversees progress of entire construction project.

CONSTRUCTION INSPECTOR observes work in progress and inspects finished construction to insure that procedure, materials and workmanship comply with local building codes, job specifications and safe working conditions; makes necessary recommendations and reinspects changes made.

SURVEYOR uses transit, level and measuring tapes; furnishes accurate information about boundaries and land features of construction site; establishes exact location and size of structure to be built.

EQUIPMENT OPERATOR operates compressors, pumps, hoists, cranes, shovels, tractors, scrapers or motor graders to excavate and grade earth, remove debris, move materials and pour concrete.

DISCUSSION TOPICS

1. What are the first structures to be erected on a construction site?
2. What utilities will be needed by the contractor during construction?
3. What does a surveyor do?
4. What is the responsibility of a construction superintendent?
5. What is salvaging?
6. List the ways by which debris is removed from the site.
7. What is OSHA?
8. How is rock loosened for removal?
9. What is a scraper?
10. What is meant by backfilling?

Chapter 38

FOOTINGS AND FOUNDATIONS

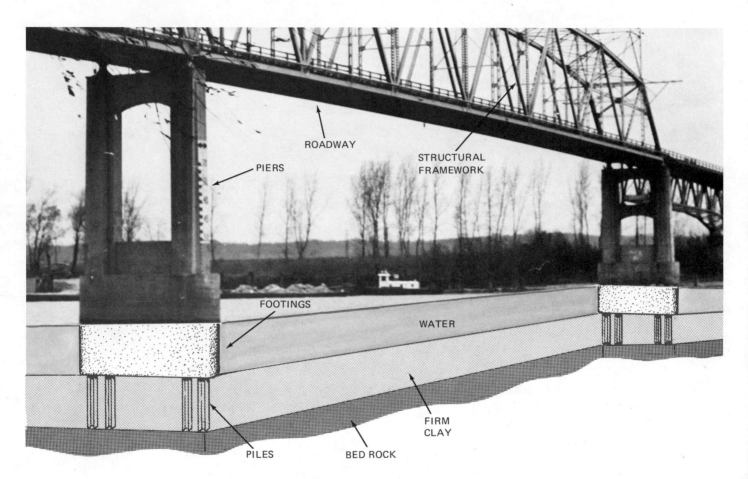

Fig. 38-1. Bridge footings (under water) and piers (above water) make up substructure of this bridge. Roadway and structural framework about the roadway are called the superstructure.

Foundations are the part of the building or structure below the surface of the earth. The weight of a structure is transferred through foundations, piers and columns to footings and from there onto the earth. Footings and foundations must be of a size and type to meet the bearing capacity of the soil.

Each structure has two parts, the substructure and the superstructure, Fig. 38-1. The substructure, generally, is that part of the building which is below ground. The superstructure rests on the substructure, or a part of the substructure located above ground.

The substructure of a house is the foundation or basement wall and footing. While part of the foundation is above ground, it remains the substructure until the type of construc- tion changes. Normally, the first floor and everything above it is the superstructure. This chapter is mostly concerned with the substructure.

THE CONTINUOUS FOUNDATION

Most residences or small buildings have a continuous foundation. A foundation wall extends above a footing slab. Fig. 38-2 shows two types of foundation walls. Top view is a poured wall. Bottom view shows foundation of concrete block. The wall provides:

1. Protection against water and frost penetration.
2. Support for lateral (sideways) pressure exerted inward by

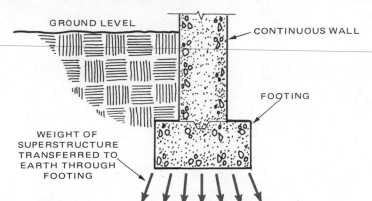

3. Distribution of vertical loads of the superstructure to the footing.

COLUMN PIERS

Concrete piers are used when rock or suitable load-bearing soil lies some distance below the surface, Fig. 38-3. Shafts, with or without metal casing, are sunk below the surface until they reach bedrock or other suitable load-bearing material, Fig. 38-4.

Fig. 38-2. Two types of foundation used in residential construction. Above. Typical continuous foundation is poured concrete substructure. Below. Similar type of dwelling may have foundation of concrete block.

Fig. 38-4. Huge auger is being readied to sink shaft for pouring of concrete pier.   (Turzillo)

When the shaft reaches the right depth, it is filled with concrete. If the shaft goes through a water-bearing soil, it may be necessary to seal the shaft. This is done with a watertight tube or chamber. It is called a caisson. Usually of steel or wood, it is carried down the shaft as the dirt is taken out.

A bell-shaped enlargement is sometimes made at the bottom of shafts dug for piers. These bells are used especially when bedrock is not reached. When concrete is poured in the shaft it fills up the bell and the pier will have a wider base. This helps to distribute the weight over a larger area. A concrete structural column is then poured on top of the pier. See Fig. 38-5.

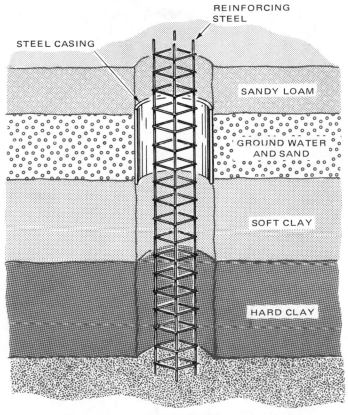

Fig. 38-3. Excavating for piers. Holes are bored many feet into the ground. Unstable or water-filled soils must be sealed out with metal cylinder (caisson).

Fig. 38-5. Structural columns, placed on top of piers sunk deep into the earth, form substructure of this bridge. Note concrete form being placed on top of poured piers for next column.

PILES

Piles are poles that are driven into the ground at a construction site. They help to support the structure that is to be built. The piles firm up the soil at the excavation site and keep the structure from settling unevenly.

Of themselves piles have no ability to hold up heavy loads. But they do transmit the loads pressing down on them to the soil around them or to bedrock underground. See Fig. 38-6.

It is easy to see how a pile driven down to solid rock could support heavy loads. It is not as easy to see how a pile can transmit the weight of a building to the soil it is driven into. In

Fig. 38-6. Steam-operated pile-driving hammer pounds piles into earth in Turkey.

such cases, the pile is tapered. As it is driven into the ground it compacts the soil around its surface. Friction between soil and pile provides support. This packed soil also resists the pressure of the weight transmitted through the pile.

Piles can be made of wood, steel or concrete. Condition of the subsoil and cost will determine which is used.

WOOD PILES

Wood piles will last for centuries if kept wet. Made from the trunks of trees they have the limbs and bark trimmed off. They are driven, small end first, into the ground. The end may be fitted with a steel point if there are rocks underground. Cedar and cyprus are used for piles because of their natural resistance to decay. Douglas fir and southern pine are also used because of their resistance to the shock of driving. Though not as decay resistant, they can be chemically treated to slow down decay, Fig. 38-7.

The heavier the load placed upon them the bigger around the pile must be. The minimum diameter of the driving point is 6 in. (15 cm). The top end is seldom less than 12 in. (30 cm).

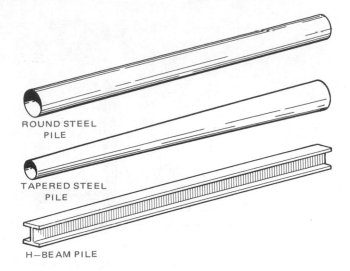

Fig. 38-8. Steel piles have different shapes.

Fig. 38-9. Steel piles are driven in rows to give continuous support to structure. (Raymond International)

Fig. 38-7. Lightweight wooden piles are supporting this structure. They are logs that have been trimmed, debarked and treated against decay. (Koppers Co., Inc.)

STEEL PILES

Steel piles, Fig. 38-8, may be round, round tapered, or flanged like an I-beam or H-beam. They are usually driven to bedrock. Many may be driven in rows to give support for a large structure, Fig. 38-9. Structural steel piles up to 200 ft. (61 m) long have been driven.

PRECAST CONCRETE PILES

Precast concrete piles reinforced with steel are widely used. They may be round, square or rectangular in shape. Fig. 38-10

Fig. 38-10. Concrete piles are usually used where soil is soft. Here they support a bridge. Piles are sunk deep into bed of river.

Fig. 38-11. Driving piles from a barge for a bridge. Precast cylindrical pier on barge awaits pile driver. Note how piles are capped in metal to keep concrete from crumbling under heavy hammer blows.

shows round or cylindrical precast piles used to support a bridge. Precast piles must be heavily reinforced to withstand the impact of driving, Fig. 38-11. Precast piles are somewhat more costly than the others.

DRIVING PILES

Driving piles takes heavy equipment. The simplest type is a drop hammer in a vertical frame. The hammer, weighs 500 to 2000 lb. (250 to 1000 kg). It is raised to the top of the frame and dropped onto the end of the pile.

Two types of steam-aided hammers are the single acting and the double acting. With the single-acting hammer, a cylinder is placed over the pile and steam is allowed to enter the enclosed chamber raising the hammer a distance of 2 to 4 ft. (60 to 80 cm). When the steam is released the hammer falls of its own

weight. The double-acting hammer or ram is raised 4 to 20 ft. (80 cm to 6 m). Instead of being allowed to fall of its own weight, steam pressure forces it downward, See Fig. 38-6.

Water is used to set some piles in sand or soft soil. A water pipe, lowered beside the pile, forces a jet stream of water through the pipe. The water washes away the sandy soil below the pile and acts as a lubricant. The pile is usually driven the last 3 ft. (90 cm) without water. This gives it a more solid permanent rest.

RETAINING WALLS

A retaining wall prevents the flow of soil down a slope or it contains the coil around a substructure, Fig. 38-12. The basement wall of a house is a retaining wall as well as a support for the building.

There are several methods of retaining soil. The type of retaining wall that will be used depends on:
1. Soil makeup.
2. The function of the structure.
3. Depth of excavation.
4. Economy of construction.

SHEET PILING

Sheet piling is used to hold back earth in dry land excavations. Or it can be used to hold out water when the excavation is in or alongside a body of water. Usually the sections of piling lock together as those shown in Fig. 38-13. The sheet piling structure used to hold out water during an excavation is called a cofferdam.

LAYING OUT FOUNDATIONS

Laying out a foundation is usually a simple job. Though easy to do, it is important that it be square and level. The

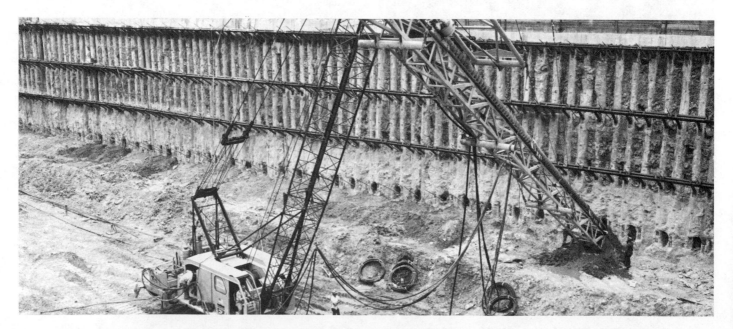

Fig. 38-12. Drilling holes for "earth anchors" to hold reinforcing steel for retaining wall. Before excavation was started, a "wall" of sheet piles was sunk to stop dirt from caving into excavation. Piling will create part of form for concrete retaining wall. It will remain in place permanently.

Fig. 38-13. Steel sheet piling driven into a lake, seals out water behind excavation for building. Note pile-driving hammer on top of piling. Sheet piling is stacked at right.

success of the whole project depends upon it. If a foundation is not right, the rest of the building will not be square or level. Doors will not hang right. Cabinets and paneling will not fit.

Before laying out the foundation, the construction crew must know where the foundation is to be located on the lot. It must conform to all local restrictions such as setbacks from streets and lots next to it.

When these are known, the next step is to lay out the foundation lines, Fig. 38-14. To do this they will locate each outside corner of the foundation wall and drive stakes into the

ground at these points. They will drive a nail into the top of each stake at the exact point of the outside corner. Then it is necessary to check for squareness by measuring from opposite corner to opposite corner. The distances will be equal if the layout is square. They may also check for squareness of any corner by measuring down one side 6 ft. and the other side 8 ft. The length of the diagonal line across these two points should be 10 ft., Fig. 38-15. Batter boards are then located about 3 to 4 ft. outside the finished foundation line. Workers place a string or chalk line across the tops of opposite batter

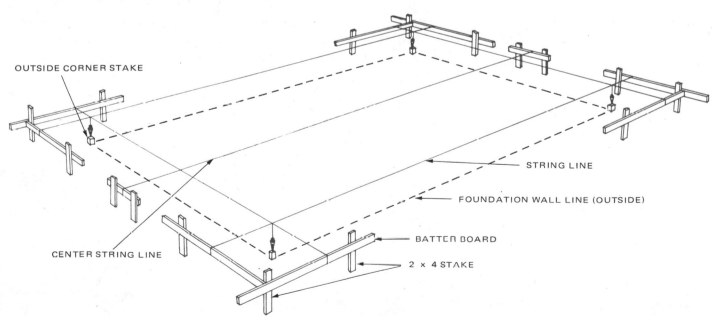

OUTSIDE CORNER STAKE

STRING LINE

FOUNDATION WALL LINE (OUTSIDE)

CENTER STRING LINE

BATTER BOARD

2 x 4 STAKE

Fig. 38-14. Batter boards and string line are necessary equipment for simplest of foundation layouts. Outside corners of foundation are at points where lines cross.

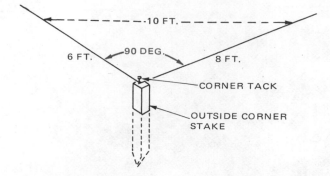

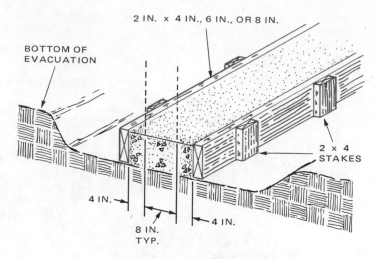

Fig. 38-15. Carpenters use the 6-8-10″ rule to make sure that batter board and string line layouts are square. A corner is square (a true 90 deg.) if one side measures 6 ft., the other side 8 ft. and the diagonal line 10 ft. If corner is more than 90 deg., diagonal will measure more than 10 ft. Diagonal will measure less than 10 ft. if corner is less than 90 deg.

Fig. 38-16. A typical form for a footing to support basement wall.

boards. Using a plumb bob, they adjust the lines so that the intersecting points are exactly over the nails in the corner stakes. A saw kerf about 1/8 in. deep cut in the batter board where the line touches holds lines in correct position.

THE FOOTING

When the foundation layout is completed, forms are placed so that a footing of concrete can be poured, Fig. 38-16. Such footings are commonly used for houses. Footings prevent settling or cracking of building walls. They must be completely level and should extend at least 12 in. below frost line. See Fig. 38-17 for frost depths for different regions of the U.S.

SUMMARY

Every bridge, road, dam or building can be easily divided into two general parts, the superstructure and the sub-

Fig. 38-17. Frost depth lines for the United States.

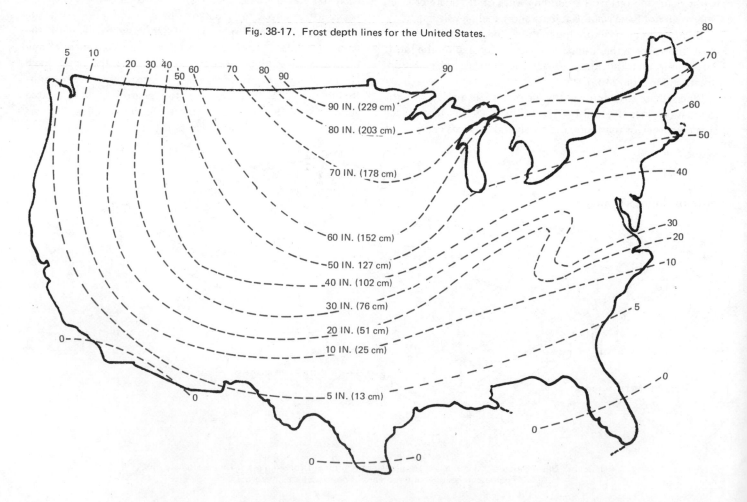

structure. The substructure or foundation transfers the immense weight of the building to the earth. The weight of the building must be distributed uniformly. Otherwise uneven settling can cause cracks and even collapse of the structure. Foundations will vary with the type and size of structure and the earth strata beneath it. Residences may have poured concrete walls set on a poured concrete footing. This footing must be located on firm earth 12 in. below the frost line of the region. Large structures may rest on piles driven deep into the earth. Some piles rest directly on bedrock.

CAREERS RELATED TO CONSTRUCTION

PILE DRIVER operates huge power diven hammer to sink piles into ground or bed of lakes or rivers usually to provide solid foundation for buildings, bridges and dams.

FORM SETTER assembles and installs prefabricated wooden forms to mold concrete columns, walls, girders, beams and floor slabs. Lifts sections of forms into place. Uses fasteners and braces to hold them in place. Aligns and levels. When he builds the forms from lumber he is called a FORM BUILDER.

DISCUSSION TOPICS

1. Of what does the substructure/superstructure consist? In the house or apartment where you live can you tell where the substructure ends and the superstructure begins? What is the substructure of a bridge, dam or highway?
2. Identify one example of a continuous foundation.
3. Describe the difference between a pier and a pile.
4. How are piles installed? What materials are used for piles?
5. What is the purpose of retaining walls?
6. What is steel piling used for?
7. Describe the process of laying out a foundation.
8. What is a plumb?
9. Why are batter boards used?
10. What is a footing? Where is the footing located?

Chapter 39

POURED IN PLACE CONCRETE

Poured concrete is used in structures large and small. In homes and small construction it serves as footings, foundations, slabs, steps and walls. Other areas where loads are placed on the structure might also call for poured concrete.

There are many reasons why concrete is used. It resists decay, moisture and corrosion from acids. But, more important is the fact that, before it hardens it can be molded to any shape. That is to say, it takes the shape of the form it is poured into.

Concrete can be reinforced (made stronger) with steel rods or wire mesh. This material is buried in the concrete while it is soft. The strength of concrete and steel together is greater than when either is used alone. Fig. 39-1 shows good use of concrete.

Fig. 39-2. Ready-mixed concrete is delivered by special truck to site. A trough places it directly into the form.
(Challenge-Cook Bros.)

DELIVERY OF CONCRETE

Concrete is prepared by two methods:
1. Mixed on the site with material stockpiled nearby.
2. Ready mixed in central plants and transported to the building site in huge mixer trucks.

JOB MIXED CONCRETE

Mixing concrete on the job site is less popular than it once was. It is not recommended unless the job is large enough to warrant expensive equipment for measuring and handling the material. Stockpiling and mixing takes a lot of room.

READY-MIXED CONCRETE

Ready-mixed concrete is purchased from a central plant. These plants are equipped to furnish concrete that will meet exact needs. It will be mixed as required and is guaranteed to meet strength requirements. Ready-mixed concrete is delivered by a special truck designed for this purpose, Fig. 39-2. The

Fig. 39-1. Reinforced concrete withstands forces of compression and tension. (Compression tends to push the concrete into its support. Tension tends to pull it apart by stretching.) When combined with the design principle of the arch, such concrete is capable of bridging large distances. (Howard, Needles, Tammen, and Bergendoff)

mixing of dry ingredients is done at the central plant. Water is added and mixed while the truck is driven to the site.

PLACEMENT OF CONCRETE

Good concrete work depends on proper pouring, finishing and curing. For best results, these operations should be directed by an experienced construction supervisor. To get a strong, long-lasting concrete structure, several steps are necessary:

1. It must be poured uniformly.
2. The surface must be compact (free of honeycombing).
3. Poured concrete must be cured so that few surface cracks develop.

METHODS OF PLACEMENT

If concrete cannot be placed (poured) directly where it is needed by the mixer chute, it must be carried there by other means. Several methods are used:

1. Pumping.

Fig. 39-4. Flexible hose allows workers to place concrete exactly where it is needed.   (Whiteman)

2. Conveyor belts.
3. Boom and bucket.
4. Concrete buggie.
5. Wheelbarrow.

Concrete is always placed in horizontal layers. It is never allowed to flow or free-fall for distances greater than 3 or 4 ft. (.92 or 1.2 m). The aggregates tend to separate if it falls further. The heavier particles collect at the bottom leaving the fine aggregates on top.

Concrete is pumped through pipes to lift it to great heights. The boom, Fig. 39-3, can be swung to left or right for delivery into forms set high above the ground. See Fig. 39-4.

Pumping equipment is of two kinds:

1. Trailer units that have their own power, Fig. 39-5.

Fig. 39-5. Large engine powers trailer-mounted concrete pump.

Fig. 39-3. Special booms, pipes and pumps lift concrete high into the air and place it in forms.

2. Units mounted on a truck, Fig. 39-6.

The concrete is dumped into the hopper of the pump from the ready-mix truck, Fig. 39-7. Pressure forces the concrete up the pipe. This pressure is supplied by a double-acting piston pump, Fig. 39-8. The pump is run by a powerful diesel engine.

Fig. 39-6. Truck body type pumping unit has separate engine for pump. (Whiteman)

Fig. 39-7. Concrete goes directly from the ready-mix truck to the concrete pump. (Challenge-Cook Bros.)

The boom can reach as high as 110 ft. The pump truck can pump 60 to 80 cu. yd. (46 m³ to 61 m³) of concrete an hour.

Sometimes air pressure is used to spray layers of concrete on a wall. This is called pneumatic placement. This method has been used for years. Air pressure forces a dry mixture of cement and sand through the hose. Water is mixed at the nozzle. This mixture is called gunite, sprayed concrete, or, more commonly shotcrete. Fig. 39-9 shows shotcrete being applied.

Boom and bucket method of placing concrete is used to transport concrete to high places, Fig. 39-10. Boom and bucket method is slower than pump. Placement is very accurate.

CONSOLIDATION OF CONCRETE

Concrete thicker than 12 in. must be agitated to consolidate it. Consolidation does three things: settles the concrete;

works out air pockets; forces aggregates closer together.

The simplest tool for agitating is a rod which is plunged into soft concrete to push aggregates around and fill air pockets. The vibrator, which is the tool most often used, vibrates as the worker moves it through concrete. Consolidation of concrete for thin slabs, like sidewalks, driveways and floor slabs is accomplished by a simple process called "jitterbuging." Jitterbuging uses an open screen device in a tamping action on the surface of freshly placed concrete to consolidate the mix.

LEVELING

When all the concrete has been placed it should fill the form. To level the concrete, we can use a straightedge, Fig. 39-11. The long straightedge is rested on the screeds and worked back and forth in a sawing motion to strike off (remove) excess concrete.

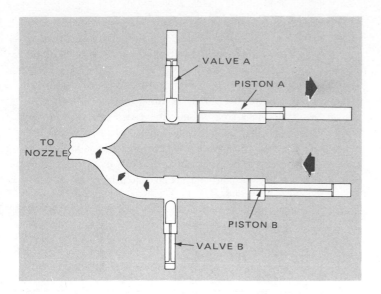

Fig. 39-8. Valve-piston system of concrete pumps moves concrete in a steady flow by maintaining constant pressure. As piston A, above, begins its back stroke, valve A closes to stop concrete from flowing back with the piston. As piston A begins its back stroke, piston B is beginning its forward or pressure stroke. As it begins to move forward, valve B opens.

Fig. 39-10. Concrete is placed in a large bucket and lifted by crane to where it will be poured. When bucket is exactly over the spot to be poured a trap door in bottom of bucket is opened to dump concrete. (American Plywood Assoc.)

Fig. 39-9. Equipment has been developed which will spray concrete with aggregates as large as 3/4 in. (Wire Reinforcement Institute)

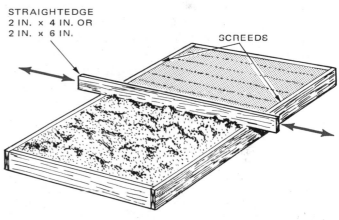

Fig. 39-11. Straightedge screed method is probably the simplest way to level concrete surface.

FLOATING

Soon after leveling the concrete, workers will float it. This must be done before it hardens. *Floating pushes aggregate below the surface and brings finer particles to the top.* A cement finisher uses a wooden or aluminum float which is moved back and forth over the leveled surface. Floating continues until no coarse particles can be seen, Fig. 39-12. Floating insures that there will be a smooth mixture of cement and sand on top for finishing.

FINISHING

Concrete slabs like floors and sidewalks are worked until the top has the proper finish. A smooth finish is called a steel trowel finish. A rough finish may be called a brushed finish.

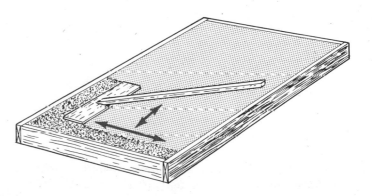

Fig. 39-12. Floating concrete surface with wood float.

Fig. 39-13. Finishing machine above has three or four steel trowels fastened to wheel. Wheel is turned by gasoline engine. It can finish a large surface in a short time. Note finisher in background using hand trowel. (Master Builders)

The cement finisher makes a smooth surface by pressing and moving a steel trowel back and forth over the surface. A finishing machine like the one in Fig. 39-13 may be used to produce this finish.

CURING

Concrete needs proper curing if it is to be strong and free from defects. Its strength improves rapidly when it is first placed. Improvement continues, but more slowly, as long as conditions are favorable. It must not be allowed to dry too fast. Chemicals in the cement need water to act. Temperature should be between 60 F and 80 F (16 C and 27 C).

In hot weather, when temperatures are above 80 F, the drying process should be slowed down. Several methods are used:

1. Sprinkling with water.
2. Covering with moist burlap.
3. Covering with waterproof paper.
4. Covering with plastic sheets.

Almost no chemical action will take place with temperatures below 30 F (−1 C). Freezing ruins freshly poured concrete.

REINFORCEMENT OF CONCRETE

Steel rods or wire are placed in concrete to make it stronger. Such materials are called reinforcement. Concrete withstands compression (squeezing together) very well. Steel has great tensile strength. (This means it resists stretching force.) With reinforced concrete, you combine these properties and get extra strength. Reinforced concrete, in effect, has the properties of both concrete and steel.

The location of the steel within the concrete is very important. When a vertical load is placed on a beam or slab that rests on upright supports, the beam or slab tends to sag in the center between the supports.

During this bending action, a squeezing force is created on the top of the beam. At the same time, a stretching force is exerted on the lower side. Where the beam passes over the supported column a shear (cutting) force is present. Where the beam rests on the column, the forces are reversed. Tension force is on the top and compression force is underneath. See Fig. 39-14. Reinforcement rods can equalize these natural

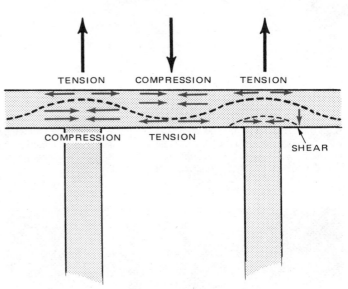

Fig. 39-14. Forces created within a concrete slab or beam over vertical supports.

forces, Fig. 39-15. To reinforce against stretching forces, the steel rods are generally located low in the concrete beam but are spaced high enough in the form to allow concrete to completely surround the rod. Refer again to Fig. 39-15. See how the rod is near the bottom of beam. Fig. 39-16 shows reinforcing in place for a beam. Forms will be placed around the reinforcing in the next construction step. Saddles are used, as in Fig. 39-17, to hold reinforcement bars above the bottom of the form.

Wire mesh reinforcements are used in continuous slabs, Fig. 39-18. The most common is the right angle mesh.

SLAB FLOORS

Slab floors are continuously supported by beams and columns. They are like slabs on the ground. They are only 4 to 6 in. thick. Floors poured between vertical support columns need to be thick. However, to reduce the amount of concrete necessary to fill the form and to reduce the total weight, tubes and tubs may be set in the floor of the form before concrete is poured.

In the tube slab system metal or paper tubes are spaced evenly in the form. Reinforced concrete is poured around them, Fig. 39-19.

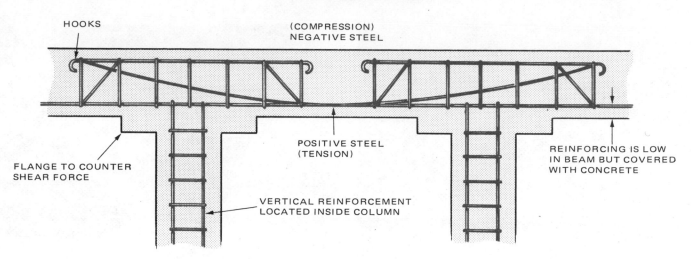

HOOKS

(COMPRESSION) NEGATIVE STEEL

FLANGE TO COUNTER SHEAR FORCE

POSITIVE STEEL (TENSION)

REINFORCING IS LOW IN BEAM BUT COVERED WITH CONCRETE

VERTICAL REINFORCEMENT LOCATED INSIDE COLUMN

Fig. 39-15. How reinforcement is "buried" in a beam and in vertical supports to strengthen concrete against tension and shear forces. Note how reinforcing for beam is tied into reinforcing for columns.

Fig. 39-16. Beams in this stadium complex will carry heavy weights. The cage-like reinforcing will be buried in concrete when beam is poured. This section is ready for workers who will build forms. Then it will be ready for pouring of concrete. Note how reinforcing is tied to supporting steel columns.

Fig. 39-17. Arrow points to patented saddles which support reinforcement bars.

The tubes "float" in the slab so that the concrete can form completely around them. The thickness of slab, the diameter of the tube, and the spacing between them can change. It depends on two things:

1. How much weight must be supported by the floor.

Fig. 39-18. Workers place and tie wire mesh into form. Saddles hold mesh above form. (Wire Reinforcement Institute)

Fig. 39-19. Tube slabs range from 10 in. to 30 in. thick and may span from 18 to 80 ft.   (Bergan Built)

Fig. 39-20. Tubs sit on floor of form. Reinforcing rods will be laid in network between them. Foreground shows reinforcing rod sticking out of a column. Floor slab reinforcing will be tied to them.
(Bergan Built)

2. Distance between supports.

In the grid or tub floor system, plastic or paper pans are used as temporary forms. They are turned upside down on the floor of the slab form, Fig. 39-20. Reinforcement steel and concrete are placed between them. The ceiling of the level below will have cavities left by the tubs. See Fig. 39-21.

EXPANSION JOINTS

All large slabs of concrete will expand or contract as temperatures change. To protect the concrete against this expansion and contraction, spaces are placed between slabs. This space is filled with special sealants. These materials remain flexible. They can be compressed without damage. Yet, while joint or space widens they will expand to fill and seal the space. This helps to keep water from leaking through the

Fig. 39-21. Underside of construction pictured in Fig. 39-20. Tub forms leave cavities between beams to cut down weight of structure.

space. Expansion joint materials are manufactured in three types:
1. Premolded type including vinyl, neoprene, cork, rubber and asphalt-impregnated fibrous material.
2. Mastic type including epoxy, asphalt, synthetic rubber and silicones.
3. Metal type including a bellows-like strip of copper or stainless steel.

SUMMARY

Concrete can be used in so many ways to build structures. Because of the compressive strengths of concrete and the tension strength of steel rods, reinforced concrete is an extremely strong building material. Knowing where the greatest forces are exerted on a structure, we can position reinforcing for greatest strength. At the same time beams and slabs can be made lighter. This saves on material and cuts down total weight of the structure.

CAREERS RELATED TO CONSTRUCTION

CONCRETE FINISHER (also called CEMENT MASON) smooths and finishes surfaces of poured concrete floors, walls, sidewalks or curbs to textures needed. Also spreads concrete to workable consistency and specified depths. Finishes vertical concrete surfaces by rubbing with abrasive stone and by applying coating of cement compounds. May supervise subgrading, mixing, form setting and mixing of concrete.

CONCRETE MIXER OPERATOR tends mixing machine at ready-mix batch plant. Mixes sand, cement and aggregates. Places this mixture in ready-mix truck. Adds water in right amounts. May work at smaller mixer on building site. Duties there may include wheeling cement to forms in two-wheeled cart or wheelbarrow.

CONCRETE PUMP OPERATOR tends pump to force concrete from mixer or truck into concrete forms. Must position pump and snap pipes together. Starts pump, gives signals to concrete mixer-truck operator. Turns valves to regulate flow of concrete.

READY-MIX TRUCK OPERATOR drives truck equipped with cement mixer between bulk plant and construction site. Cleans truck after delivery.

DISCUSSION TOPICS

1. Why are steel rods or wire mesh placed in concrete?
2. Explain compression and tension forces.
3. How is a concrete slab leveled?
4. How does floating make concrete better?
5. List the methods of lifting and placing concrete.
6. What is meant by ready-mix concrete?
7. Why should concrete cure slowly? At what temperature does concrete cure best?
8. Discuss the placement of tubes and tubs in concrete slab floor systems.
9. Explain why concrete is agitated after it is placed in the form?
10. Why provide for expansion of concrete?

Chapter 40

WATERPROOFING
AND DAMPPROOFING

Waterproofing and dampproofing refer to methods of construction that prevent moisture from entering a structure by hydrostatic pressure. In this regard, hydrostatic pressure is pressure exerted by or existing within a liquid at rest with respect to adjacent bodies.

The deeper a structure is set into the earth, the greater the hydrostatic pressure will be. It will force water into the pores

WATERPROOFING

To effectively prevent water from entering any portion of the structure that is below grade, the walls must be designed to withstand the lateral (side) pressure of the back fill. Hydrostatic pressure also exists, especially below the ground water level (or below the water level of rivers, lakes or seas). See Fig.

Fig. 40-1. Hydrostatic pressure is present in all soils during the rainy season. However, on sites near large bodies of water, hydrostatic pressure is ever present. (Port of New York Authority)

of masonry or concrete surfaces and assist in a more rapid deterioration of the wall of the foundation. Hydrostatic pressure also can cause water seepage through pores, cracks and construction joints.

The difference between dampproofing and waterproofing is that dampproofing is done at or above ground, while waterproofing is a heavier build of material below ground to prevent water from entering by hydrostatic pressure.

40-1. The waterproofing membrane must be strong enough to prevent this water from entering through pores and voids in the wall.

Whenever possible, drainage systems must be devised to reduce this pressure by carrying away any excess water. See Fig. 40-2. This may be either clay tile with open joints or a perforated plastic tube to allow the free passage of water. In either case, the tile or tube is covered with gravel to speed the

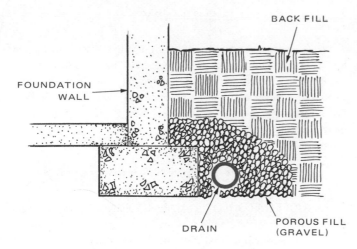

Fig. 40-2. Proper drain placement at base of foundation will carry away excess ground water.

absorption of moisture.

Waterproofing membranes are built up by means of hot asphalts and felts, glass or rags in extreme water problem areas. In areas having less of a problem (residence on high ground, for example), only a single application of hot or cold emulsified asphalt or coal tar is applied.

Plastic and synthetic rubber membranes and similar waterproofing materials also are available. The thin sheets are held tightly against the wall surface with special adhesive. Fig. 40-3.

Prefabricated membranes are produced with a plastic or glass fiber core between two layers of asphalt-impregnated felt.

The outer surface is coated and finished asphalt. This product is particularly effective under conditions of excessive moisture and vapor. Another type of waterproofing membrane is made by simultaneously spraying cut-up glass fibers and asphalt to form a monolithic membrane.

Sheet lead joined by welding is another material used. It is placed under the wall, then turned up the wall to make a continuous waterproofing pan. Lead is particularly effective under outdoor pools, but it is considerably more expensive than asphalt-base membranes.

DAMPPROOFING

Many different dampproofing systems are used to prevent surface moisture from penetrating through masonry or concrete walls. The materials used must penetrate and fill the pores of the wall surface, and also must be sufficiently elastic to resist minor expansion and contractions.

The dampproofing materials are applied by brush, roller or spray. These include hot applied and emulsified coal tar, hot and cold applied asphalt, and emulsified asphalt combined with glass fibers. Other materials used for dampproofing are vinyls, epoxy resins and cement-based aggregates.

One of the latest developments in dampproofing is a metallic material called hydrolithic or ferrous waterproofing. This consists of pulverized iron aggregates mixed with an oxidizing agent. When the iron particles oxidize or rust, they expand four and one-half times their original size. This wedges them into the pores of the wall to create a relatively dense barrier that is impervious to water.

Fig. 40-3. Plastic membranes placed under a wood foundation can be adhered to the wall high enough to reach above grade level. (Koppers Co., Inc.)

DAMPPROOFING SLABS

Slabs placed on grade (not subject to hydrostatic pressure) may be placed over a granular fill of 4 in. to 5 in. (10 cm to 13 cm), Fig. 40-4. The granular fill tends to prevent water from being drawn up through the bottom of the slab by capillary action. To complete this water stop, a polyethylene, polyvinyl or butyl-rubber sheet covers the granular fill and under the slab.

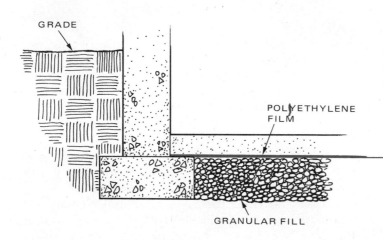

Fig. 40-4. Granular fill is placed under slab to prevent water from entering slab by capillary action.

In dampproofing slabs with granular fill and sheeting, there is always the danger of puncturing the vapor barrier sheet by the irregularities in the fill. Therefore, a different technique is necessary where it is imperative that the structure remain dampproof, or even where it may be subjected to hydrostatic pressure. A structural slab is placed directly on the granular fill, finished and allowed to cure. Then, a waterproof membrane is placed or built up over the slab in much the same manner as for basement walls. Finally, 1 1/2 in. to 2 in. (3.8 cm to 5 cm) of finished concrete is poured over the membrane.

SUMMARY

Waterproofing and dampproofing are extremely important for structures built deep into the earth. These treatments are especially necessary in locations having high ground water levels (structures built along river fronts, lakes or seashores). For a residence to have a "dry basement," proven techniques must be used to insure that moisture does not gain entry to the structure.

CAREERS RELATED TO CONSTRUCTION

OWNER/OPERATOR OF DAMPPROOFING FIRM oversees or actually performs dampproofing operation; brushes, rolls or sprays dampproofing material on wood, brick, concrete or other masonry surfaces to seal pores.

LEAD BURNER joins sheet lead by welding to form a waterproofing pan for placement under and alongside of wall to protect it from moisture penetration.

EMULSION SPRAYER produces prefabricated waterproofing membranes by spraying various asphalt-based materials on a plastic or glass fiber core.

DISCUSSION TOPICS

1. Describe the difference between waterproofing and dampproofing.
2. What is hydrostatic pressure?
3. How much hydrostatic pressure is present 20 ft. (6 m) below the ground level?
4. Why should waterproofing be done?
5. How would a drainage system be installed?
6. What methods are used to apply dampproofing materials?
7. Describe the hydrolithic or ferrous system of dampproofing.
8. How are slabs at grade level dampproofed?
9. What is capillary action?
10. Describe a build-up system of dampproofing for slab construction.

Chapter 41

MASONRY

Fig. 41-1. Masonry units find many exterior and interior uses, for load bearing and non-load bearing walls, curtain walls, fire walls, chimneys and fireplaces. (Patent Scaffolding Co.)

Fig. 41-2. A good masonry mortar must remain workable long enough to permit the mason to position the units.

Masonry is the art of building with small units bonded together with mortar, Fig. 41-1. Bricklaying, block laying, stone setting and hollow clay tile setting are masonry construction methods.

Building structures by the masonry method is made possible by the bond that develops between the mortar and the building units (brick, stone, block or tile). The strength of masonry walls and their ability to resist water penetration depends on the completeness of the mortar bond.

MORTAR BOND

The water content of the mortar mix must be just right to obtain a strong bond between mortar and the brick. As a rule of thumb, the mortar should be mixed with the minimum amount of water possible and still maintain plasticity, Fig. 41-2. If mixed mortar loses water by evaporation and stiffens, additional water can be added and the mortar remixed. All mortar should be used within two hours after mixing.

Most building codes recognize four types of mortar, each recommended for a specific use. Each type is identified by a letter designation: M, S, N and O.

TYPE M MORTAR

Type M mortar is suitable for general use, and it is recommended specifically for unreinforced masonry below grade and in contact with the earth. Type M consists of 1 part portland cement, 1/4 part hydrated lime and 3 parts sand. Another acceptable M mix is 1 part portland cement, 1 part masonry cement and 6 parts sand by volume.

TYPE S MORTAR

Type S mortar is a general purpose mortar bond recommended where high resistance to lateral force is required. Type S consists of 1 part portland cement, 1/2 part hydrated lime and 4 1/2 parts sand. Another acceptable S mix is 1 part masonry cement and 4 1/2 parts sand by volume.

TYPE N MORTAR

Type N mortar is suitable in exposed masonry above grade, and it is recommended specifically for exterior walls subjected to severe exposure. Type N consists of 1 part portland cement, 1 part hydrated lime and 6 parts sand. Another acceptable N mix is 1 part masonry cement and 3 parts sand by volume.

TYPE O MORTAR

Type O mortar is recommended for load-bearing walls of solid units where the compressive stresses do not exceed 100 pounds per square inch (7 kg per cm^2). Another consideration is that the masonry will not be subjected to freezing temperatures and heavy moisture. Type O consists of 1 part portland cement, 2 parts hydrated lime and 9 parts sand. Another acceptable O mix is 1 part masonry cement and 3 parts sand by volume.

MORTAR JOINTS

The size, texture and color of mortar joints affect the strength and water resistance of a masonry wall. The joint also enhances the appearance of the wall. A wide mortar joint in a contrasting color may be used to develop a pattern. A wide mortar joint cut back from the surface can be used to create shadow lines, Fig. 41-3.

The size of masonry joints depends on the type and size of masonry unit and on effect desired. The joint usually used in brick is 1/2 in. (12 mm). Joints may vary from 1/4 in. to 1 in. (6 mm to 25 mm) in width. However, joints over 1/2 in. (12 mm) are more difficult to lay, and they are more expensive.

MASONRY VENEERED WALLS

Veneering is a protective or ornamental facing used in conjunction with frame buildings. The wood frames carry the necessary loads, while the veneering is supported by the foundation and secured to the wood frames, Fig. 41-4. Masonry veneering is a means of obtaining the appearance of a solid brick wall. Other advantages include:

1. Economy of construction.
2. Better insulating qualities against moisture and extreme

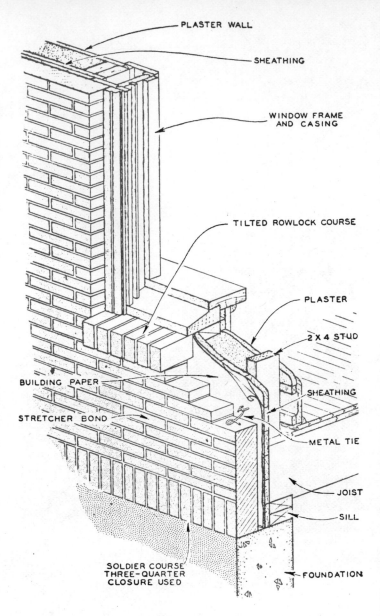

Fig. 41-4. A cross-sectional view of brick veneer over a wood frame structure.

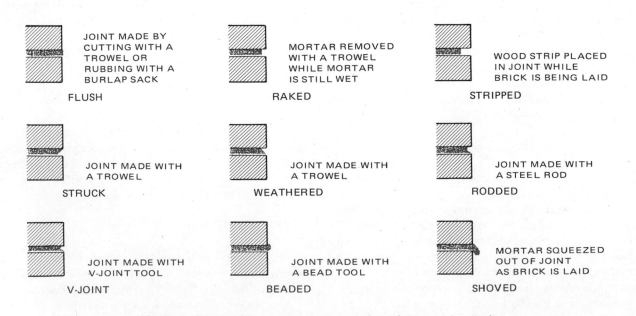

JOINT MADE BY CUTTING WITH A TROWEL OR RUBBING WITH A BURLAP SACK **FLUSH**	MORTAR REMOVED WITH A TROWEL WHILE MORTAR IS STILL WET **RAKED**	WOOD STRIP PLACED IN JOINT WHILE BRICK IS BEING LAID **STRIPPED**
JOINT MADE WITH A TROWEL **STRUCK**	JOINT MADE WITH A TROWEL **WEATHERED**	JOINT MADE WITH A STEEL ROD **RODDED**
JOINT MADE WITH V-JOINT TOOL **V-JOINT**	JOINT MADE WITH A BEAD TOOL **BEADED**	MORTAR SQUEEZED OUT OF JOINT AS BRICK IS LAID **SHOVED**

Fig. 41-3. Common joints used in masonry work are shown and described.

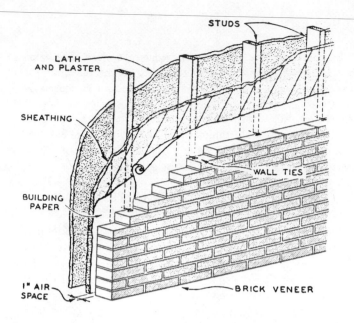

Fig. 41-5. Brick veneer is "tied into" studs of wood frame structure. Note ties.

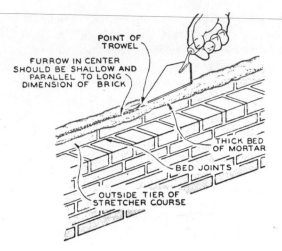

Fig. 41-7. Technique of furrowing a mortar bed helps to insure satisfactory bond between mortar and bricks.

temperatures.

Veneering is *not* simply an addition to the regular structure as a siding material. The masonry must be tied to the structure, Fig. 41-5, and one inch of air space must separate the sheathing of a wood frame structure and the masonry wall. For the general residential construction, brick veneer is used. Commercial buildings, apartment buildings and other larger structures are veneered over concrete blocks or poured concrete walls. See Fig. 41-6.

LAYING BRICK

Bricklaying is the act of setting small building units in a bed of mortar to form a strong, cohesive block or mass. The mortar in horizontal and vertical joints tends to tie all units together. The process of bricklaying is called bonding. Good bonding is accomplished by lapping one brick across or over at least two others in the course below it. This method of lapping is generally referred to as stretcher bond. See Fig. 41-4.

BEDDING JOINTS

Bedding joints (horizontal spread of mortar) are important from the standpoint of bonding bricks together, creating equal pressure throughout the wall. Mortar for bedding joints should be spread at least one-inch thick. Usually, the bricklayer will run the point of the trowel along the middle of the mortar bed to make a shallow furrow, Fig. 41-7.

It is not advisable to spread mortar more than the distance equal to the length of four or five bricks in advance of laying. This is especially true in hot and dry weather. By using this technique, the mortar remains soft and plastic, allowing the bricks to be laid (bedded) easily and properly, Fig. 41-8.

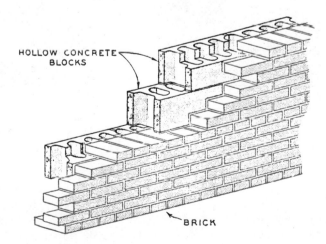

Fig. 41-6. Brick veneer over concrete block frame. Note that blocks are cut away to tie in with bricks.

Fig. 41-8. Bricklaying is being completed by a group of masons, using a stretcher bond. Note that scaffolding is maintained at a convenient height for the bricklayers. (Patent Scaffolding Co.)

Head joints (vertical spread of mortar), like bedding joints, must be completely filled with mortar. Plenty of mortar should be thrown on the end of each brick to be placed. The bricks should be layed in such a way that the mortar squeezes out between the joint. Excess mortar is scraped off by means of a trowel from the bottom edge or end of each brick, Fig. 41-9.

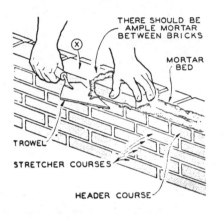

Fig. 41-9. Bricklayer pushes brick into bed and head joints until excess mortar squeezes out. Excess mortar is trimmed with the trowel.

KEEPING COURSES LEVEL

Constant care must be exercised in checking the level of each course. A spirit level is used to check the level of each corner, as well as the plumbness (vertical accuracy) of the rising corner. These checks are especially important because the corners serve a guide for the wall between. In order to lay the face wall, a cord or line is stretched between the corners to serve as a guide for each course. The line is placed in proper position for laying the first course between corners. Next, nails

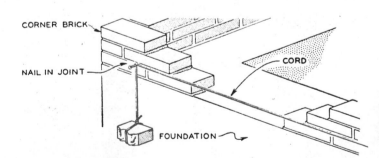

Fig. 41-10. Corners are worked first, then a line is strung between corners as a guide in laying the courses.

are driven into the mortar joints, Fig. 41-10. Then, the line is wound around these nails with a small weight attached to hold the line tight and in proper position. The cord should not touch the outer corner of the brick.

For each succeeding course, the line should be moved up one course. When the mortar can be indented with the thumb without too much pressure, the joints should be tooled.

CLEANING

Efflorescence is a soft white powder that often appears on the face of brickwork. It is particularly noticeable after the bricks have been exposed to water. This action is caused by salts (such as sodium or magnesium) in the brick or mortar, brought to the surface by the action of the water. Efflorescence can be removed from masonry walls with dilute muriatic acid, followed by a clean water rinse. Masonry supply firms also market a specially prepared rinse solution.

FIREPLACE CONSTRUCTION

A fireplace adds much to the attractiveness of a room and to the joy of living in a house, Fig. 41-11. Besides providing the pleasure of having an open fire, a fireplace can serve as a source of heat to supplement or substitute for normal means of heating. Materials in common use for facing of fireplaces include brick, marble, slate, stone, tile and glass. Wood trim and paneling may be used to complete the elevation design.

Fig. 41-11. A fireplace provides a hub for grouping furniture and creates a special interest area in the home.
(Western Wood Products Assoc.)

It is good planning to build the fireplace on an inside wall. This location reduces heat losses to the outside and keeps the chimney warm, thereby helping the fireplace to draw better. The fireplace may be built flush with the interior wall or project into the room. If it projects, a mantel can be installed and/or cabinets can be built alongside the fireplace.

BUILDING FIREPLACES

After the fireplace floor and trimmer arches have been completed, the jambs, end walls, fireplace side walls, and the vertical back wall are laid up at the same time, course by course, Fig. 41-12. Side walls and vertical back walls are laid up with firebricks (made of clays having a high percentage of alumina or silica and flint) to withstand the extreme heat. The inclined portion of the back wall is laid course by course with the other sections, so that it can be supported by the side walls.

The angle steel lintel (horizontal member spanning fireplace opening) should be placed where the jambs are up to the maximum opening height. Each end of the lintel should have a bearing surface of at least 3 in. (7.6 cm). The smoke shelf is made of mortar in the form of a curved or dished surface. Curved smoke shelves or smoke chambers help create an updraft in the chimney.

Mantels are made of brick, stone or wood and installed above the fireplace opening. Mantels may be surrounded by woodwork installed by a carpenter.

SUMMARY

Masonry is the bonding together of brick, block, stone or tile to construct walls, partitions, fireplaces, chimneys and other structural members. Masonry units must be placed in such a manner that they tie together to form a cohesive block or mass. The mortar in the horizontal bedding joints and vertical head joints tends to tie all bricks together. Unless the individual masonry unit is placed and bonded properly, a wall will not have enough strength to support heavy loads. The process of tying a masonry wall or unit together is called bonding. Good bonding is accomplished only by lapping one unit across or over at least two other units in the course below.

Generally, fireplaces are built of masonry material lined with firebrick to withstand the extreme heat.

CAREERS RELATED TO CONSTRUCTION

BRICKLAYER lays out work and calculates angles and courses for building brick walls, partitions, arches, fireplaces and ornamental brick work; mixes and spreads mortar; lays brick in prescribed alignment, using chalk lines, plumb bobs, tapes, squares and levels.

STONE MASON builds stone structures such as walls, fireplaces, piers, abutments and walks; shapes stone before setting; mixes and spreads mortar; sets and aligns stone in residences, office buildings, hotels, churches and restaurants.

TILE SETTER installs hollow clay tile walls, floors and walks, using mortar and following design specifications.

DISCUSSION TOPICS

1. What are the advantages and disadvantages of masonry construction?
2. What are the characteristics of a good masonry mortar?
3. What are the basic ingredients of a mortar?
4. List and identify common joints used in masonry work.
5. Explain masonry veneering.
6. What is bedding?
7. Describe how a bricklayer keeps each course of brick level.
8. What is efflorescence?
9. What is a lintel?
10. What is firebrick?

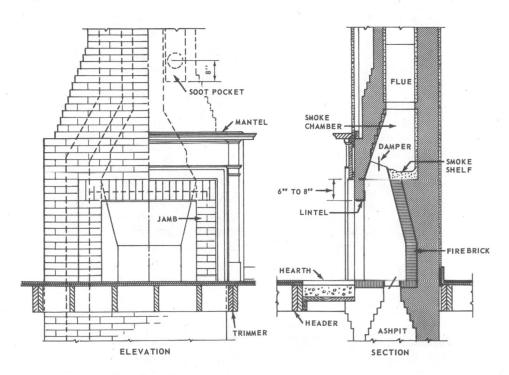

Fig. 41-12. Typical brick fireplace, illustrating details of construction.

Chapter 42

FLOOR FRAMING

Floor framing is the system of sills, girders, joists and subflooring installed in a wood-framed structure.

Floors are constructed in many different ways, using a variety of materials and techniques. Two types of floor surfaces are used in residential construction: concrete and wood. Houses without a basement usually have a concrete slab-on-earth subfloor. Houses with a basement generally have wood-framed subfloors and a concrete slab-on-earth basement floor.

Each type of floor surface receives a final covering by a variety of materials.

POSTS AND GIRDERS

Posts and girders are the basic structural members that support the floor joists (horizontal supporting members) along the center line of the house, Fig. 42-1. Posts and girders may be of wood, metal or a combination of both, Fig. 42-2. The house plan will specify post type and spacing; also girder span length and type. The long dimension of the house dictates the overall length of the girder. For example, a 48-ft. (15 m) long house will require three girder sections approximately 16-ft. (4.9 m) long.

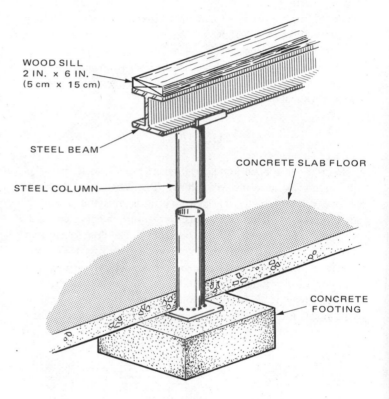

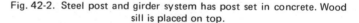

Fig. 42-2. Steel post and girder system has post set in concrete. Wood sill is placed on top.

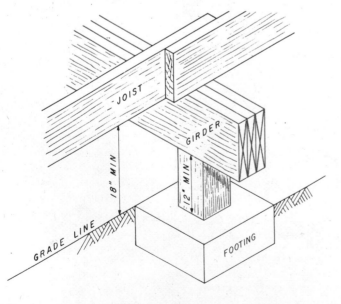

Fig. 42-1. Posts and girders are used to support floor joists at the center lap.

JOIST SYSTEMS

Joists are the main supporting members of the floor, Fig. 42-3. Each joist rests on the sill plate at the outside end and on the girder at the inner end. In residential construction, joists are 2-in. (5 mm) stock placed on edge.

The most common spacing for joists is 16-in. (41 cm) on center (OC). Spacing of 24-in. (61 cm) OC or wider also may be used.

Any wood joist having a slight bow edgewise should be placed so that the bow or crown is on top. A crowned joist will tend to straighten out when subfloor and finish floor loads are applied. Also, straight joists having large knots close to the edge should be positioned so that the knots will be located on the top edge. Joists are stronger in this position.

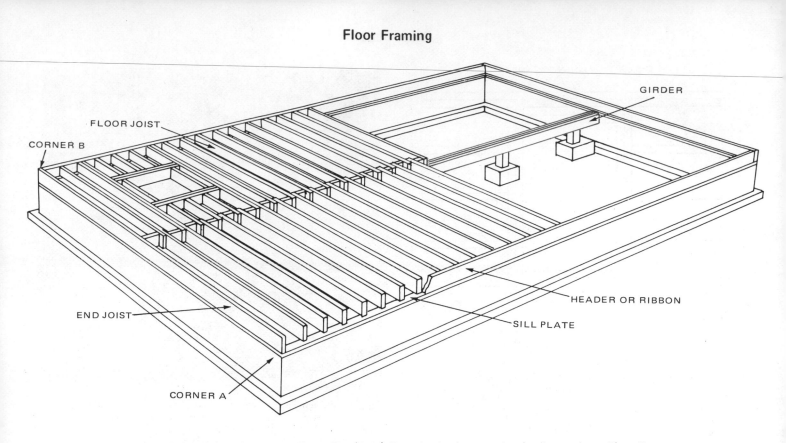

Fig. 42-3. Wood floor joists generally are 2-in. (5 cm) dimension lumber spanning the distance from sill to sill.

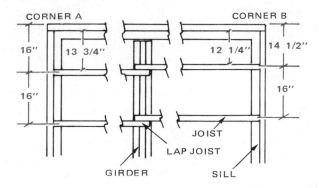

Fig. 42-4. Layout of joists. Note that 16-in. OC measurements between joists begin at starting edge, corner A to corner B.

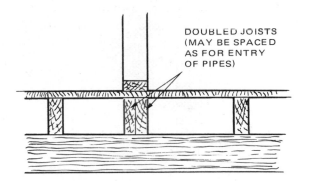

Fig. 42-5. Doubling joists helps support additional weight of a wall. Regular joist can serve as one of doubling joists.

JOIST LAYOUT

Floor joist positions are laid out by marking joist positions on the sill plates. Two marks are used, one for each side of the joist. The end joist location is marked on the sill, even with the outside edge of the sill, Fig. 42-4. Then, the location of the first interior joist is marked, so that spacing between joists is 13 3/4 in. (35 cm) and 16 in. (41 cm) from the outside of the sill to the center line of the first interior joist. Next, the positions of other joists are marked, spacing them 16-in. (41 cm) OC, continuing to the end of the house. In addition, the location of the end joist at this end of the house should be marked on the sill.

This marking process is repeated on the opposite sill, except that spacing between the end and the first interior joist is 12 1/4 in. (32 cm) and 14 1/2 in. (37 cm) from the outside of the wall to the center line of the first interior joist. This allows for joist overlap over the girder beam at the center of the house.

The floor plan should be checked for partition walls running parallel to the joists. At this point, a double joist generally is required to support the load of the partition, Fig. 42-5.

The location of any opening (stairwells, chimneys, etc.) should be marked. Regular joists are installed on each side of these openings. Special framing joists are left out until the regular joists have been placed, Fig. 42-6.

SETTING JOISTS

Floor joists should be selected from straight pieces of material. For example, No. 2 dense, yellow pine is excellent for floor joists to be used in residential construction.

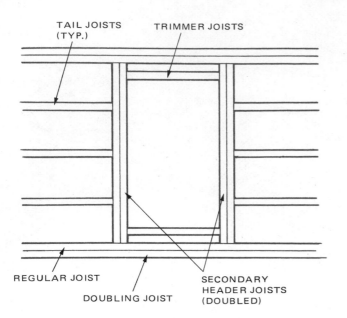

Fig. 42-6. Floor openings require special framing joists to maintain floor strength.

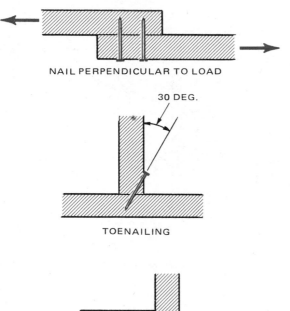

NAIL PERPENDICULAR TO LOAD

TOENAILING

NAIL IN WITHDRAWAL

Fig. 42-7. Various nail loading techniques are shown, each is designed to best support a given load.

The joists are cut and nailed in place, Fig. 42-7. Using 16d common or box nails, the header joist is toenailed into the sill, with the nails spaced about 16-in. (41 cm) apart. The corners of the header joist are installed with 16d nails, using a minimum of three nails.

By using a steel framing square, vertical alignment of each joist can be marked on the inside face of header joist. The

Fig. 42-8. Portable nailer can be used when working in a tight area where it would be difficult to swing a hammer. (Duo-Fast Fastener Corp.)

horizontal lines on the sill are continued to the inside face of the header.

The joists are cut to length, if necessary. Lap of the joists over the center girder should be a minimum of 4 in. (10 cm), but no more than 28 in. (65 cm).

The joists are placed at the marks and a minimum of two 16d nails are driven into the end of each floor joist from the outside of the header joist. Then, the joists are toenailed to the girder, using 16d nails. The lapped joists are nailed together over the girder, using two or three 10d common or box nails. See Fig. 42-8.

JOIST SPANS

Safe spans for joists under live loads are indicated in Fig. 42-9. Joists must be strong enough to carry the load that rests on them and stiff enough to prevent eventual bending or vibration.

Most building codes specify that the deflection (downward bending of joist) must not exceed 1/360th of the span under normal live load, Fig. 42-10. Normal load is considered to be 40 lb. per sq. ft. (195 kg per m^2). This load would result in a deflection of 1/2 in. (1.27 cm) for a 15-ft. (4.6 m) span.

BRIDGING

Bridging is a stiffening process designed to eliminate lateral (side) deflection of the floor joists. Bridging consists of rows of small diagonal braces nailed between the regular floor joists, Fig. 42-11. These braces distribute any load on the floor over a wide area. This places the load on many joists, rather than one or two. In addition, briding stiffens the floor and tends to keep the joists in line by preventing warping.

NOMINAL SIZE		SPACING OC		40 lb./sq. ft. (195 kg/m^2) LIVE LOAD GRADE NO. 2 DENSE SPAN		30 lb./sq. ft. (146 kg/m^2) LIVE LOAD GRADE NO. 2 DENSE SPAN	
In.	cm	In.	cm	Ft.	m	Ft.	m
2 x 6	5 x 15	12	30	11—5	3.5	12—5	3.8
		16	41	10—5	3.2	11—4	3.4
		24	61	8—11	2.7	10—0	3.1
2 x 8	5 x 20	12	30	14—9	4.5	16—1	4.9
		16	41	13—6	4.1	14—9	4.5
		24	61	11—11	3.6	13—1	4.0
2 x 10	5 x 25	12	30	18—3	5.6	19—11	6.1
		16	41	16—9	5.1	18—3	5.6
		24	61	14—10	4.5	16—2	4.9
2 x 12	5 x 30	12	30	21—9	6.6	23—9	7.2
		16	41	19—11	6.1	21—9	6.6
		24	61	17—7	5.4	19—2	5.8
3 x 8	7.5 x 20	12	30	17—0	5.2	18—7	5.6
		16	41	15—7	4.8	17—0	5.2
		24	61	13—10	4.2	15—1	4.6
3 x 10	7.5 x 25	12	30	21—1	6.4	23—1	7.0
		16	41	19—4	5.9	21—2	6.4
		24	61	17—1	5.2	18—8	5.7

PERMISSIBLE SPANS FOR SOUTHERN PINE FLOOR JOISTS

Fig. 42-9. Permissible spans are charted for southern pine floor joists.

Fig. 42-10. Worker levels a steel joist system by adjusting individual screw jack supports. (Dickson-Basford, Inc)

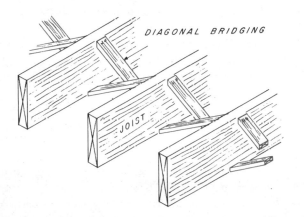

Fig. 42-11. Diagonal or herringbone bridging between joists provides stiffening effect.

Bridging should be installed in rows from one side of the house to the other. Normally these rows are spaced less than 10-ft. (3 m) apart. Typically, bridging members are cut from 1-in. by 4-in. (25 mm by 100 mm) stock.

Another good method of bridging is called "solid bridging," Fig. 42-12. Solid bridging is made by cutting full size pieces

Fig. 42-12. Solid bridging lends added strength to floor framing.

(same dimension as floor joist stock) and nailing them between the regular joists. This type of bridging may be used for stiffening joists that must carry above-normal loads. Solid bridging, for example, would be used in framing the floor to support a floor overlayment of heavy tile set in a concrete bed.

The subfloor should be installed before the bridging is nailed tightly in place. Usually, the top ends of the bridging are nailed to the joists before the subfloor is laid. The lower ends are fastened *after* the joists have adjusted to the subfloor.

Fig. 42-13. Plywood subfloor is installed with a coil nailer, using 2 3/8-in. ring shank nails for added holding power. (Duo-Fast Fastener Corp.)

SUBFLOOR

Today, most subflooring is plywood. However, solid wood may be used. The subfloor forms a working surface over the floor joists to serve as a platform, or base, for finish flooring, Fig. 42-13. In this example, a two-layer floor system is assumed. The subfloor is made of plywood and a separate layer of underlayment, then the floor covering is added.

Plywood makes an excellent material for subflooring for these reasons:
1. Convenient sheet size.
2. Ease of application.
3. Smoothness and uniformity of finished surface.

Plywood can be obtained in a number of grades and thicknesses suitable for use as subflooring. However, C-D interior grade plywood is commonly used. Panels of this grade carry an identification index marking on the back that gives allowable spacing of joists for various thicknesses of plywood.

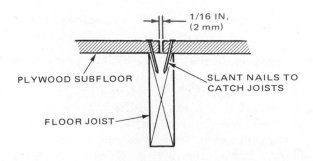

Fig. 42-14. Plywood subflooring must have clearance between sheets to allow for expansion. Space also helps to eliminate squeaking of subfloor.

Identification index 32/16 indicates that support spacing of 32 in. is allowable for roof rafters, 16 in. for floor joists.

When applying plywood subflooring, the sheet is placed on the joists so that the C grade (better quality) is up. C-D grade plywood has C grade veneer on the face and D grade on the back.

Plywood panels should be spaced 1/8-in. (3 mm) apart at edge joints and 1/16-in. (1.5 mm) apart at end joints over joists, Fig. 42-14. In wet or humid areas, this spacing must be doubled.

SUMMARY

Floor framing consists of sills, girders, joists and subflooring. All members are tied together to support the loads expected on the floor and to give lateral support to the exterior wall.

Sills which rest on continuous foundation walls usually consist of one thickness of nominal 2-in. (5 cm) lumber. The sills are positioned on the walls to provide full and even bearing. Usually, they are anchored to the foundation walls with 1/2-in. (12 mm) bolts spaced approximately 8-ft. (2.4 m) apart.

Joists should have a bearing surface on the sill of at least 1 1/2 in. (38 mm). Wood joists should be placed so that the top edge provides an even plane for installation of the subfloor and finished floor.

Under normal conditions, subflooring that is properly nailed to the joists will prevent lateral movement of the top edges of the joists. However, to prevent the joist from warping or turning over, bridging is installed. Bridging distributes the load evenly over several joists.

CAREERS RELATED TO CONSTRUCTION

FRAMING CARPENTER studies building plans; selects specified lumber or other materials; installs posts and girders, sills, joists, bridging and subfloor; erects wood framework.

DISCUSSION TOPICS

1. What is meant by "OC"?
2. What is a header joist?
3. Sketch the location of a sill as it relates to the foundation, joists, and header joists.
4. If a board to be used as a joist has a bow, how should it be installed?
5. Where is the first interior joist located?
6. What size of nails are needed to secure joists to the girder and header?
7. What is meant by live load?
8. What is the maximum amount of downward deflection allowed by most building codes?
9. Describe the application of subflooring.
10. What is the purpose of bridging?

Fig. 43-1. Certain basic shapes of homes have become generally accepted and are often known by names associated with regions where they were first developed. From top: Cape Cod, Ranch, Colonial, and French Provincial.

Chapter 43

WALL FRAMING

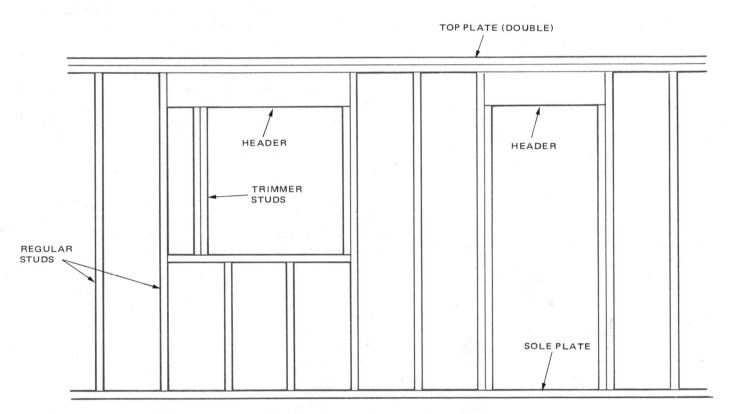

Fig. 43-2. Wall framing is the "skeleton" of the wall. It supports insulation, dry wall and other materials attached to it.

Humans first lived in natural caves or rock ledge overhangs. As they learned to use tools and make shelters they left the caves and moved about from place to place. They carried with them moveable houses made of animal hides stretched over a wood frame.

But when they began to grow food they were forced to build more permanent housing. Building materials were logs from trees, stone, straw, mud and any material which the local region would yield.

Today, wall framing is very often done in wood. The frame is covered with a variety of materials. Some are natural materials. Some are manufactured.

The American home buyer can choose from many styles. See Fig. 43-1. *A house style is the outward appearance of the structure. Usually a style is a traditional way of constructing homes in a region.*

The Cape Cod region of Massachusetts developed the small white clapboard cottage of one or two stories which has become known as the Cape Cod Colonial cottage. In open prairie areas people built low, long houses. Today we call that style the ranch house.

The style and character of a house is set largely by the wall framing and the shape of the roof. In this unit we are going to talk about how we assemble the wall frame.

PLATE AND STUD SYSTEM

Wall framing is a term that includes the vertical members called studs, and the horizontal members called plates. The bottom plate next to the subfloor is called the sole plate. The one at the top is the top plate. Both exterior walls and interior partitions are constructed alike, Fig. 43-2.

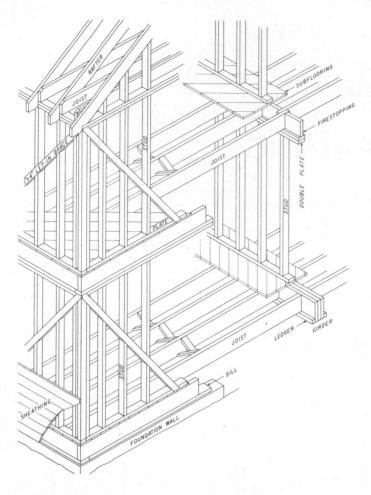

Fig. 43-3. Skeleton of a house built by the platform frame method.

PLATFORM FRAMING

The platform method is the most common type of framing. *In platform construction, the subfloor extends to the outside edges of the building and provides a base.* Exterior walls and interior partitions are erected on this base as shown in Fig. 43-3.

Platform construction is most generally used in one-story houses. However, it is also the method most used in units having many stories. See Fig. 43-4. In some parts of the country almost all residential building uses the platform system.

Platform construction is easier. It provides a flat surface at each floor level on which to work. It is also most easily adapted to prefabrication. (In prefabrication, whole sections of a building are made up in a factory.)

With a platform framing system, it is common practice to assemble the wall frame on the floor. Then the entire unit is tilted up into place.

Precut and prefabricated houses normally follow the platform framing system of construction. The precut house is one where most of the material is cut to size at a central manufacturing plant. It is then transported by truck to the building site. Carpenters erect it piece by piece. But they do little, if any, cutting.

In the prefabricated house the major sections are constructed at a factory. The modules or sections are then moved to the building site and assembled. Some housing contractors use both methods along with the cutting and sizing of stock pieces at the site.

BALLOON FRAMING

In balloon framing both studs and first floor joists rest on the sill plate, Fig. 43-5. But the second floor joists rest on a 1 by 4-in. ribbon which has been let (notched) into the inside edges of the studs.

Material for wall framing is generally of 2 by 4 in. (5 cm by 10 cm) lumber. The support headers, over the windows and doors are usually heavier material. Headers for windows are normally 2 by 6 in. (5 cm by 15 cm) stock or larger. Two are nailed together for strength.

Fig. 43-4. Wood framed four-story apartment building uses the platform method.   (Western Wood Products Assoc.)

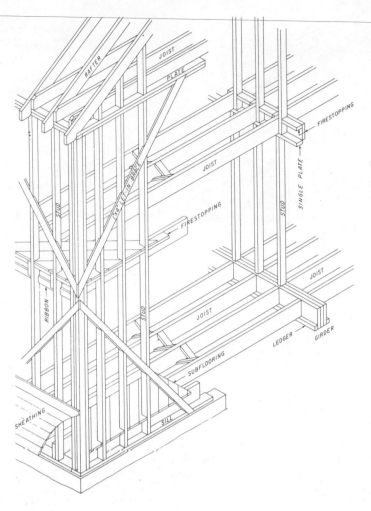

Fig. 43-5. Balloon frame construction. Compare it with Fig. 43-3. See how studs go all the way from sill to roof.

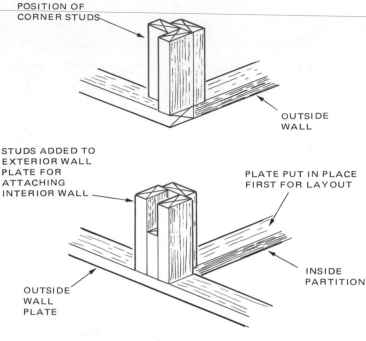

Fig. 43-6. Special framing is necessary for corners and mid-wall intersections. Extra studs must be placed at such points so that there will be support for nailing on drywall or lath.

Balloon framing is preferred for two-story buildings where the exterior covering is of brick, stone, veneer, or stucco. There is less likelihood of movement between the wood framing and the masonry veneer. However, it is more costly to build this way. Then too, it is hard to get longer stock for studs. This method of framing is being used less often now.

FRAMING LAYOUT

Let us suppose that we are looking on as carpenters lay out and place the walls. They will refer often to the blueprints for the dwelling. First they will lay out and mark on the subfloor the location of all outside walls and inside partitions. Next they select 2 by 4 in. bottom plates for all outside walls and cut them to length. These plates will be nailed in place temporarily. The outside edge will be flush to the edge of the platform. Eight penny box nails are used since they permit easy withdrawal.

Their next task is to position all the bottom plates for partitions. These, too are nailed in place temporarily. Laying out the walls this way helps the builder identify all the wall intersections. This knowledge is needed to determine the special framing pieces necessary for corners or mid-wall

intersections. See Fig. 43-6.)

While the bottom plates are still nailed in the place the carpenters will mark the location of all major openings on all plates, Fig. 43-7. Then all stud locations are marked on the top

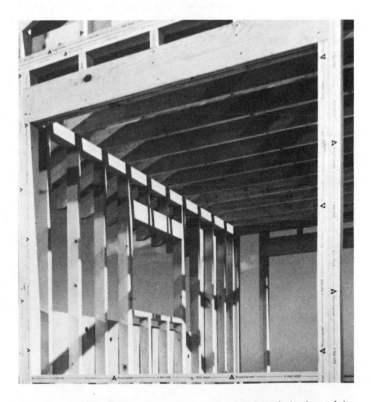

Fig. 43-7. Rough opening is made ready for window. It is about 1 in. larger than window unit. (Weyerhaeuser)

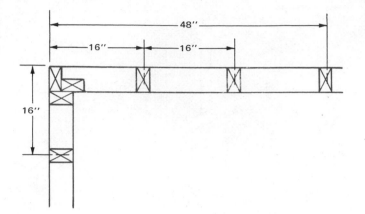

Fig. 43-8. Starting at the corner, all studs are layed out on 16-in. centers. Remember that the exterior sheathing is usually 48 x 96 in. Therefore, dimensions of the plywood sheet and the spacing of the studs come out even for nailing. Note different method of placing corner studs. Some carpenters prefer this to method shown in Fig. 43-6.

and bottom plates. Studs are spaced 16 in. apart on centers. See Fig. 43-8. (On centers means that the stud is centered over the pencil mark made at each 16-in. point.) Extra studs are needed at corners and partition intersections. Extras are also required to carry special loads at door openings. These are noted and marked also.

After all bottom plates are marked, carpenter will cut a second set of plates exactly like the first. These will be marked exactly like the bottom plates. These pieces will become the lower members of the double top plate.

Now all wall height stud members are prepared. To the floor-to-ceiling height dimension the carpenters will add about 1 in. for floor underlayment (plywood) and ceiling material thickness. They will subtract 4 1/2 in. for three 2 by 4-in. plate thicknesses. Usually precut studs can be bought for standard heights of ceilings. Finally, all trimmers, sills, cripples and headers are cut, Fig. 43-9.

Assembly of long-wall, outside sections is about to begin. Workers move the interior partition plates out of the way to clear the floor area for a working space. They turn the bottom plate on edge, Fig. 43-10, and toenail it occasionally to the subfloor. This temporarily holds it in place while they build the wall frame section. They move the lower top plate into position and distribute the regular studs into position.

Plates and studs are fastened with 16 penny common or box nails, Fig. 43-11. Studs for special wall intersections are nailed into the frame at this time. Next to be added are all cripples, headers, rough sills and trimmers. See Fig. 43-11.

Before raising the wall or attaching sheathing the workers will check for squareness. If the wall frame is square, the two diagonal measurements from corner to corner will be equal in length. Temporary bracing may be nailed on to hold it square until sheathing is applied.

SETTING UP WALL FRAMES

When wall sections are ready for raising, carpenters cut several 1 by 4 in. or 1 by 6 in. braces. Once the wall is upright,

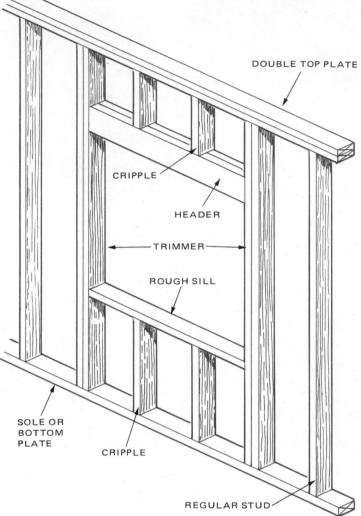

Fig. 43-9. Typical section of exterior wall with window opening.

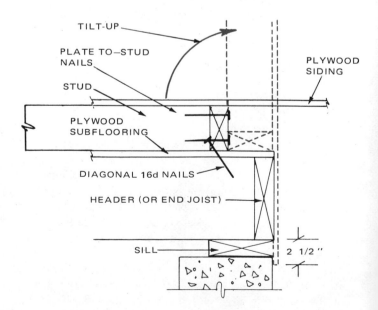

Fig. 43-10. Wall section, built laying on the subfloor will be tilted upward and positioned easily.

SPAN		HEADER SIZE	
ENGLISH	METRIC	ENGLISH	METRIC
3 1/2 FT.	106.6 cm	2 x 6 IN.	5 x 15 cm
5 FT.	152 cm	2 x 8 IN.	5 x 20 cm
6 1/2 FT.	198 cm	2 x 10 IN.	5 x 25 cm
8 FT.	243.8 cm	2 x 12 IN.	5 x 30 cm

Fig. 43-11. Chart gives correct sizes of headers to carry weight of building across window openings.

Fig. 43-13. Carpenters tilt up wall partition. Note placement of top plate and door openings.   (Weyerhaeuser)

these braces can be used to attach the wall framing temporarily to the subfloor. See Fig. 43-12. This prevents the wall from falling over while the next section is being raised.

The carpenters erect the longer exterior wall first. They then build interior partitions and raise them into place, Fig. 43-13. When in position they are nailed to the subfloor.

Interior partitions are laid out exactly as were the exterior walls, except no sheathing is attached. Interior sections are diagonally braced and nailed to the intersection stud of the intersecting wall. (See detail in Fig. 43-6.)

When workers have all wall sections in place, a second top plate is applied as shown in Fig. 43-14. Carpenters will make sure the joints in the two layers are offset. This insures a proper tie for all intersecting walls so they will not pull apart later. It also makes the wall stronger.

PLUMBING THE WALL

Immediately after the wall section is in a vertical position and diagonal braces are in place, final adjustments are made in the position of the sole plate. It has to be straight before it is nailed to the floor frame with 20 penny common or box nails. These are driven through the subfloor and into the floor joists.

Then workers loosen the braces one at a time and carefully plumb (position perfectly upright) the corners and mid-wall.

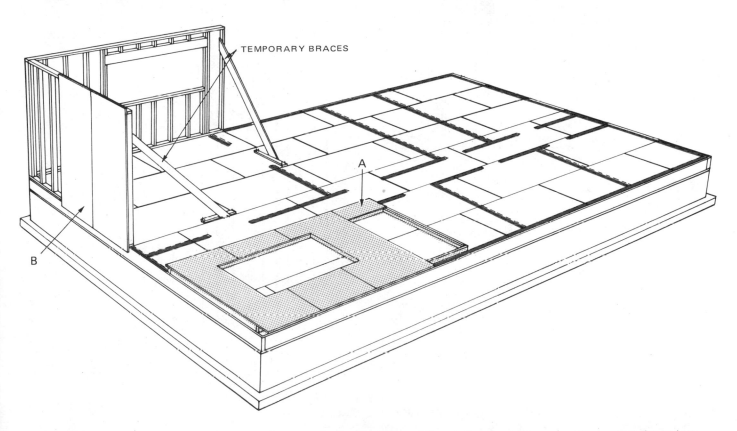

TEMPORARY BRACES

A

B

Fig. 43-12. Newly raised wall sections are braced temporarily so they will not fall over. Note brown lines on subfloor. They show where other walls and partitions are yet to be placed. A—Completed wall section ready for raising. Sometimes carpenters attach sheathing before raising the section. B—Sheathing attached to wall section after raising.

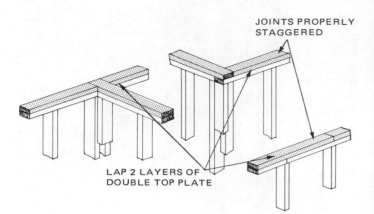

Fig. 43-14. Method of lapping second top plate to lock all wall sections together.

Fig. 43-16. Carpenter secures studs to sole plate swiftly with this pneumatic hammer. (Duo-Fast Fastener Corp.)

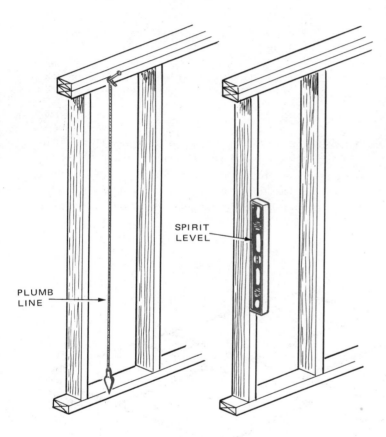

Fig. 43-15. Methods of plumbing a wall.

This can be done with a plumb line or carpenter's spirit level, Fig. 43-15. After the wall is plumb they will again secure the temporary diagonal bracing. If they have not done so before this point, the workers will toenail all studs. This job can also be done with a power nailer as in Fig. 43-16.

SHEATHING

Conventional exterior wall sheathing covering consists of a sheathing material. This is overlaid with siding. Sheet material in both layers provides strength, stiffness and durability, Fig. 43-17. Siding provides a pleasing appearance, Fig. 43-18.

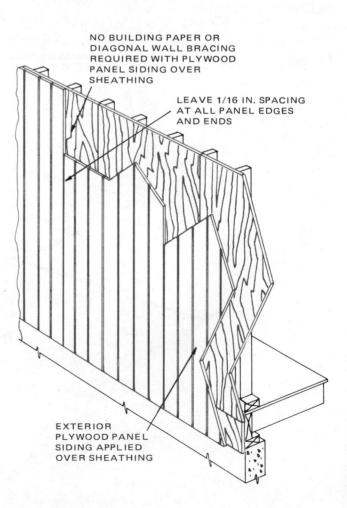

Fig. 43-17. A typical wall sheathing with siding covering the sheathing.

Fig. 43-18. Siding, the finish wall covering for the exterior side, looks nice and protects wall from weather.
(Western Wood Products Assoc.)

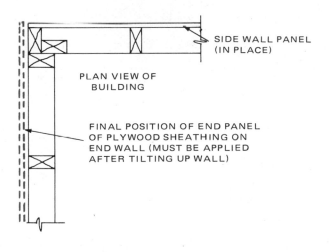

SIDE WALL PANEL
(IN PLACE)

PLAN VIEW OF
BUILDING

FINAL POSITION OF END PANEL
OF PLYWOOD SHEATHING ON
END WALL (MUST BE APPLIED
AFTER TILTING UP WALL)

Fig. 43-19. Corner section of sheathing is nailed on last to finish out
the corner.

Sheathing is nailed to the wall frame with 6 penny common or box nails spaced about 6 in. apart at the panel edges and about 12 in. apart over the intermediate studs.

When the longer side walls are sheathed the carpenters may apply sheathing to the shorter end sections before tilting them up. They place the first panel at the end (corner) and temporarily nail it into place as in Fig. 43-19. They will allow the first panel to overhang the framing on the end. This overhang is measured so the panel will cover the end of the side wall when the wall is in place. All remaining panels will be nailed down to the wall frame. But the last panel at the other end of wall will be left off.

Then the carpenters will remove the temporarily fastened panel from the wall. The open framing at either end of the wall allows the end wall sections to be tilted up past the side walls.

After the wall is tilted into position and nailed to the side walls and floor, the corner panels are nailed on to complete the sheathing operation.

SUMMARY

House framing started as a means of satisfying a need for shelter. The American home of today is usually wood framed. Simply put, it consists of pieces of wood covered with a skin of plywood. This frame is generally layed out on the subflooring. All wall openings, windows and doors are positioned by a special framing technique. Included in this framing are the header, trimmers, rough sill and cripples. After each of the wall sections are assembled, they are erected by tilting up and then nailed to the floor joist. Secondly, the interior partitions are assembled and erected. Sheathing is generally applied to the exterior wall frame before tilting up.

CAREERS RELATED TO CONSTRUCTION

FRAMING CARPENTER OR ROUGH CARPENTER does the framing work on home construction. Work includes installing studs, joists, sheathing, subflooring, partitions and rafters. Will also build forms for concrete work.

DISCUSSION TOPICS

1. What were the first type of constructed shelters like?
2. What is a house style? List as many styles as you can.

3. How does platform framing differ from balloon framing?
4. Describe the difference between site assembly, precut and prefabricated methods of construction.
5. What does the abbreviation, OC, stand for?
6. What is the standard ceiling height?
7. What are headers, trimmers, rough sills and cripples?

8. What are the steps for building and erecting wall and partition sections?
9. What is meant by "plumb"? Explain how a wall section is plumbed.
10. Sketch the method by which wall intersections are constructed.

Chapter 44

CEILING FRAMING

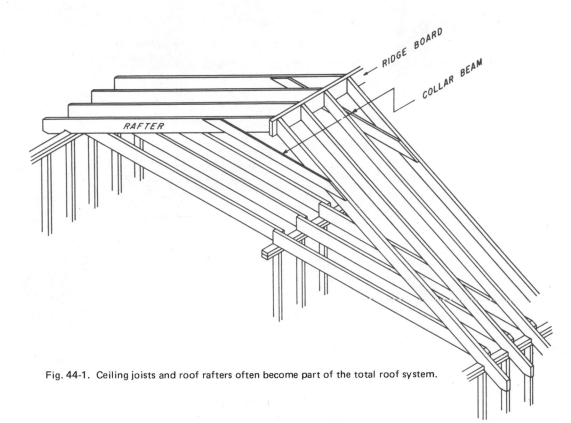

Fig. 44-1. Ceiling joists and roof rafters often become part of the total roof system.

Ceiling framing is the system of top plates, joists and ceiling support members installed in a wood-framed structure.

Ceiling framing has three main purposes:

1. Ties together opposite walls and roof rafters to resist outward pressure on walls imposed by pitched roof, Fig. 44-1.
2. Supports finished ceiling, Fig. 44-2.
3. Supports second story floor or attic storage area.

Basic ceiling framing construction is similar to floor framing. However, individual members are lighter, and header joists are not included around the outside of the ceiling joists. Smaller live loads are carried by the joists. Therefore, ceiling joist load capacity does not need to be as great as floor joist capacity.

The joists are the main ceiling framing members. As in floor framing, joist size is determined by spacing and length of span, Fig. 44-3. Like all house framing, ceiling framing is 2-in. (5 cm) stock. Local building codes must be checked for minimum specification requirements.

Ceiling joists and roof rafters should be *laid out* at the same time. They should be equally spaced.

The joists are *installed* before the rafters to provide a working platform for building the roof. Ceiling joists and roof rafters must lap, and the rafters must be securely fastened to the joists. It is much easier to use the same spacing for joists and rafters, than to stagger the joists.

With plywood sheathing and siding and the double top plate, it is not necessary to position the joists directly over the wall studs. In some cases, wall studs may be spaced 16-in. (41 cm) OC, while the ceiling joists may be located on 24-in. (61 cm) centers.

If the load-bearing interior wall is not continuous, a beam is needed to carry the inside ends of the ceiling joists.

CEILING JOIST LAYOUT

The position of the ceiling joists and roof rafters should be laid out in approximately the same manner as in floor joist layout. Each location is marked on both outside plates and the interior wall partitions. At this point, the two rafters on opposite sides of the house will frame opposite each other at the center. However, the ceiling joists will lap the rafters, so an

303

Fig. 44-2. Ceiling framing provides surface support for finished ceiling and its various decorative systems. Ceiling uses a grid of wood mouldings that is coordinated with wall mouldings. (Western Wood Moulding and Millwork Assoc.)

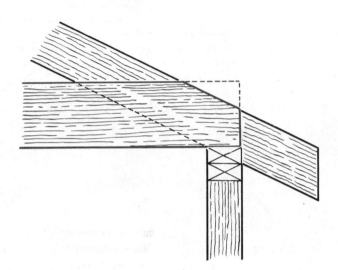

Fig. 44-4. Use of filler blocks will position ceiling joist to allow roof rafters to run in direct alignment.

Fig. 44-5. Corner of ceiling joist must be cut to match slope of rafter.

alignment problem exists.

The easiest method for making the various framing members line up is to use a filler block over the interior bearing partition, Fig. 44-4. Then, the two ceiling joists meeting over the center partition can be placed on each side of the roof rafter. The fetter block occupies the same space as a roof rafter at the center point.

CUTTING CEILING JOISTS

The corners of the ceiling joists at the outer walls must be trimmed to match the rafter slope, Fig. 44-5. The ceiling joists

NOMINAL SIZE		SPACING OC		20 lb./sq. ft. (98 kg/m²) LIVE LOAD GRADE NO. 2 DENSE		NO ATTIC STORAGE GRADE NO. 2 DENSE	
In.	cm	In.	cm	Ft.	m	Ft.	m
2 x 4	5 x 10	12	30	9–5	2.9	11–10	3.6
		16	41	8–7	2.6	10–9	3.3
		24	61	7–5	2.3	9–5	2.9
2 x 6	5 x 15	12	30	14–7	4.5	18–5	5.6
		16	41	13–3	4.0	16–9	5.2
		24	61	11–6	3.5	14–7	4.5
2 x 8	5 x 20	12	30	19–6	5.9	24–7	7.5
		16	41	17–9	5.4	22–4	6.4
		24	61	15–5	4.7	19–6	5.9
2 x 10	5 x 25	12	30	24–9	7.6	31–2	9.5
		16	41	22–6	6.9	28–3	8.6
		24	61	19–6	5.9	24–9	7.6

PERMISSIBLE SPANS FOR SOUTHERN PINE CEILING JOISTS

Fig. 44-3. Permissible spans are given for southern pine ceiling joists.

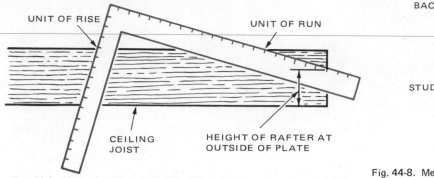

Fig. 44-6. Method is illustrated for using a framing square to establish slope of rafter.

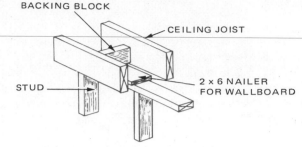

Fig. 44-8. Method is shown for fastening ceiling joists and wall partitions which run parallel to each other.

must be cut to proper length, with the outer end flush with the outside of the wall. At least a 4-in. (10 cm) overlay should be allowed.

To lay out the pattern for the slope cut, a framing square must be used to establish the rise and run, Fig. 44-6. If this corner is left on low-pitched roofs, it will interfere with the roof sheathing.

INSTALLING CEILING JOISTS

Ceiling joists are toenailed to the top plate of the exterior walls with two 10d common or box nails on each side. At the center lap, the joists are nailed to the fetter block, then toenailed to the plate of the load-bearing wall. Since the ceiling joists supply the tie across the building, they must be well nailed. If an individual joist has a bow in it, the crown of the bow should be placed on top.

Any attic access opening or other opening must be included in the ceiling frame to provide sufficient structural strength. Fire regulations and local building codes usually list minimum size requirements for an attic access.

The architectural plan usually includes access opening size, description and location. The opening is framed in much the same manner as a floor opening, Fig. 44-7. Doubling of the headers and joists may or may not be required.

ANCHORING PARTITIONS TO CEILING JOISTS

Ceiling joists give the builder a good opportunity to anchor the top of the wall partition in place. The joists also prevent any lateral (side) movement of walls or long interior partitions. Where the ceiling crosses the partition, two 10d common or box nails must be toenailed into the double plate.

Ceiling joists and non-load bearing partitions which run parallel are harder to anchor together. However, by doing some minor framing, they also may be secured, Fig. 44-8. A 2-in. by 4-in. (5 cm by 10 cm) backing block is nailed between

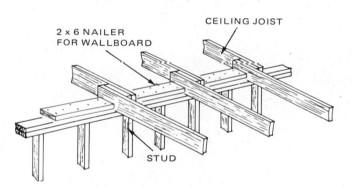

Fig. 44-9. A "nailer" is installed above each double plate of non-load bearing wall to provide for later fastening of interior ceiling finishing materials.

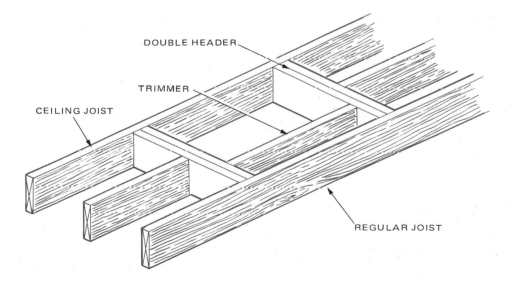

Fig. 44-7. Framing an attic access opening calls for changes in basic ceiling framing system.

the ceiling joists, running across the double plate which, in turn, is toenailed to the double plate.

Finally, a 2-in. by 6-in. (5 cm by 15 cm) nailer is nailed on top of each non-load bearing wall, Fig. 44-9. This board will extend off both edges of the double plate to provide a nailing ledge for fastening ceiling finishing materials.

PLANK AND BEAM FRAMING

The plank and beam (or post and beam) system of construction has been used for many years in the construction of public and commercial buildings. The adaptation of this system to residential construction has given designers and builders great new ideas when it comes to the placement of interior wall partitions, Fig. 44-10.

Fig. 44-10. Exposed beams recall early building techniques when native materials were used for homes on the frontier. (Koppers Co., Inc.)

The plank and beam system of framing offers many advantages:
1. Distinctive architectural effect, Fig. 44-11.
2. Added height in living areas.
3. Serves as finished ceiling when roof planks are selected for appearance, Fig. 44-12.
4. No further ceiling treatment is needed, except for the application of stain, sealer or paint.
5. Savings in cost of construction.

Well planned plank and beam framing also permits substantial savings in labor. The pieces are larger and fewer than in conventional framing, Fig. 44-13. Cross bridging of joists is eliminated. Larger and fewer nails are required.

Locating the electrical distribution system may present a problem because of the lack of concealing spaces in the ceiling.

Fig. 44-11. Beams eliminate the need for ceiling joists. Finished ceiling is underside of roof planks. (California Redwood Assoc.)

However, this may be resolved by using "spaced" beams, Fig. 44-14. Spaced beams are made of several pieces of 2-in. (5 cm) stock separated by short blocking. This setup provides a space for the electrical system, plumbing pipes and other mechanical services.

CONSTRUCTION DETAILS

The plank and beam system is essentially a skeleton framework. Planks are designed to support a moderate load, uniformly distributed. This load is carried to the beams which, in turn, transmit their loads to posts supported by the foundation, Fig. 44-15. Where heavy concentrated loads occur in places other than over main beams or posts, supplemental beams are needed.

SUMMARY

Ceiling framing is closely associated with roof framing. To more fully understand ceiling-roof framing technique, study the next chapter.

The ceiling joist typically serves to:
1. Tie exterior walls together, preventing an outward movement caused by downward thrust of roof.
2. Help maintain the position of the interior wall positions.
3. Provide support for finished ceiling by making available ready nailers.
4. Provide floor joists for addition of second story or attic storage area.

Plank and beam system of construction eliminates the need for ceiling joists, thereby reducing building time and cost of construction. The roof planks become the finished ceiling.

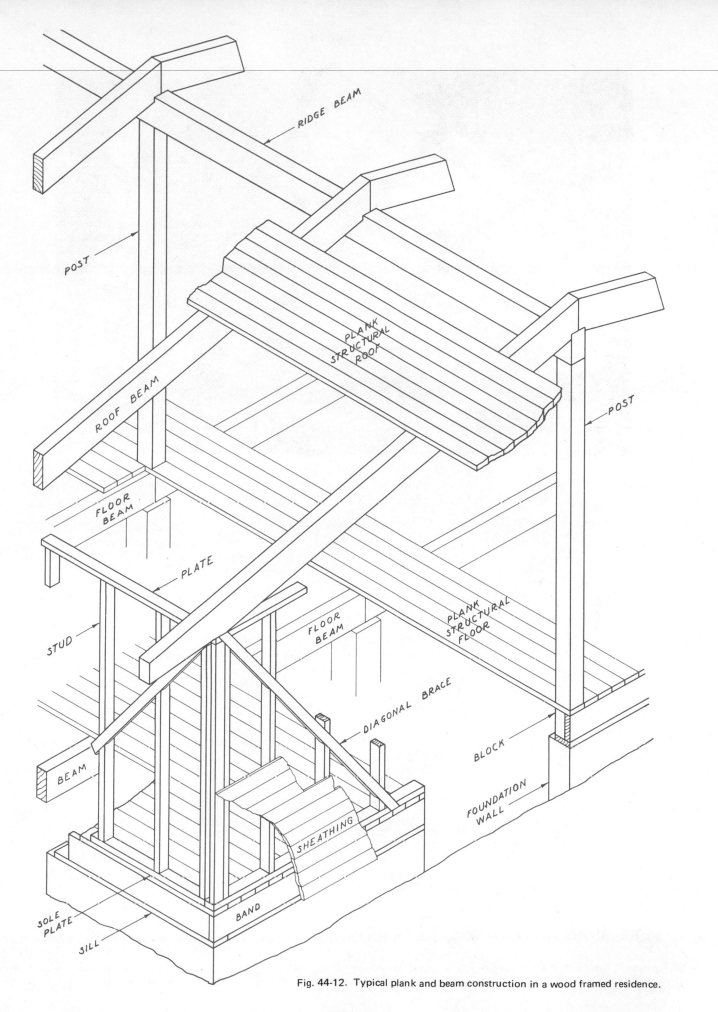

RIDGE BEAM

POST

ROOF BEAM

PLANK STRUCTURAL ROOF

POST

FLOOR BEAM

PLATE

STUD

FLOOR BEAM

PLANK STRUCTURAL FLOOR

DIAGONAL BRACE

BLOCK

BEAM

FOUNDATION WALL

SHEATHING

SOLE PLATE

SILL

BAND

Fig. 44-12. Typical plank and beam construction in a wood framed residence.

Fig. 44-13. Architectural styles vary greatly with plank and beam system of construction. Individual pieces are larger and spaced wider than in stud-plate method of construction.   (California Redwood Assoc.)

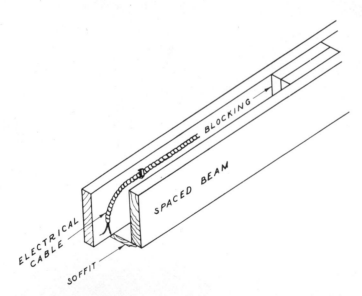

Fig. 44-14. "Spaced" beams provide a space for concealing electrical cables and plumbing pipes.

Construction materials are larger and spaced differently from those of the stud-plate system.

CAREERS RELATED TO CONSTRUCTION

GENERAL FRAMING CARPENTER erects wood framework in buildings; lays subflooring, builds stairs and installs

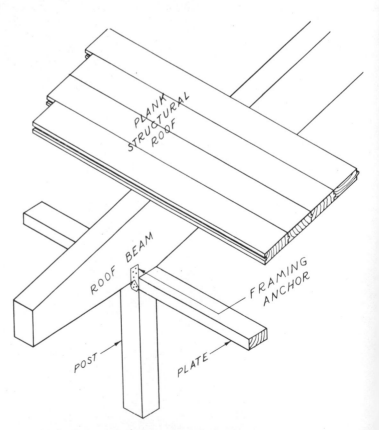

Fig. 44-15. Plank and beam system of construction transmits weight through well-placed posts.

partitions; erects scaffolding and temporary buildings at construction site; builds forms and chutes for pouring concrete.

CARPENTER'S APPRENTICE works at and learns various carpentry techniques in a four-year, on-the-job training program; completes a minimum of 144 hours of related classroom instruction per year; learns elementary structural design, laying out, framing and finishing.

DISCUSSION TOPICS

1. Why are ceiling joists necessary for the stud-plate system?
2. Why should the builder be concerned about ceiling joist size?
3. What is the purpose of filler blocks placed between the ceiling joist at the point of lap?
4. What does the slope of the roof have in common with the ceiling joist?
5. Describe "bow," and how to handle a bowed ceiling joist.
6. How and where would you make an attic access?
7. What is a backing block?
8. Why is a 2-in. by 6-in. (5 cm by 15 cm) nailer placed on top of the double plate of each non-load bearing partition?
9. What is the difference between a wall and a partition?
10. How does plank and beam construction differ from the stud-plate system?
11. How can ceiling joists be eliminated with the plank and beam system?

Chapter 45

ROOF FRAMING

Fig. 45-1. Flat roof of lightweight concrete is being poured over ribbed galvanized metal form. Top coating of hot asphalt and aggregate (gravel or stone) is applied to make watertight seal.   (Perlite Assoc.)

The roof of any building is a watertight skin. This skin covers the top of the building. Roofs may be flat or pitched (sloped).

Flat roofs usually are made of metal and concrete or plastic. The built-up type is most popular. In this type, layers of roofing are secured by layers of sticky bituminous sealant. Aggregate such as fine stone or gravel is bedded into a thick top layer of sealant. See Fig. 45-1.

The pitched roof may be covered with metal sheets, rolls of asphalt paper or shingles. Shingles are made of wood or asphalt, Fig. 45-2 shows three types of roofs.

Roofs are usually framed and finished before exterior walls are sided. This is because residential structures are usually made of wood. The roof goes on as quickly as possible to protect the wood from moisture.

Large high-rise buildings are exceptions to this rule. Since floors and walls may be of poured concrete, these are finished first. The roof goes on last.

GABLE ROOF   HIP ROOF   FLAT ROOF

Fig. 45-2. Basic roof types.

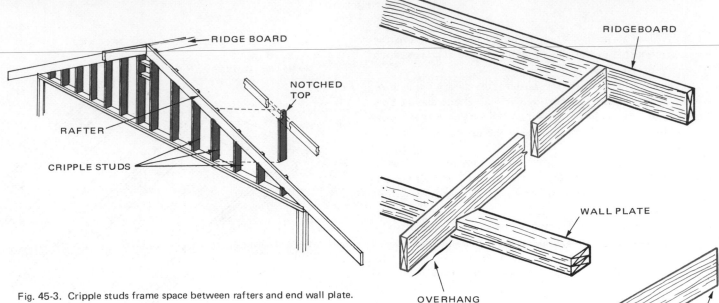

Fig. 45-3. Cripple studs frame space between rafters and end wall plate.

RAFTER SYSTEMS

Roof framing for the pitched or sloped roof includes several parts. They are the rafters, ridgeboard, collar beams and cripple studs. (Cripple studs are the short 2 by 4 in. pieces that run vertically from the end wall plate to the rafter. See Fig. 45-3.) Collar beams are shown in Fig. 45-11.

CUTTING THE COMMON RAFTER

In gable roof construction, all rafters are the same length and shape. They are all cut from one pattern. This pattern is

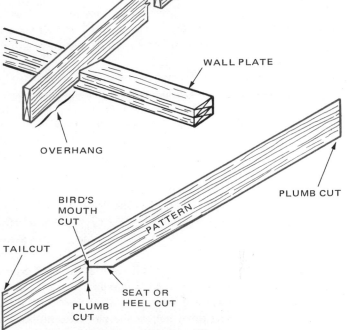

Fig. 45-4. Rafters should be carefully cut. Angles must be right for good fit at peak where rafter meets ridgeboard and at plate. Above. Sketch of properly fitted rafter shows name of cuts. Below. General shape of a pattern for a common rafter.

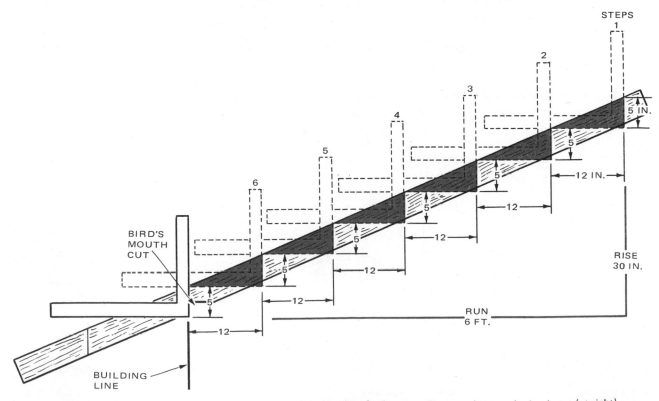

Fig. 45-5. Step-off method for finding rafter length and angles of rafter cuts. First step is to mark plumb cut (at right).

Construction

made full size as in Fig. 45-4. In producing the patterns the carpenters must find the exact length and the angles at which the cuts must be made.

In view A of Fig. 45-4, the various cuts are shown as they will fit at the plate and at the ridge. In view B you can see how the rafter pattern will look when all the cuts have been made. The pattern is made from a piece of the lumber intended for rafters.

Finding the length of a rafter can be done in several ways. All ways are based on one fact: You can find the length of the hypotenuse of a right angle triangle if you know the length of the other two sides.

A carpenter may use the step-off method, or length can be found by using tables on the side of the framing square.

Fig. 45-5 shows the step-off method. In this figure, the rafter has a slope of 5 in 12. This means that for every 12 in. of run the roof slopes or drops 5 in. In this illustration the run is 6 ft. The vertical distance from lowest point to highest is 30 in. This is called rise.

You will see that in Fig. 45-5 the slope per foot (5 in.) is marked on the short side (tongue) of the square and the run (1 ft.) is marked on the long side (blade). By stepping off the slope and run six times on the rafter's edge, the length is easily found.

Look again at Fig. 45-5. You will see that the angle on the tongue of the square is correct for the plumb cut at the peak. The angle of the blade is correct for half of the notch called the bird's mouth. The other half of the bird's mouth has the same angle as the plumb cut.

You can see how the step-off method works by noting that there are six brown triangles along the rafter in Fig. 45-5. They are miniatures of the rise and run. Each is one-sixth of the actual triangle described by the rise, run and rafter. Since the hypotenuse of each small triangle is marked right on the rafter, their sum must equal the length of the rafter.

In the third method, a mathematical formula is used:

Hypotenuse squared = altitude squared plus base squared

$$(H^2 = A^2 + B^2)$$

In geometry you will hear it explained as: The square of the hypotenuse of a right angle triangle is equal to the sum of the square of the other two sides.

In a rafter problem, the base is the run and the altitude is the rise. The rafter is the hypotenuse. See Fig. 45-6.

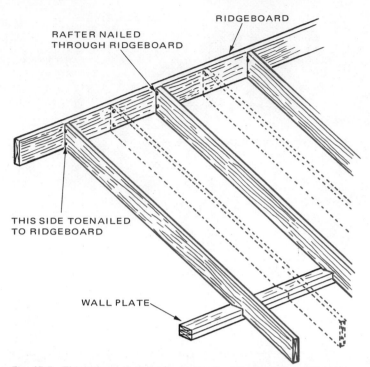

Fig. 45-7. This is how partially framed roof will look. A few rafters are shown nailed in place. Note that opposing rafters are directly across from each other. First rafter can be nailed through face of ridgeboard. Mating rafter must be toenailed.

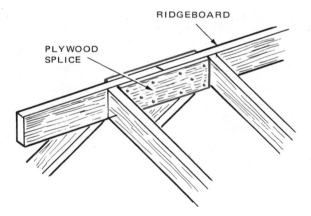

Fig. 45-8. Method of splicing ridgeboard by nailing on scabs.

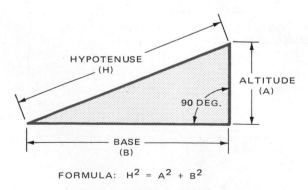

Fig. 45-6. Mathematical method of figuring rafter length is the same as figuring out the hypotenuse of a right angle triangle.

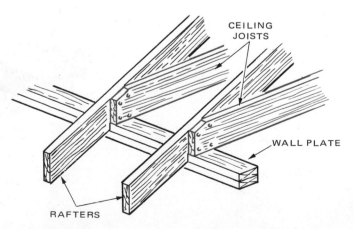

Fig. 45-9. Ceiling joists are nailed to rafter ends for added strength.

ATTACHING RAFTERS

Each pair of rafters is fastened at the top to a ridgeboard. A 2 by 8 in. board is commonly used with 2 by 6 in. rafters.

The ridgeboard supports the rafter ends. At the same time it keeps them evenly spaced and provides a better nailing surface. See Fig. 45-7.

An extra piece of ridgeboard is allowed to hang over the gable end. This is needed to support fascia rafters. (See Fig. 45-11.) On dwellings of any great size the ridgeboard must be spliced. The splice should be made between two rafters, as shown in Fig. 45-8.

When rafters are properly located there will be one alongside each ceiling joist. Rafters will be spaced the same as the joists, Fig. 45-9.

Workers will place the first pair of rafters flush with the outside edge of the end wall. The next rafter will be 16 or 24 in. (41 or 61 cm) away. Measurement is from the end of the building to the center of the rafter. All other rafters are located the same way. They will always be located at the side of each ceiling joist.

SELECTING THE RAFTER

A rafter must be capable of carrying the normal live load placed upon it. There is also the dead load of its own weight. As the length of the rafter increases, the total load has more effect upon it. See tables in Fig. 45-10.

The run of a rafter is the clear distance from the two points supporting the rafter. For the common rafter, this is from the exterior wall plate to the ridgeboard.

From the pattern they have made, the carpenters cut all rafters they will need to complete the house. In addition, they cut two pairs of fascia rafters for the ends of the ridgeboard forming the gable overhang, Fig. 45-11.

NOMINAL SIZE		SPACING OC		HEAVY LOAD 30 lb./sq. ft. (147 kg/m^2) LIVE LOAD GRADE NO. 2 DENSE		LIGHT LOAD 15 lb./sq. ft. (73.5 kg/m^2) LIVE LOAD GRADE NO. 2 DENSE	
In.	cm	In.	cm	Ft.	m	Ft.	m
		12	30	10–6	3.2	12–3	3.7
2 x 4	5 x 10	16	41	9–1	2.8	10–7	3.2
		24	61	7–5	2.3	8–8	2.6
		12	30	16–4	5.0	19–1	5.8
2 x 6	5 x 15	16	41	14–2	4.3	13–6	4.1
		24	61	11–6	3.5	13–6	4.1
		12	30	21–9	6.6	25–6	7.8
2 x 8	5 x 20	16	41	18–10	5.7	22–0	6.7
		24	61	13–5	4.1	17–11	5.4
		12	30	27–7	8.4	32–3	9.8
2 x 10	5 x 25	16	41	23–10	7.3	17–10	5.4
		24	61	19–6	6.0	22–9	6.9

Fig. 45-10. Permissible runs for southern pine rafters.

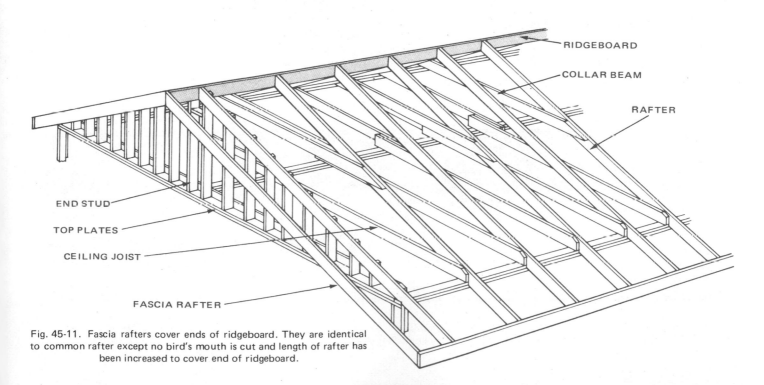

Fig. 45-11. Fascia rafters cover ends of ridgeboard. They are identical to common rafter except no bird's mouth is cut and length of rafter has been increased to cover end of ridgeboard.

Fig. 45-12. This photograph shows jack rafter system used to form roof without gable end. It is known as a hip roof.
(California Redwood Assoc.)

THE JACK RAFTER

Besides the common rafter, there is the jack rafter, Fig. 45-12. The jack rafter is usually shorter than the common rafter. It does not extend from the ridgeboard to the outside wall. There are three types of jack rafters:

1. The hip jack.
2. The valley jack.
3. The cripple jack.

The hip jack extends from the double plate to a jack rafter. The valley jack extends from the ridgeboard to the valley rafter. The cripple jack is used to frame in an opening in the roof. For example, it reaches from the double plate to the frame of a flue opening.

ERECTION OF THE RAFTER

Carpenters will test rafters at ground level to insure the accuracy of the cuts. Erection is much easier if there are at least two helpers. A considerable amount of temporary bracing will be required if only one or two carpenters are on the job.

First they erect the ridgeboard and the rafters nearest its ends. If the ridge of the house is longer than the individual pieces of ridgeboard, they will find it easier to erect each piece separately. It is difficult to splice the ridgeboard full length without rafters for support. The first piece of ridgeboard is supported at both ends with temporary props. The first pair of rafters will be securely toenailed to the ridgeboard. Then, rafters will be nailed to the double plate. Rafters on the other side of the roof will be nailed on in the same manner. The crew will continue until all rafters are in place. For added strength the workers will cut and install 1 by 6 in. collar beams at every other pair of rafters. See Fig. 45-11. Collar beams should be in the upper one-third of the attic crawl space.

To finish off the lower ends of the rafters, a fascia board is cut and installed. It is the same length as the ridgeboard. The top edge is trimmed to the same plane as the roof slope. (See Fig. 45-11.) The fascia board is nailed to rafter ends. The fascia rafters are then attached.

Fig. 45-13. With steel tubular trusses on the outside, this exhibit hall needs no heavy supports on the inside. Trusses like this provide great strength and give structure dramatic appearance. (C. E. Murphy Assoc.)

ROOF TRUSSES

Roof trusses are prefabricated lightweight rafters. They serve both as rafters and as ceiling joists.

Usually roof trusses are made to order for each job. They can be supplied in lengths from 20 to 36 ft. (6.1 m to 10.9 m) or even longer. The truss system follows the same principle as the triangle. The triangle is a rigid self-supporting shape.

Because a truss will span from one outside wall to the other, no interior load-bearing wall is needed. In Fig. 45-13, three trusses support the roof of a large building. Truss systems are used in some residential construction. It works best in houses which are simple rectangles. These require only one type of truss.

TRUSS DESIGN

There are many types and shapes of roof trusses. See Fig. 45-14. One type, Fig. 45-15, uses a rafter shaped like a single arc.

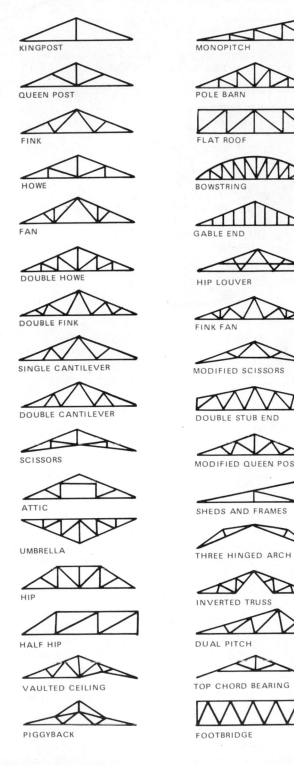

KINGPOST

QUEEN POST

FINK

HOWE

FAN

DOUBLE HOWE

DOUBLE FINK

SINGLE CANTILEVER

DOUBLE CANTILEVER

SCISSORS

ATTIC

UMBRELLA

HIP

HALF HIP

VAULTED CEILING

PIGGYBACK

MONOPITCH

POLE BARN

FLAT ROOF

BOWSTRING

GABLE END

HIP LOUVER

FINK FAN

MODIFIED SCISSORS

DOUBLE STUB END

MODIFIED QUEEN POST

SHEDS AND FRAMES

THREE HINGED ARCH

INVERTED TRUSS

DUAL PITCH

TOP CHORD BEARING

FOOTBRIDGE

Fig. 45-14. Typical truss forms. The most common for dwellings is the fink truss. It is also called the "W" truss. Note how bracing forms triangles in each truss.

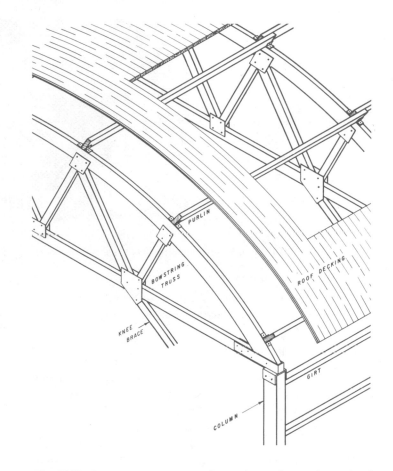

Fig. 45-15. Bow strung truss erected on column. It may be made of wood or steel.

Trusses should always be built according to designs developed from engineering data. The "W" or fink truss, Fig. 45-16, is designed according to a carefully worked out formula. The bottom cord is divided into three equal portions, marking the location of the joining of the webs. The top cord (rafters) is divided into four equal spaces. The upper location of the tension web is always at the ridge area of the truss. This arrangement gives the truss the maximum strength for the material used.

Gussets are used at the joints to fasten members of a wood truss together as in Fig. 45-17. The gussets can become part of the ceiling decoration.

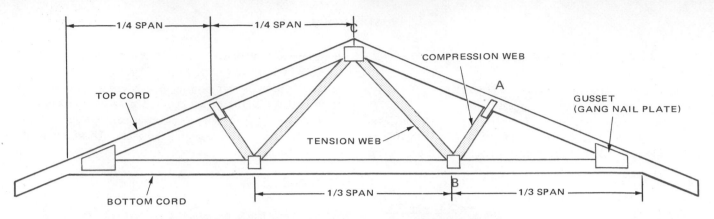

Fig. 45-16. The common "W" or fink truss design used in residential construction. Force applied at any point is passed along to other part of truss by webs. For example, a weight at A will push down at point B. But point B is stopped from sagging by web extending up to point C.

Fig. 45-17. Decorative exposed trusses of this building are reinforced by heavy steel gussets bolted to the individual members.

Plywood gussets are applied with glue and nails. Metal truss plates are attached with mechanical nailers. Heavy metal truss plates are bolted on.

When a truss is in position and "loaded," there will be a slight sag. To compensate for this sagging of the bottom cord, the truss is arched slightly during fabrication. This adjustment is called camber. Camber is measured at the midpoint of the span. A standard truss with a 24-ft. (7.3 m) span, will usually require about 1/2 in. (12.7 mm) of camber.

Most trusses for residential structures can be erected without special equipment. Each truss is simply placed upside

Fig. 45-18. Unusual roof line of this structure is possible with glued-laminated wood rafters. (California Redwood Assoc.)

down on the walls at the midpoint of installation. Then the peak is swung upward into position.

GLUED-LAMINATED SYSTEM

Laminated wood beams and arches are built in many shapes and sizes. They are used in both residential and commercial construction, Fig. 45-18. Glued-laminated structural members are custom built in special plants. Units are usually large and heavy. Special equipment is needed to place them. The beam is usually a foundation-to-foundation arch. Purlins, as shown in Fig. 45-19, are used to tie all the arches together. Roof sheathing is attached directly to these purlins.

ROOF SHEATHING

Roof sheathing is the covering over rafters, trusses or the beam-purlin system. Sheathing provides strength and rigidity. It makes a solid base for fastening the roofing material.

SHEATHING LAYOUT

Before starting with the roof framing, the carpenters lay out the complete sheathing plan. This layout will show panel sizes and where they will be placed. It will also indicate quantities needed. See Fig. 45-20. If the diagram should show

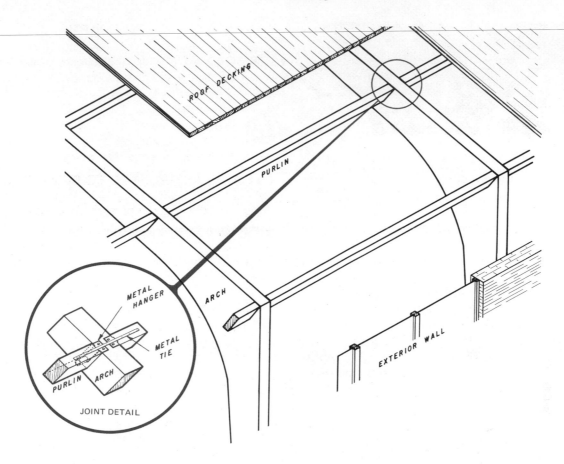

Fig. 45-19. Each laminated beam is carefully spaced and held in line by purlins.

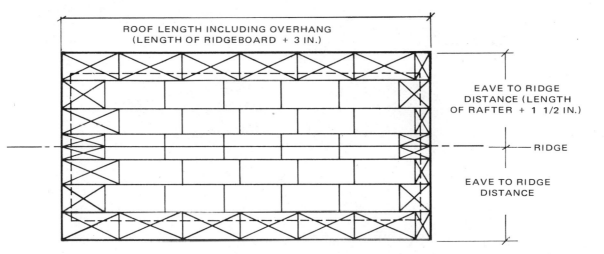

Fig. 45-20. Roof sheathing layout. Broken line is for the soffit and gable overhang. Diagonal lines in edge panels show the worker how edge panels are pieced in to fill all spaces. Small "X's" indicate partial panels.

that they will have too much cutting, they may be able to reduce scrap by shortening the rafter overhang at the eave or gable overhang.

PLACEMENT OF SHEATHING

Nailing on of sheathing starts at any corner of the roof. Workers will fasten the first course of panels to the roof framing members with 6 penny common or box nails. For additional holding power a ring shank nail might be used.

As in applying subflooring panels, sheathing panels will have 1/16 in. expansion room at the ends and 1/8 in. along the sides. In high humidity regions, twice this much spacing will be allowed. End joints are staggered on different rafters as shown in Fig. 45-20. Note that the broken line in this layout drawing indicates soffit and gable overhang.

Fig. 45-21. An open soffit with exposed beam and rafter construction. These rafters are used with post and beam construction. (Koppers Co., Inc.)

SOFFITS

Soffits are the underside of the roof overhang at the eave and gable areas. Fig. 45-21 shows an open soffit design. To keep the roofing nails from showing through the underneath side, these exposed soffit panels must be at least 1/2-in. (12 mm) thick. Many of the textured, finished plywoods of 1/2-in. (12 mm) and 5/8-in. (15 mm) thickness can be used with the textured side down to provide attractive soffits. Closed soffits must be framed to hold up the soffit material as shown in Fig. 45-22.

SUMMARY

Roof framing provides the support members to lay on a watertight coating. Principal systems are:
1. Rafter/ridge method.
2. The truss system.
3. The beam/purlin system.
Rafter length can be determined by the step-off method or by finding the hypotenuse of a right triangle.

CAREERS RELATED TO CONSTRUCTION

ROOF FRAMING CARPENTER is the rough carpenter who lays out rafter systems, cuts the rafters and other parts of the system and then installs them. Two or more will work together. In small crews, this work will be done by the same carpenters who frame the rest of the structure.

TRUSS ASSEMBLER is a production worker in a factory which manufactures roof trusses. Work involves placing precut parts into jigs and fastening the truss together.

TRUSS PLANT SUPERVISOR directs factory workers in laying out, cutting and assembling trusses.

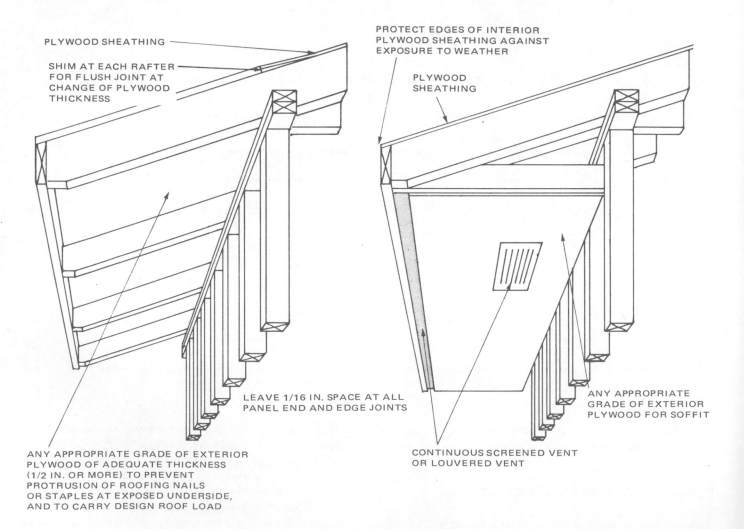

PLYWOOD SHEATHING

SHIM AT EACH RAFTER FOR FLUSH JOINT AT CHANGE OF PLYWOOD THICKNESS

ANY APPROPRIATE GRADE OF EXTERIOR PLYWOOD OF ADEQUATE THICKNESS (1/2 IN. OR MORE) TO PREVENT PROTRUSION OF ROOFING NAILS OR STAPLES AT EXPOSED UNDERSIDE, AND TO CARRY DESIGN ROOF LOAD

LEAVE 1/16 IN. SPACE AT ALL PANEL END AND EDGE JOINTS

PROTECT EDGES OF INTERIOR PLYWOOD SHEATHING AGAINST EXPOSURE TO WEATHER

PLYWOOD SHEATHING

CONTINUOUS SCREENED VENT OR LOUVERED VENT

ANY APPROPRIATE GRADE OF EXTERIOR PLYWOOD FOR SOFFIT

Fig. 45-22. Construction technique for the open and closed soffit. Vents in closed soffit allow air to circulate.

TRUSS ERECTOR works on a crew which attaches truss rafters to walls of partially completed structures. May also work on erection of laminated beams and rafters.

DISCUSSION TOPICS

1. How is a ridgeboard used in the rafter system?
2. What is a plumb cut and a bird's mouth cut?
3. How would you splice the ridgeboard?
4. Explain how the hypotenuse of a right triangle is used to identify the length of a rafter.
5. What is a fascia rafter? A jack rafter?
6. What is a collar beam?
7. Explain the compression and tension forces of a roof truss. What is camber?
8. What is a gusset? Why is it used?
9. How is roof sheathing laid out?
10. What is a soffit?

Chapter 46

SPECIAL FRAMING

Fig. 46-1. Classic winding stairs gradually change direction in rising from one level to another. (Seery-Hill Associates)

Fig. 46-2. U-shaped open stairway is bounded on one side by a wooden railing; bounded on the other side by a wall. (California Redwood Assoc.)

Several areas of residential and commercial construction require techniques having particular standards for their completion. Stairways, cornice returns and roof dormers are a few of the many with special framing requirements. Several framing methods have been developed to accomplish the installation of these features.

STAIR FRAMING

Stairs are a series of steps that enable people to pass from one floor to another. See Figs. 46-1 and 46-2. When the series of steps is continuous (without breaks formed by landings or other construction), it is referred to as a flight of stairs. Stairway and staircase are other terms used to describe a series of steps. Basically, stairs are classified as either open, closed or a combination of open and closed. The open stair is bounded on one or both sides by a railing. The closed stair is bounded on both sides by solid walls.

Stairs are listed as straight run, platform or winding, Fig. 46-3. *The straight run stairway is continuous from one level to another level, without any breaks in the progression of steps.* The straight run is the easiest to build. It requires a total run (distance covered at floor level) of approximately 12 ft., 3 in. (3.7 m) for a total rise (height between two levels) of 8 ft. (2.4 m).

Platform stair design includes a landing, where the direction of the stair run usually changes. Platform design utilizes limited space. Straight run stair design requires a long stairwell, which often presents problems in smaller homes.

Winding stairs are circular or elliptical, and they gradually change direction in their ascent from one level to another. Usually, the geometric change is 90 deg. from the starting

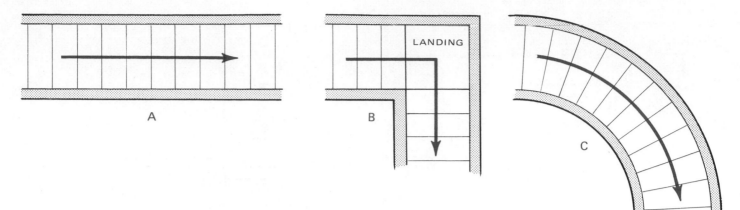

Fig. 46-3. Types of stairways. A—Straight run. B—Platform. C—Winding.

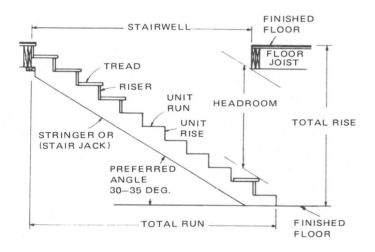

Fig. 46-4. Basic descriptive terms accompanying stair construction are called out.

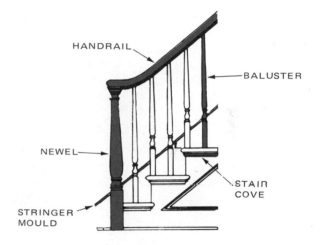

Fig. 46-5. Railings for open stairways provide safety in stair use.

point. A corresponding curved wall is required. Winding stairs are seldom installed because of the high cost and large amount of space required.

IDENTIFYING STAIR PARTS

Stairway terminology is basic and unique. Each part is identified in Fig. 46-4. Note that stairs consist of treads and risers, supported by stringers. Treads are the horizontal steps of a stairway. Risers are the vertical boards between the steps. Stringers are the sides of the steps.

The width of each tread is called "unit run." The height of each riser is called "unit rise." The sum of all the treads is "total run." The sum of all the risers is "total rise."

The stairwell is the framed opening through the floor that receives the stairs. The stairwell must be large enough to allow ample height (headroom) between the double header and the tread underneath to allow freedom of passage. The Federal Housing Administration requires a minimum of 6 ft., 8 in. (2.0 m) for main stairs and 6 ft., 4 in. (1.9 m) for service stairs.

Open stairs must have a railing for safety purposes, Fig. 46-5. The railing consists of:

1. A newel, or upright post, for the start of the railing (many styles are used).
2. The handrail, which is a continuous rail for security.
3. The balusters, or individual vertical pieces that fill the spaces between the newel posts and the handrail-tread area. The general appearance of a stairway is greatly affected by railing design.

STAIR DESIGN

In any stair run, it is important to maintain the same height for each riser and the same width for each tread. *Unequal risers and/or treads can cause a fall and possibly serious bodily injury.* Therefore, the proper combination of measurements is important.

Uniformity of measurement of home stairs, steps leading to public buildings and stairs of commercial structures help to prevent or minimize accidents. Inconsistent stair measurement can cause stair climbing to be tiring, resulting in extra strain on leg muscles. If the tread is too narrow, it will cause the riser to be kicked each time. If the tread is too wide, or if the riser is too high, it can cause tripping.

Certain general rules are used to establish a correct

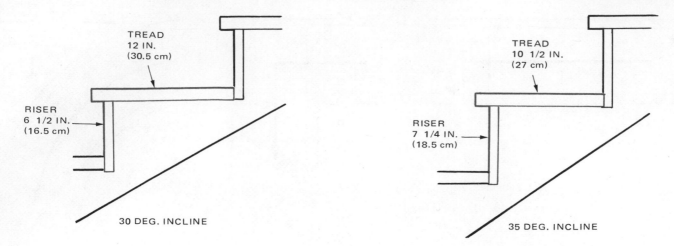

Fig. 46-6. Tread and riser measurements of 30 deg. incline stairs and 35 deg. incline stairs are compared.

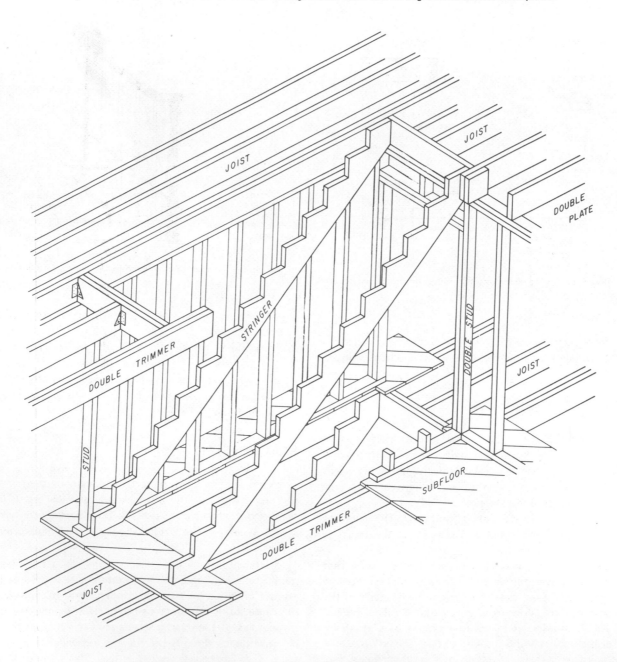

Fig. 46-7. Basic structure of an interior straight run stairway. Note stringers.

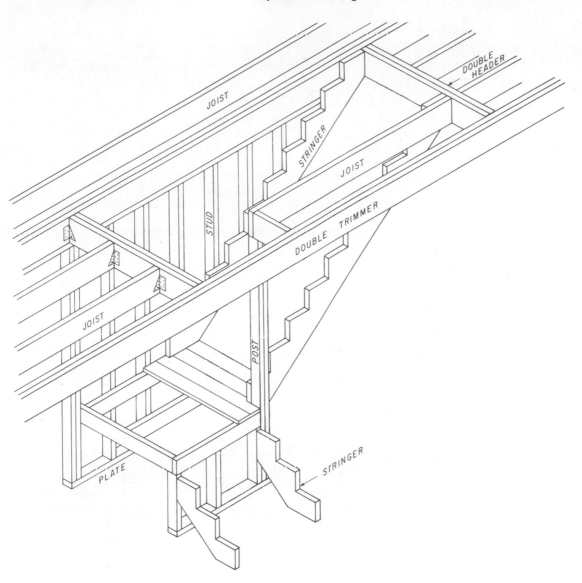

Fig. 46-8. Basic structure of a platform stairway. Width of landing should equal width of stairs.

tread-to-riser relationship. The ratio of inches of tread to inches of rise should result in a 30 to 35 degree angle of incline, Fig. 46-6. For a safe and comfortable ratio, these three generally accepted rules apply:

1. The sum of one tread and two risers should equal 25 in. (64 cm).
2. The sum of one tread and one riser should equal 17 to 18 in. (43 to 46 cm).
3. The product obtained by multiplying the width of a tread by the height of a riser should be approximately 75 in. (190 cm).

STAIR STRUCTURE

The structuring of wood stairs is based on the design of the stringers, Figs. 46-7 and 46-8. The stringer, sometimes called the carriage, is made from a piece of No. 2 dense yellow pine, selected from 2-in. by 10-in. or 2-in. by 12-in. (5 by 25 cm or 5 by 30 cm) straight stock.

The layout is started by placing the pine board across a pair of saw horses for convenience. Using a framing square, the location of the treads and risers is laid out, Fig. 46-9. With the location lines established, the thickness of the riser and tread boards must be taken into account. The stair tread is either

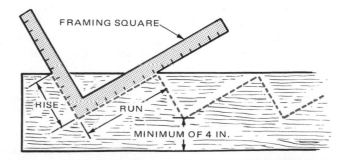

Fig. 46-9. When the number of treads and risers has been determined, the stringers can be laid out as shown. Space must be allowed for thickness of riser and tread boards.

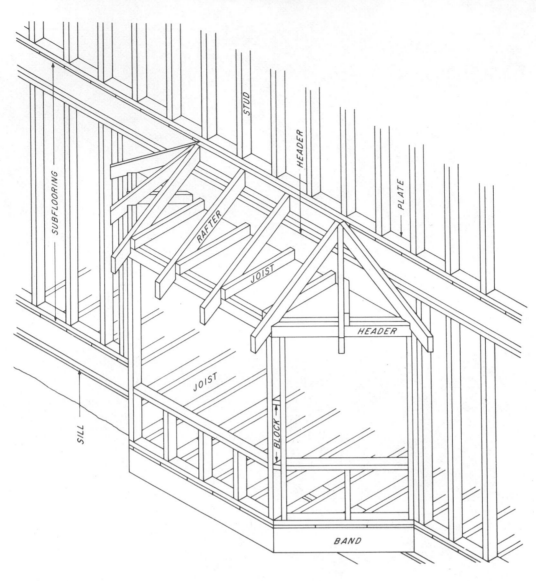

Fig. 46-10. A bay window is a special framing feature, requiring extra studs, headers, joists and rafters.

1 1/16-in. or 1 1/8-in. (26 mm or 28 mm) thick. The riser usually is 3/4-in. (19 mm) thick.

FRAMING A BAY WINDOW

The bay window is an extension of the structure beyond the consistent framework of the unit. At the point of opening in the wall, provision must be made to carry the vertical load, Fig. 46-10. A header of adequate size is required, and the ends of the header are supported on the studs. If the opening exceeds 6 ft. (1.8 m), in width, it is good practice to use triple studs with the end of the header resting on two studs.

PREPARATIONS FOR FRAMING A FIREPLACE

In fireplace and chimney construction, wood framing should be separated from the firebox by masonry. All wood trim should have proper clearance from fireplace openings, Fig. 46-11. All headers, beams, joists and studs should be at least 2 in. (5 cm) from the outside face of fireplace and

chimney masonry.

A minimum of 8 in. (20 cm) of masonry should separate the firebox and any wood framing. All wood trim mantles and other wood trim should be kept at least 6 in. (15 cm) from the fireplace opening. Parts of the mantle which project outward more than 1 1/2 in. (38 mm) from the fireplace should have additional clearance equal to the amount of projection.

FRAMING A ROOF DORMER

A roof dormer is used to allow light and ventilation into the attic space. It is framed with standard framing material, Fig. 46-12. Roof rafters are doubled to support the weight of the dormer walls and roof.

The roof dormer opening is squared off by installing double headers between the double roof rafters. The dormer walls are framed by the typical stud-plate method. Roof framing of a gable dormer is completed by installing a ridgeboard and a valley rafter. See Fig. 46-12. A flat roof dormer does not have a valley, but it must be pitched slightly for drainage.

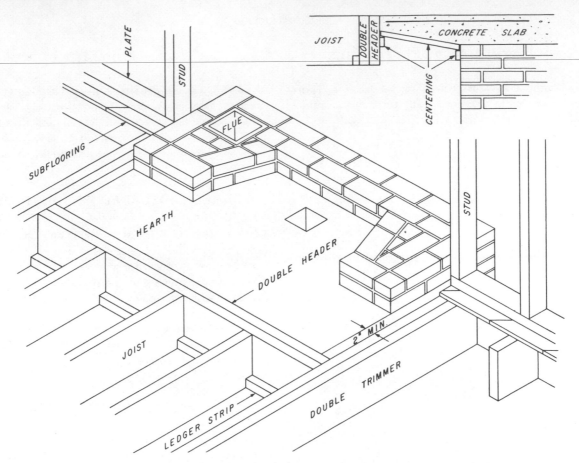

Fig. 46-11. Framing the opening for a fireplace. Note that wood framing is kept clear of firebox.

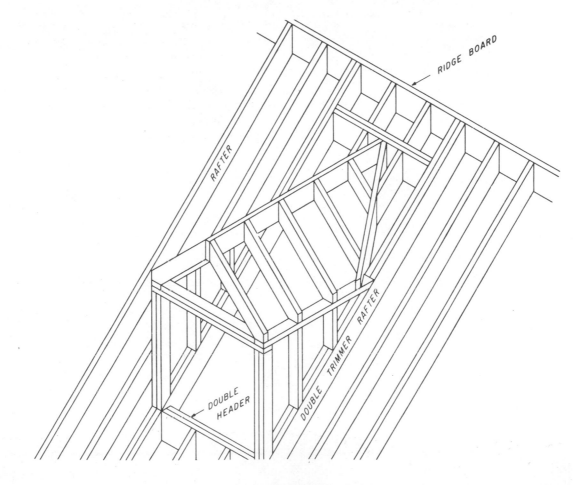

Fig. 46-12. Method of framing a roof dormer calls for double rafters and double headers.

FRAMING FOR PLUMBING FIXTURES

Certain plumbing fixtures require special framing before installation. For example, certain bath tubs and lavatories require frame supports, Fig. 46-13.

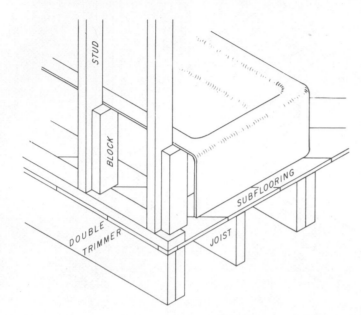

Fig. 46-13. Cast iron bathtubs, because of their weight and design, often need special framing features.

A wall-mounted wash basin may be supported by nailing a 2-in. (5 cm) member between the studs. Depending on the size and weight of the basin, this support may be 2 in. by 8 in. (5 cm by 20 cm), 2 in. by 10 in. (5 cm by 25 cm) or 2 in. by 12 in. (5 cm by 30 cm).

SUMMARY

Special framing departs somewhat from standard framing practice. Special framing features usually vary with each project. However, when basic framing practices are followed, these variations can be adapted to any project with the confidence the building will be structurally sound and safe for the occupants.

CAREERS RELATED TO CONSTRUCTION

ROUGH CARPENTER studies building plans or receives oral orders to construct concrete forms and pouring chutes; builds scaffolding and ladders for assembling structure above ground level; erects temporary frame structures on construction site; assists carpenters to rough-in wood framework of structure.

FRAMING CARPENTER studies building plans, selects materials and prepares framing layout; shapes, assembles and fastens materials to erect wood framework; lays subflooring and stairs.

DISCUSSION TOPICS

1. What is the inclined angle suggested for stairs?
2. Why should stairs be of standard size?
3. What is the tread-riser ratio?
4. Describe the method of laying out a stringer.
5. What is an open stair? What is a closed stair?
6. Why should the wood-framed fireplace opening be at least 2 in. larger than the masonry?
7. Who is responsible for framing the special features?
8. Describe the gable roof dormer framing feature.
9. What is a newel post? What is a baluster?
10. In constructing stairs, what is meant by total run?
11. What is meant by total rise?

Chapter 47

METAL FRAMING

Fig. 47-1. In structural steel-framed buildings, vertical columns support weight of building. Here rigger directs derrick operator by telephone to hoist steel wall framing section. (Port of New York Authority)

Large, commercial buildings, as contrasted to home construction, have frames of structural steel rather than wood. See Fig. 47-1. This metal framework provides the skeleton that supports the walls, floors and roof in structures of this type. It also supports the weight of the building.

STRUCTURAL STEEL FRAME

A tall, multi-story building with a steel frame is often referred to as skeleton construction. All live loads and dead loads (including walls) are supported by this steel skeleton.

The exterior walls are non-load bearing or curtain walls.

Exterior and interior columns are spaced to provide support for the beams running between them. There is no limit to the area of floor or roof that can be supported by adding another row of columns and beams.

The horizontal members and beams connecting the exterior columns are called "spandrel beams." They are equipped with shelf brackets to support the masonry spandrel over window and door openings.

Large industrial buildings, auditoriums, sports arenas, exhibit halls, bridges Fig. 47-2, and similar structures may require steel as the material for framing. In addition, many structures require that great distances be spanned. To carry the load, girders, trusses, arches or other types of framing systems may be used, Fig. 47-3.

The built-up girder consists of plates, angles and other steel shapes necessary to obtain the strength needed. The top and bottom flanges resist bending. The web acts as a spacer to keep the two flanges apart and give vertical strength. The individual parts may be assembled by welding or riveting, Fig. 47-4.

FASTENING METAL FRAMEWORK

The principal methods used today to fasten metal framework are welding, riveting and bolting. The selection of the right fastening system is based on many requirements:
1. Local building code.
2. Fabricator's preference.
3. Economic consideration.

The common shop methods of fabrication are welding and riveting. However, high strength bolts are replacing rivets in many shops. All three methods are used in the field. In fact, it is not uncommon to find one method used in the shop and another used in the field. For example: a subassembly may be welded in the shop and delivered to the job site; then erected with bolts or rivets, Fig. 47-5.

WELDING

Today, welding is the principal method of joining steel building frames in the shop. It ranks equally with high-strength structural bolts as the two principal methods of erecting structural steel. Welding has the basic advantage of fusing the metals to be joined, thereby eliminating the need for reinforcing angles and cover plates. It is particularly well suited for use

Fig. 47-2. Structural steel is used to make large trusses for bridges and skeletal frame for large structures. (Missouri Public Service Co.)

Fig. 47-3. Only steel could be used to frame this structure. Sections are welded, riveted or bolted together. (City of Tempe, Arizona)

Fig. 47-4. A welder assembles steel columns of a structure. Note V-groove for filler material at center; finished weld at right. (Port of New York Authority)

in fastening the ridge frame and architecturally exposed structural steel members, Fig. 47-6.

Welding is accomplished by heating two solid pieces of metal to the melting point, then flowing the molten portions together, making them one.

Two common methods of welding are done in the shop and in the field: arc welding (electric) and gas welding (oxygen acetylene gases). There are no set rules which govern the type of welding to be used for a particular job. In general, the controlling factors are:

1. Kinds of metal to be joined.
2. Cost involved.
3. Nature of the products to be fabricated.
4. Production techniques.

Some jobs are easily accomplished by gas welding; others are more easily done by the arc welding method.

Fig. 47-5. Subassemblies of this steel structure were welded in the shop, delivered to the construction site, lifted in place and fastened by bolts. (Stran-Steel Corp.)

Fig. 47-6. In expressway construction, columns support girders and structural steel framework that span great distances. (Missouri Dept. of Highways)

329

Fig. 47-7. Large assemblies are welded together at the construction site. Welding is a fast, economical and permanent method of fastening metal framing members. (Corps of Engineers, Kansas City District)

Gas welding is used in all metal fabrication shops and in the field. Because of its flexibility and mobility, gas welding is widely used at the construction job site.

The chief advantage of arc welding is its speed. A high quality weld can be made rapidly at relatively low cost. Arc welding is considered ideal for fastening metal members of structural steel buildings and bridges, Fig. 47-7.

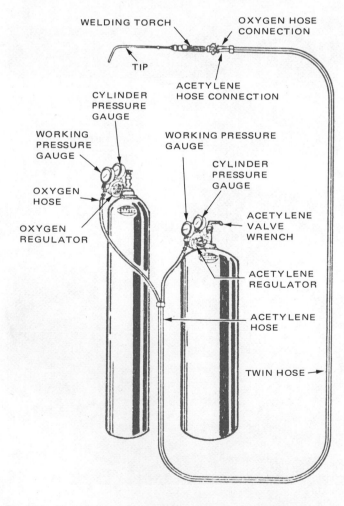

Fig. 47-8. Components of a typical oxyacetylene welding setup are called out.

GAS WELDING PROCEDURE

Many gases will burn and generate sufficient heat to weld metal. At present, the oxygen-acetylene combination is most adaptable to the construction trades, Fig. 47-8. Temperature of this burning gas mixture will reach in excess of 6000 deg. F (3282 deg. C).

To light the gas torch:

1. The separate oxygen and acetylene cylinder valves are set to working pressures to correspond with the size tip used, Fig. 47-9. NOTE: Operator must not stand directly in front of regulator when opening the cylinder valve.

TIP SIZE AND SUGGESTED PRESSURE RATINGS

TIP SIZE	THICKNESS OF METAL		OXYGEN PRESSURE		ACETYLENE PRESSURE	
	In.	mm	lb./sq. in.	kg/cm^2	lb./sq. in.	kg/cm^2
00	1/64	0.4	1	0.07	1	0.07
0	1/32	0.8	1	0.07	1	0.07
1	1/16	1.6	1	0.07	1	0.07
2	3/32	2.4	2	0.14	2	0.14
3	1/8	3.2	3	0.21	3	0.21
4	3/16	4.8	4	0.28	4	0.28
5	1/4	6.4	5	0.35	5	0.35
6	5/16	7.9	6	0.42	6	0.42
7	3/8	9.5	7	0.49	7	0.49
8	1/2	12.7	7	0.49	7	0.49
9	5/8	15.9	7 1/2	0.53	7 1/2	0.53
10	3/4 and up	19.0	9	0.63	9	0.63

Fig. 47-9. Recommended oxyacetylene torch tip sizes and pressure ratings are given for metals of various thicknesses.

2. Next, the torch acetylene valve is opened approximately three-fourths of a turn.
3. A spark lighter is held about one inch from the end of the tip, and the acetylene is ignited. CAUTION: Never use a match to light the torch.
4. Then, the acetylene flame is adjusted to bring it to peak efficiency, or just until the yellow flame begins to emit black smoke.
5. Finally, the oxygen valve is opened and adjusted until a well-defined white cone appears near the tip, surrounded by a second bluish cone that is faintly luminous.

This "neutral flame," Fig. 47-10, is burning approximately equal portions of oxygen and acetylene, making it chemically neutral. A neutral flame is used for most welding operations.

Selection of the appropriate torch tip size is all-important to the job of gas welding. Tip selection must be based on:

1. Experience of the welder.
2. Type of material being welded.
3. Thickness of material being welded.

If a tip is too large for the job being performed, the base metal may burn and create a weak weld. If the tip is too small, the welding job will take longer due to lack of heat in the base metal. By varying gas pressure, any given tip is capable of a wide range of heat.

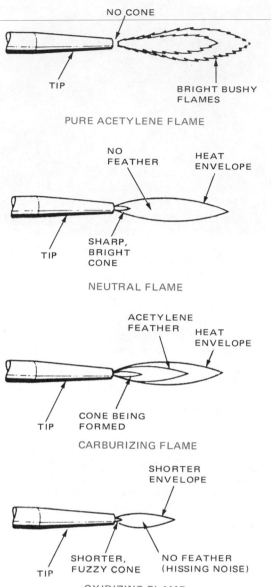

Fig. 47-10. Types of oxyacetylene flames are identified.

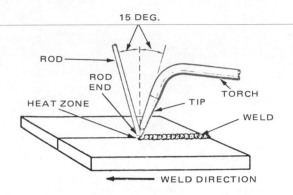

Fig. 47-11. The 15 deg. positions of welding torch tip and filler rod usually give good results in laying a bead.

Fig. 47-12. Flame cutting in the field is a fast and efficient means of sizing metal components. (Exxon)

BASIC GAS WELDING TECHNIQUE

The most popular welding technique is called "forehand welding." In forehand welding, the torch is inclined at an angle of about 15 deg. from a vertical position in the direction of travel, Fig. 47-11. At this angle, the torch will efficiently preheat the work to be welded. While preheating, a filler rod is dipped in the puddles, also at an angle of about 15 deg., but in the opposite direction. The rod is withdrawn slightly, allowing molten metal to drip from the rod into the puddle. This fills the space between the two pieces of base metal.

Experienced welders may vary the angle of the torch and the filler rod to suit the job they are doing. If the torch is sloped away from the vertical at an angle over 15 deg., the puddle can be forced along the joint from the torch pressures. This also helps to eliminate the chance of "burn through" on light gauge metal.

FLAME CUTTING OF METAL

Flame cutting is a process whereby an oxygen and gas flame is utilized to cut metal, Fig. 47-12. Flame cutting is used in steel fabrication, construction and demolition. In this process, a very hot flame makes the cut by removing some of the metal.

When the flame-cutting process is used to cut metals containing iron there is a chemical reaction. The reaction occurs because oxygen has a chemical attraction to ferrous metal when the metal is heated above the kindling (burning) temperature. During the cutting process, the welder always uses more oxygen than acetylene gas.

When molten iron is exposed to the extra oxygen used for cutting, the molten metal will be removed by a chemical reaction known as oxidation. When iron oxidizes, it simply burns up and leaves ashes known as slag. The added energy of the cutting oxygen stream aids in removing the metal by washing away as much as one third of the molten metal.

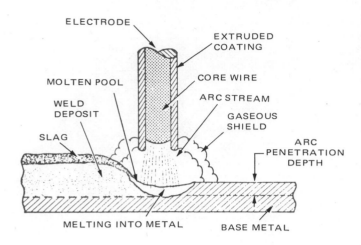

Fig. 47-13. The arc welding process utilizes an arc stream between tip of filler rod and metal being welded.

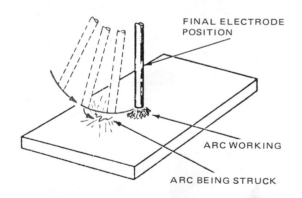

Fig. 47-15. Arc welding can be damaging to eyes if left unprotected. Note helmet and leather gloves.　(Exxon)

ARC WELDING PROCEDURE

Arc welding is a method of melting two pieces of metal into one solid piece. The edges of the pieces to be joined are heated to a temperature high enough to melt the faces together. The high temperature required is generated by an electric arc, which is made between the tip of the electrode (filler rod) and the metal being welded. The arc takes place in an atmosphere of ionized gas, produced when the current is shorted, Fig. 47-13.

A very high arc temperature, about 6000 deg. F (3282 deg. C), is produced between the tip of the electrode and the base metal. This high temperature arc cuts into the metal, forming either a crater or a pool of molten metal (depending on intensity of arc's force). The arc's force is determined by the amount of current from the power source, Fig. 47-14.

important part of the arc welding process. Electrodes are designed for specific amperage flows according to their diameter. Each electrode diameter has a minimum and a maximum amperage range listed in the manufacturer's suggested rating guide.

STRIKING THE ARC

The electrode holder and electrode are held vertical to the base metal and tilted to about a 15 deg. angle in the direction of the travel. Next, the electrode is dragged across the base metal, Fig. 47-16. As it touches the base metal, the arc will start. To maintain the arc flow, the tip of the electrode is held about the distance of its diameter above the base metal.

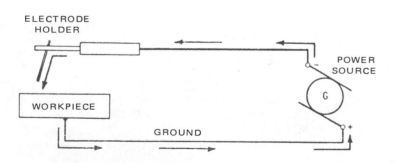

Fig. 47-14. Basic arc welding circuit requires a very high electrical current for satisfactory operation.

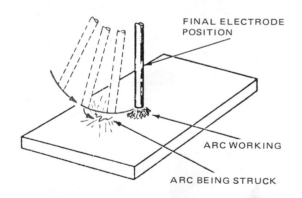

Fig. 47-16. Striking the arc is done by scratching electrode across surface of work.

Before practicing arc welding, the operator should review the protective equipment needed during the welding process. Arc flashes can cause blindness. Therefore, helmets with dark-colored vision lenses should be used to protect the eyes from the arc flash, Fig. 47-15. Heavy leather gloves prevent burns from molten metal blown away from the weld.

Electrodes are "filler rods" fastened into an electrode holder. Choosing the right electrode for a given job is an

RUNNING THE BEAD

When the arc is struck, the base metal starts to melt just below the tip of the electrode. At the same time, metal droplets from the hot electrode fall into and mix with the molten pool in the base metal. This mixing of melted electrode and melted base metal is referred to as "fusion."

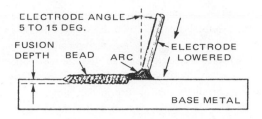

Fig. 47-17. Running a bead strengthens the weld.

Fusion takes place constantly as the electrode moves across the base metal, causing a metal buildup known as a "bead," Fig. 47-17. The bead gives a slight raised portion to the base metal at the point of union, thereby strengthening the weld.

As the bead forms, a glass-like crust called "slag" forms over it. Slag is created by impurities that float out of the molten metal. It also forms a protective coating to prevent oxidation during the welding process, Fig. 47-18.

Fig. 47-18. Chipping away slag prepares weld for inspection. (Exxon)

RIVETS

Rivets are soft, annealed, carbon steel fasteners. They usually consist of a rounded button head and a cylindrical shank. The shank is long enough to extend through the parts to be fastened and form a head at the other end when set. This second head may be a button head, flattened head, countersunk head or a countersunk chipped head, Fig. 47-19. The button head is most common.

Rivets are heated to a cherry red color and placed in holes slightly larger than the shank diameter. The rivet head is formed with a pressure type impact riveter. The riveter is a heavy C-shaped frame suspended from a crane, capable of holding the head firm while striking the opposite side. The second rivet head is formed in one stroke with a force of as much as 80 tons (72 metric tons). As the hot rivet cools, it

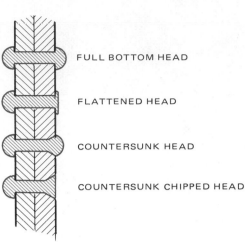

Fig. 47-19. Various types of rivet heads are identified.

contracts and exerts a strong clamping force on the metal parts between the rivet heads.

HIGH STRENGTH BOLTS

One of the newest fasteners in steel construction is the high-strength structural bolt, Fig. 47-20. Advantages include: speed of construction, strength, simplicity and safety. Allowable carrying capacity of the bolts is equal to that of rivets of the same size. Local building codes generally give bolts the same rating as rivets.

VERTICAL TRANSPORTATION

Steel frames usually are erected at the construction site, one piece or one section at a time. Derricks are used to hoist

Fig. 47-20. Structural bolts are installed, using a pneumatic wrench. (American Iron & Steel Institute)

Fig. 47-21. Derrick is mounted on top of structural steel framework. This equipment is dismantled and moved upward as height of building increases. (Port of New York Authority)

these large sections into place, Fig. 47-21. This is called "vertical transporation."

Derricks have two main parts: a mast and a boom. The mast is the tall vertical assembly held upright by cables and guy wires. The boom is the movable main horizontal portion which is designed to pivot around the mast.

Fig. 47-22. Rigging operation is attaching structural members for safe lift and arrival top side in correct position for fastening. (U.S. Dept. of Interior)

Winches (drum-shaped machines on which cable is coiled) are used to lower cables over the outside edge of the structure. Below, a rigger safely attaches the structural members to the cable end. Rigging must be done so the piece will arrive at the top in correct position to be fastened in place, Fig. 47-22. Hoisting is accomplished by winding the cable around a drum at the bottom of the mast section of the derrick. A series of pulleys and blocks give the derrick sufficient power to lift extremely heavy weights, Fig. 47-23.

Fig. 47-23. Derrick has a counterbalanced boom and tremendous lifting power. Note operator's cabin under mast. (Canadian National Railways)

Guiding the derrick's load into position can be something of a problem. Many times the operator cannot see the rigging area, the load or the positioning point. In this case, a signalman uses a series of hand signals to guide the load into position. If the derrick operator is out of range of hearing and vision of the construction superintendent, radio or telephone communications are used. See Fig. 47-24.

Smaller projects erected at lower levels may use lift systems of a portable nature. The crane, for example, is a mobile lift system set on a wheel or crawler type vehicle with an outrigger to give the crane stability. See Fig. 47-25.

The crane operator constantly must be aware of the angle, or degree of inclination, of the boom. If the angle should become too low, the weight of the load could cause the crane to overturn. In the case of an extra long boom, too low an angle might cause the boom to buckle and collapse.

Fig. 47-24. Construction superintendent uses a 2-way radio to direct derrick operator and iron workers to position and bolt sections of bridge truss 120 ft. above river bed. (U.S. Steel Corp.)

Fig. 47-25. Mobile crane is used to hoist steel and other materials to heights limited only by length of boom.

SUMMARY

Steel frames are assembled by a group of construction workers called "iron workers." Fastening of steel members is completed by one of three methods: welding, riveting or bolting. Two general methods of welding are used: gas flame and electric arc. Metal framing is either assembled in single pieces or prefabricated sections. Prefabricated sections may be assembled in the shop and transported to the site, or assembled at the site.

Vertical transportation is the term used to describe the hoisting and placing of structural sections "top side." Tall structures have mounted derricks with boom to reach over the side. Cables drop to ground level where riggers carefully attach sections to the cable end. The derrick is moved up, level by level, until the structure is "topped out."

CAREERS RELATED TO CONSTRUCTION

SKILLED WELDER sets up equipment for gas welding, arc welding, brazing or flame cutting and grinding metal plates and structural shapes; positions, fits and welds fabricated, cast or forged components to assemble structural forms as specified by work orders or oral instructions. Welder also may assemble parts by bolting or riveting.

WELDING INSPECTOR examines welds for bead size and penetration; tests welds to insure that joints meet requirements, using X-rays, radiographics, magnetic testing machines and other apparatus.

WELDING SUPERVISOR studies work orders, layouts, diagrams and blueprints for welding requirements on the construction job; determines needs and assigns jobs to welders, then directs and coordinates their activities. Supervisor also examines X-rays of welds, watching for defects.

IRON WORKER erects steel framework of buildings, bridges or other structures; pushes, pulls and pries steel beams and girders into proper position; temporarily connects all steel members with bolts; uses plumb bobs and levels to align the structure, then welds, bolts or rivets units together.

RIGGER splices ropes and metal cables to form hoisting slings for lifting and moving structural beams and columns, wall and floor slabs, and other materials and equipment. Rigger attaches load with grappling devices (loops, wires or chains) to crane or derrick hook; gives directions to crane or derrick operator, using hand signals, loud speaker or telephone.

HOISTING ENGINEER operates a crane or derrick to lift, move and place material and equipment, using attachments such as a sling, electromagnet, grappling hook or bucket. Operator responds to signals given by hand, bell, whistle, telephone or loud speaker; also disassembles crane or derrick for move to different level of structure, then reassembles it at new location.

DISCUSSION TOPICS

1. What is a girder?
2. Describe skeleton framing.

335

3. What are the requirements for selecting a fastening method for metal framework?

4. Give the procedure for lighting a gas welding torch.

5. What is a neutral flame?

6. What is the temperature of a gas welding torch flame?

7. List the safety equipment for arc welding.

8. What is an electrode?

9. How are rivets installed?

10. What is the advantage of bolts over rivets for fastening structural members together?

11. What is meant by vertical transportation?

12. What is a mast and boom?

Chapter 48

PLUMBING

Plumbing systems in residences contain two major parts:
1. *Large pipe for waste disposal.*
2. *Smaller pipe for fresh water distribution of both cold and hot water.*

Plumbing systems are covered by codes such as the "Uniform Plumbing Code," published by the International Association of Plumbing and Mechanical Officials. Services of a competent plumber are needed to install the many individual pieces correctly. Systems for commercial buildings are even more complex. See Fig. 48-1. The plumber will also help solve difficulties that arise after a building is under construction.

Fig. 48-1. Complex drainage system is made up of small parts that must be carefully measured and installed so they will not leak.

FRESH WATER SUPPLY

Water supply piping and fittings are made of brass, copper, iron, galvanized iron, lead or plastic. All fittings used in a water supply system must be of like material.

Valves or other similar devices can be a different material. Valves of less than 2 in. (5 cm) must be brass. If larger than 2 in. they must have a cast brass body.

GALVANIZED PIPE AND FITTINGS

Galvanized steel pipe has great strength, dimensional stability, (does not stretch or shrink much) and lasts long. The strength of steel provides safety with economy. Galvanized pipe is joined with threaded fittings as shown in Fig. 48-2.

Fig. 48-2. Screwed fittings for drainage systems. A—Elbow. B—Tee. C—Union. Iron fittings are often galvanized.

Fittings are threaded on the inside. Pipes are threaded on the outside.

The American National Standards Institute (ANSI) has two classes of pipe threads. One is tapered; the other is straight. The normal type employs a tapered internal and a tapered external thread. The taper is 1/16 in. per inch. Because of the taper, a pipe screwed into a fitting will make a watertight joint. It will not leak even under pressure.

Plumbers prefer to use a tapered external thread with a straight internal thread. They feel that the metal is soft enough to allow the threads to adjust themselves to the taper. You can be sure all pipe threads are tapered unless otherwise marked. See Fig. 48-3.

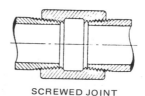

SCREWED JOINT

Fig. 48-3. Cutaway of tapered pipe connection. Unless fitting has a marking on the outside it is a tapered thread.

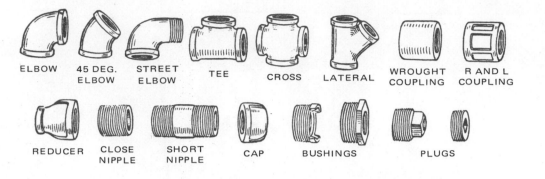

Fig. 48-4. Basic screwed fittings.

Threaded fittings are available in many shapes and bends, Fig. 48-4. They include:

1. A 90 deg. elbow, for changing in the direction of flow by 90 deg. Both ends have internal threads.
2. A 45 deg. elbow for changing the direction of flow by 45 deg. Both ends have internal threads.
3. A street elbow of 90 deg. One end has external threads.
4. Tees for connecting three pipes. One at a 90 deg. angle to the other two.
5. Crosses for connecting four pipes. The pipe connections may be of different sizes.
6. Laterals which are fittings for joining a third pipe at a 45 deg. or a 60 deg. to the main pipe.
7. Couplings designed to join straight sections of pipe or to extend a straight run. Both ends have internal threads.
8. Reducers similar in appearance to couplings. Both ends have internal threads. One end is for smaller pipe than the other. Reducers are used to reduce pipeline size.
9. Nipples which are short pieces of pipe threaded on both ends. They are available in many sizes from 1 in. to 24 in.
10. Caps to close the end of a pipe.
11. Plugs used to close the opening in a fitting.
12. Bushings also used to reduce the size of pipe runs.
13. Unions used to close systems and to connect pipes that are to be taken apart occasionally. The screwed union is composed of three pieces. Two pieces are screwed firmly on the ends of the pipes to be connected. The third piece draws them together forming a pressure-tight joint.
14. Valves used to control flow through the system. Many types of valves are used — gate valve, globe valve, swing valve, plug valve, and butterfly valve, to name a few. Valves are specified by giving the nominal size, material and type.

COPPER TUBING

Copper tubing is a soft, flexible pipe. It can carry many different liquids. Tubing is generally fitted together with flared connectors.

Copper tubing is generally used to make the final installation between the fixture or appliance and the piping. Copper tubing is soft and flexible. This helps in making the last connection. Copper does not corrode and remains free of calcium deposits.

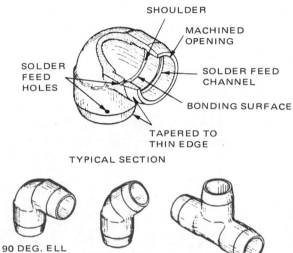

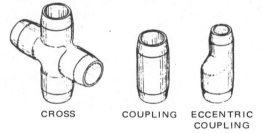

Fig. 48-5. Copper fittings for soldered connections.

COPPER PIPE

Copper pipe is hard and rigid where copper tubing is soft and flexible. Soldered joint fittings, Fig. 48-5, are used with copper pipe. Soldered joints may also be used with brass pipe.

In soldering, two or more metal surfaces are joined by a third metal which is applied in a molten state. Fig. 48-6 shows several sweat-soldered joints. The metal is an alloy (mixture) of tin and lead. Usually the solder is half tin and half lead. This 50/50 mixture melts at 415 F (210 C).

In order to do a good job of soldering, these steps must be followed:

1. Know the kind or kinds of metal being soldered.
2. Clean the surfaces thoroughly.

Fig. 48-6. Copper tubing with several tees and elbow joints sweat soldered in place.

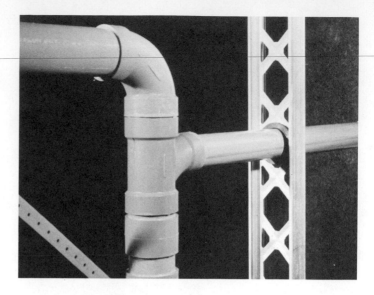

Fig. 48-7. Molded plastic fittings are cemented to plastic pipe making a quick bond.

3. Know the fluxes and how to use them properly.
4. Understand the capillary action of liquids.
5. Be able to apply the proper amounts of heat.

To complete a soldered joint on copper or brass pipe it is necessary to:

1. Clean the inside of fitting and outside of pipe with fine abrasive paper.
2. "Tin" both surfaces. (Heat and coat the surface with a thin coating of solder.)
3. Dip pipe end into a rosin type flux. The residue of rosin flux is a noncorrosive material.
4. Heat both fitting and pipe until the "tinned" coating is again molten (melted), then immediately push fitting over the pipe until the end rests firmly on the fitting shoulder.
5. While solder is still molten, wipe with a clean rag to remove excess solder.
6. Allow solder to cool and solidify before moving.

Fluxes play a very important role in the soldering process. Metals, when heated in an atmosphere of oxygen, will unite with the oxygen. This forms an oxide film. The film stops the solder from making a tight seal with the pipe and its fittings. The flux forms a protective covering over the soldering surface. This prevents formation of oxide.

Solder type fittings for copper pipes are shaped about the same as threaded fittings. The mating surface is machined to a very close fit with the pipe. Therefore, only a very small amount of solder is needed. Soldered fittings are very strong. They are fast and cheap to install.

PLASTIC PIPE

Plastic pipe does not corrode. It also has a high resistance to many chemicals. It is used widely in place of metal pipe. Plastic is easily cut with woodworking and metalworking types of saws.

The fittings are molded of similar types of plastic, Fig. 48-7. Plastic pipe is nonmagnetic, and, for the most part, will not burn.

Fittings and pipe are fastened together by solvent bonding. Certain solvents have a softening action on the plastic. With the surfaces of both fitting and pipe in a softened condition and placed together, the solvent soon evaporates. The joint becomes a permanent bond. A solvent-bonded joint cannot be disconnected.

Plastic pipe is not structurally strong. Long spans need to be supported to prevent sags. Expansion is five times greater than steel.

The speed and ease of installation make the plastic a very economical system. It is suitable for hot and cold water distribution and also for drainage systems.

DRAINAGE SYSTEMS

Each plumbing fixture in a system has:
1. A waste soil stack.
2. A trap.
3. A vent to the outside of the building.

The vertical waste soil stack carries wastes to a building drain. In turn, the drain conveys the wastes to the city's sewer main or to a septic tank. The vent stack provides air circulation through the drainage system. The purpose of the vent is two-fold:
1. To carry away the sewer gases.
2. To equalize air pressure inside the drainage system.

The trap, Fig. 48-8, provides a water seal at each fixture to

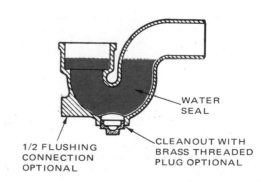

WATER SEAL

1/2 FLUSHING CONNECTION OPTIONAL

CLEANOUT WITH BRASS THREADED PLUG OPTIONAL

Fig. 48-8. Water seal created by trap prevents entrance of sewer gases into building through drains. (Josam Mfg.)

Construction

prevent the back passage of air. Under the Uniform Plumbing Code, as many as three single-compartment sinks and lavatories may be served by one trap. For a drawing of the drainage system refer back to Fig. 34-8.

CAST IRON SEWER PIPE

Cast iron pipe is made of grey iron with a bituminous coating inside and out. Cast iron pipe is fastened together in a hub and spigot design, Fig. 48-9. There are three methods of

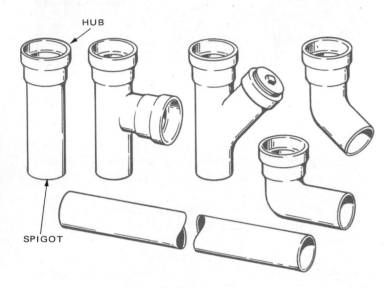

Fig. 48-9. Fittings for cast iron hub and spigot system of sewer pipe come in many shapes. Spigot end fits in hub where it is packed with a rope-like packing called oakum. Then a seal of lead is poured around it.

securing the joint:
1. The lead and oakum joint.
2. The neoprene gasket.
3. The neoprene-stainless steel sleeve, as shown in Fig. 48-10.
    The neoprene-stainless steel sleeve, often referred to as the

Fig. 48-10. Neoprene-stainless steel sleeve compression bond joint.

"no hub joint," makes a more compact joint than either of the other two types.

The joint is quickly put together. A neoprene gasket is placed on the end of a plain pipe section. Then the stainless steel shield and clamp assembly is placed on the end of the next section. The stainless steel compression clamp compresses the neoprene gasket. This makes a watertight and gastight joint that is strong and flexible. This joint can be put together in limited spaces.

COPPER DRAINAGE PIPE

Copper drainage lines are fastened together by soldering. Care must be taken to compensate for the greater coefficient of expansion of copper over steel. Copper may be coupled to iron with special adapters.

PLASTIC DRAINAGE PIPE

Plastic systems are significantly faster to install than most other materials. Each fitting is solvent welded identically similar to the system described for the installation of the water supply. Plastic has many anti-fouling qualities. This prevents the clogging problems of other systems.

CLAY SEWER PIPE

The clay used for sewer pipe is similar to that used for burned brick. Most clay pipe is of the hub and spigot type. Clay sewer pipe is usually joined with a hot poured compound. A tapered neoprene gasket is also used to seal the joint.

FUEL-GAS PIPING

Gas piping and fixtures, like other plumbing systems, are governed by local codes. Fuel gas lines must be fabricated from malleable iron or brass. The joints are usually screwed with standard pipe threaded fittings. Copper or brass fittings are secured by the flare system.

Fuel gas supply lines buried directly into the ground must be a special plastic-coated malleable iron. Fittings used underground must be wrapped with a plastic adhesive tape. Fuel gas lines may also be fastened by a welded joint, as in Fig. 48-11.

INSTALLING THE PLUMBING SYSTEM

Plumbing work can begin once the structure is closed in. It must be completed before installation of flooring, insulation or dry wall. This work is done by skilled people.

While the architect is responsible for the design of the plumbing system, the plumbing subcontractor or one of his employees prepares the working drawings. These must be approved by the architect or the general contractor.

Piping layout for a simple freshwater system is shown in Fig. 34-1. The drainage system is shown in Fig. 34-8. In addition to these two systems, the plumber may install heating pipes or ducts as well as gas fuel pipes.

Fig. 48-11. Welding a fuel line which will be buried. Note the plastic protective coating on the pipe. (Exxon)

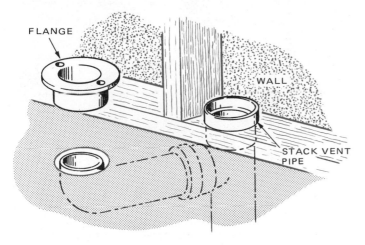

Fig. 48-12. Plumbers cut out rough flooring sections to install soil pipe for drainage systems. This section of pipe is for installation of a bathroom water closet (stool).

Using piping materials previously described, the plumber works right at the building site. Workers will measure out distance, cut out openings for pipes in flooring, Fig. 48-12, in vertical framing sections. They will cut pipes, determine placement of fittings and install them.

Many runs can be completely or partially fabricated before fastening them in place. Special supports hold the pipes to the framework of the structure. See Fig. 48-13.

When installing horizontal drainage piping, workers must be careful to slope soil pipes so waste will drain from them properly. Fresh water pipes are not sloped. The water in them is under pressure and will flow in any direction when a faucet is opened.

Fixtures are not installed and connected until after floors and walls are finished. Plumbers must be familiar with local codes for plumbing work.

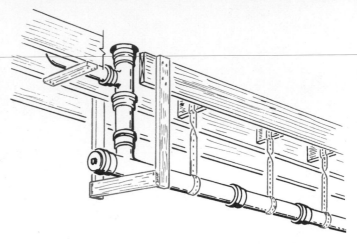

Fig. 48-13. Plumbing must be securely supported by attaching it to parts of the building. Here a long run of soil pipe is securely attached to floor joists with metal strapping.

SUMMARY

A special skill is necessary to fabricate the piping system for construction. Systems are designed around the properties of the materials used in making the pipe and fittings. Each kind of material requires special skills for sealing joints against leakage. Some materials need special support. Adapters are available to convert from one system to another. A total plumbing system is generally two or more systems. One supplies fresh water; one may also distribute heat. A third system takes away waste water. Some installations also have a system supplying fuel gas.

CAREERS RELATED TO CONSTRUCTION

PLUMBER assembles, installs and repairs pipes, fittings and fixtures of heating, water and drainage systems.

PLUMBING INSPECTOR looks at and approves new plumbing systems to assure that all codes have been followed in installation.

MECHANICAL CONTRACTOR prepares estimates, makes bids, buys materials, makes working drawings for all types of systems that supply water, heat, electricity and air conditioning to a building.

DISCUSSION TOPICS

1. Who publishes the Uniform Plumbing Code?
2. Valves are usually made of what material?
3. What is the American National Standards Institute (ANSI)?
4. List the types of screwed pipe fittings.
5. What method is used to join copper tubing?
6. What is the difference between copper tubing and copper pipe?
7. Describe the process of soldering a pipe fitting.
8. Should a trap be installed for each plumbing fixture?
9. Describe the kind of materials used for fuel gas piping.
10. Who makes working drawings for plumbing installations?

Chapter 49

CABINETRY

Fig. 49-1. Cabinets are attractively built and arranged for service and beauty. (Long-Bell)

Fig. 49-2. Many modern homes have attractive laundry storage areas. Most are built on the site by skilled carpenters. Unit pictured here is free-standing and can be moved.

Cabinets furnish specialized storage for many areas of a building. Kitchen cabinets help the homemaker keep cooking utensils orderly. They also provide counter top work area as in Fig. 49-1.

Cabinets are used throughout the home in the dressing area, dining, laundry, Fig. 49-2, serving area, Fig. 49-3, and in a variety of minor storage areas.

Cabinets must be carefully planned and well built. Functional kitchen cabinets must be adaptable to changing conditions. They must take care of different kinds of appliances. Cabinet interiors must be suitable for storing many sizes and shapes of dishes, silverware, packaged and canned goods.

Cabinetry of commercial buildings may be designed for display purposes of merchandise as in Fig. 49-4. This will change the method and materials used for attaching doors.

There are three basic methods of producing cabinets:

1. Some are built on the construction site.
2. Some are built at a mill and supplied in a knocked-down condition.
3. Some are mass-produced in a factory and delivered assembled.

STANDARD CABINET SIZES

Architectural drawings will include the location and design of cabinets for the building. Specific details concerning cabinetwork are sometimes included in the drawings. There may be elevation drawings as in Fig. 49-5. These will be drawn to scale. Dimensions will be placed on the drawings. Written specifications will describe the type of joinery as well as the kind and quality of material to be used.

Overall heights and general features of built-in cabinets are standard, Fig. 49-6. Base cabinets for the kitchen usually are 36-in. (91 cm) high and 24-in. (61 cm) deep, including the counter top. For most people, this is a comfortable working height while standing.

Most upper cabinets are 30-in. (76 cm) high and 12-in. (30 cm) deep. The distance between the base cabinet and the

Cabinetry

Fig. 49-3. Sewing center was built on site. Storage is built into wall while hinged work table folds up and stores against the wall when not in use. Support for near end of work table becomes cork bulletin board in stored position.

Fig. 49-4. Hardwood plywood display cases have an added look of elegance that serves to highlight the quality of the products shown. These cases are required to store products as well as display them. (Hardwood Plywood Manufacturers Assoc.)

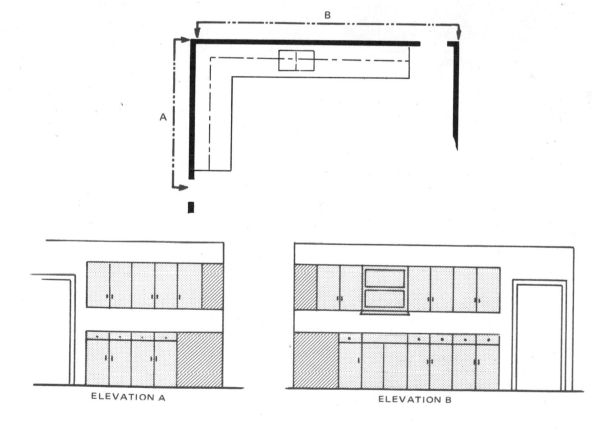

ELEVATION A   ELEVATION B

Fig. 49-5. Elevation details of cabinets found on the architectural drawing. Though not shown here, dimensions would be included.

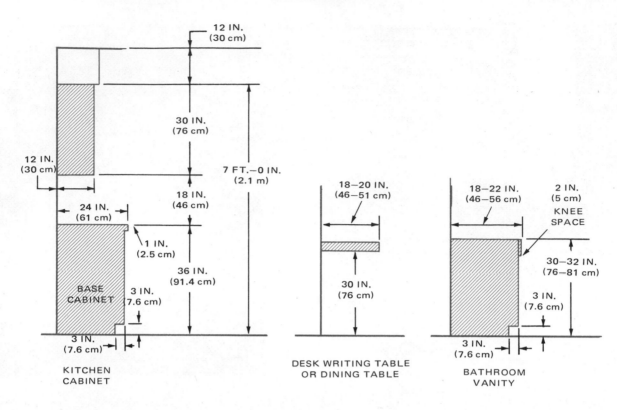

Fig. 49-6. Kitchen cabinets are built to basic standard measurements.

Fig. 49-7. The kitchen writing desk, like all desks, should be 30 in. high. (Nutone)

upper cabinet is usually 18 in. (46 cm). Writing desks should be 30 in. (76 cm) above the floor for convenient writing height. See Fig. 49-7.

In most kitchens, cabinets do not go all the way to the ceiling. The space above is closed in with rough framing and dry wall. This is called a "drop" ceiling. If there is no drop ceiling, it is better to have two sets of cabinets. Small ones are placed toward the top. Larger ones below are for storage of most-used items.

CONSTRUCTION OF CABINETS

The cabinetmaker has a choice of two methods of building cabinets on the job:
1. The frame construction.
2. The nonframe or all-panel construction.

In the frame method of construction the cabinetmaker uses solid wood framing material covered with a thin plywood end and back panel, Fig. 49-8. The only solid wood used is for the facing framing, door frame or drawer front. In the second method, heavier plywood, hardwood, or particle board is used. See Fig. 49-9.

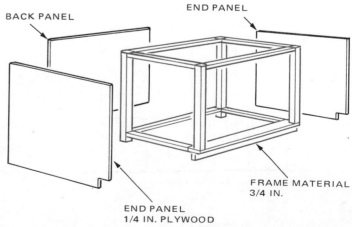

Fig. 49-8. Frame method of cabinet construction uses heavier members for frame and light panels to cover frame. Fronts are of heavier stock.

344

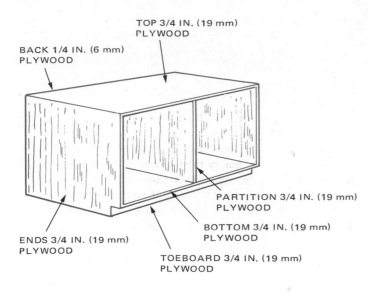

Fig. 49-9. Nonframe or panel construction for cabinets.

All base cabinets are built with a toe strip or toe space at the floor. Wall units are constructed in much the same way as base units.

During the basic framing operations, care must be taken to allow openings that are large enough for built-in appliances. Rough-in drawings are furnished by the manufacturers of the appliance units.

After the frame or panels have been assembled, facing strips are attached to the front. These facing strips frame door and

Fig. 49-10. Stock is being readied for hardwood cabinet facings. Plants specializing in ready-built cabinets stock pile various parts.
(Western Wood Moulding and Millwork Producers)

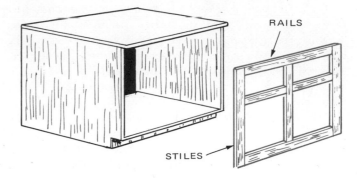

Fig. 49-11. Face frames are often assembled before being attached to cabinet front. When finished, this unit will have two drawers at top and double doors beneath.

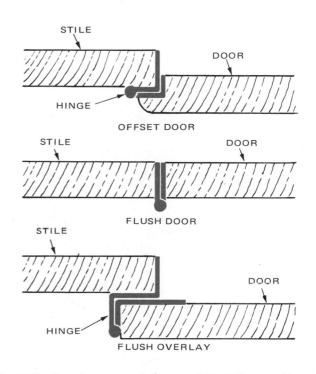

Fig. 49-12. How three types of doors would look if we could see how edges fit the frames.

drawer openings. Very good grades of hardwood are often used for this purpose. See Fig. 49-10. Often these strips are assembled into a framework before being attached to the cabinet. This is shown in Fig. 49-11. Vertical members are called stiles. Horizontal members are called rails.

DOOR CONSTRUCTION

Doors for cabinets must be attractive as well as useful. The main function of doors is to close off storage space. But doors should also add interest to the outside of the cabinet. Glass or metal grillwork is sometimes used, if the owner wishes, to show off fancy china, glassware or silverware.

Doors are fitted to the face frame in one of three ways. They are named by the method used.

1. The offset door.
2. The flush door.
3. The flush overlay door. See Fig. 49-12.

The offset door is easier to fit because its edges cover part of the face frame. Therefore, no cracks can show at the edges. A rabbet (or right-angle groove) is cut around the edges of the door. This allows part of the door to fit inside the opening. The rest covers the frame. The outer edge of the door is usually rounded or decorated by another design or edge treatment.

The flush door fits into the frame opening so that all surfaces are flush. The flush door is made about 1/16 in. smaller on each side so it will open and close without rubbing.

The flush overlay door is made 3/8 in. larger on each side than the opening. This overlay covers the entire opening.

Sliding doors are useful when a hinged door would take up too much space by opening into the room. See Fig. 49-13. Such doors are also used for safety when the doors are all glass. Sliding doors have one disadvantage. Only half of the cabinet interior is accessible at a time.

Sliding doors are made of hardboard, plywood, glass or plastic. There are several ways of installing them. The simplest is to cut grooves into the top and bottom rails. Sometimes metal or plastic track is used. This way, grooves do not need to be cut in the rails.

Doors are constructed as:

1. A solid panel.
2. A combination of frame and panel, Fig. 49-14.

The solid panel door is usually made from 3/4 in. (19 mm) plywood. The frame-panel door is made of solid wood frame. The frame requires four or more pieces. There are two vertical side pieces. These too, are called stiles. Two horizontal pieces, top and bottom, are called rails. If there is another horizontal piece in the middle of the door it is called a cross rail. Vertical divider is called a mullion or cross stile.

For most cabinet doors the thickness of the frame can be as little as 3/4 in. or as much as 1 in. Larger doors will have the thicker panels. The center panel in frame and panel doors may be constructed from 1/4-in. plywood or solid wood. A slot of some kind must be cut in the frame to hold the panel. See Fig. 49-15.

Strength of the door depends on the method of joining the corners of the frame. The mortise and tenon joint, Fig. 49-16, is most often used.

Fig. 49-13. Sliding doors are sometimes set in grooves cut in the upper and lower rails of cabinet. This illustration shows view from end of cabinet as though end of cabinet is removed.

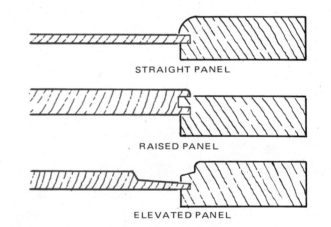

Fig. 49-15. Cutaway view shows panel shapes. Note how frames are grooved to hold panel.

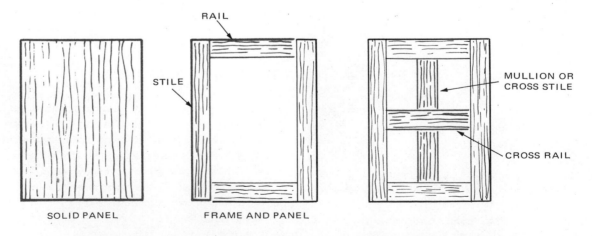

Fig. 49-14. Doors can be made of single pieces of wood or of several pieces jointed together.

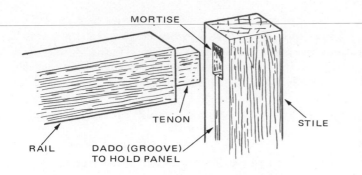

Fig. 49-16. Mortise and tenon joint is very strong.

DRAWER CONSTRUCTION

Drawers must be soundly built. In kitchen cabinets, drawers are used for storage of silverware and cooking utensils. Drawers make excellent storage space for three reasons:

1. It is easy to arrange the contents through use of dividers.
2. Drawers are relatively clean and keep contents dust free.
3. Drawers conceal items from sight until needed.

Drawer construction is often a good indication of overall furniture quality. If the drawer joints in a certain piece are good and if the drawer slides easily when pulled by the corners, then the furniture is usually of good quality.

Drawers have five parts: the front, the right side, the left side, the back and the bottom.

Drawer fronts are made of either solid wood or plywood. They are usually not less than 3/4 in. (19 mm) thick, Fig. 49-17. Stock for drawer sides is also solid wood or plywood. It is usually 3/8 in. or 1/2 in. (10 mm or 13 mm) thick.

High quality drawer sides are made from either oak or maple. Inexpensive cabinets may use pine, poplar or willow. Different kinds of joints are used for attaching sides to the front. The most difficult joint to make is the dovetail. However, the dado, dovetail dado, rabbet and others are often used. Fig. 49-18 shows some of these joints. The drawer sides have grooves, as shown in Fig. 49-17, to receive the bottom piece.

The bottom is made from 1/4-in plywood. The back piece is made from the same stock as the sides. It is joined to the

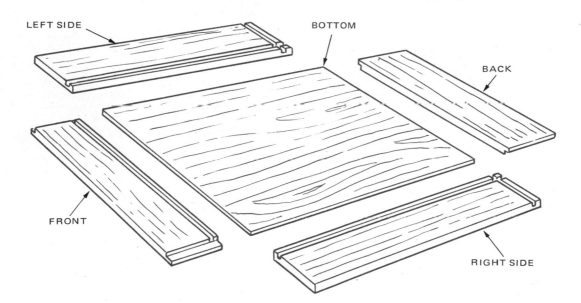

Fig. 49-17. Parts of a drawer. Note rabbet cut in front and dado cuts in side pieces.

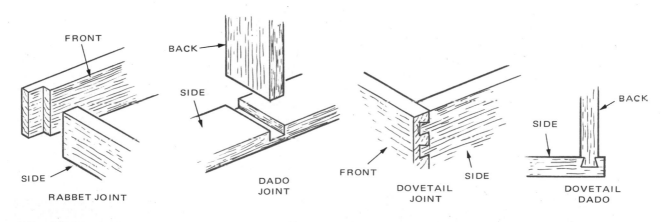

Fig. 49-18. Joints used on drawer construction. Drawer fronts on cheaper units will use rabbet joint. More expensive units will use a dovetail. Drawer backs are attached to sides with a dado joint or dovetail dado.

Construction

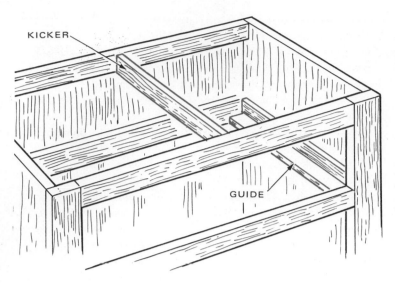

Fig. 49-19. Drawer guide and kicker are installed to steer drawer in and out of opening.

sides by a dado joint. It is butted to the bottom. Joints for fastening the drawer parts together are chosen after the cabinetmaker knows what quality of drawer is wanted. It takes more time to produce good joints and they are therefore more costly.

The drawer support system includes a guide and a kicker, Fig. 49-19. The kicker prevents the drawer from tipping when opened. The guides keep the drawer from slipping sideways.

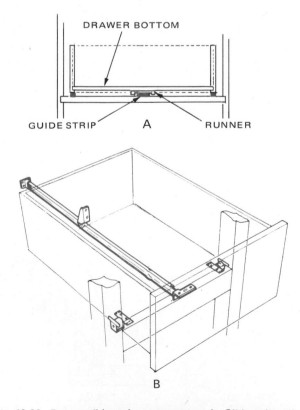

Fig. 49-20. Drawer glide and runner system. A—Glide strip and runner as seen from the back. Glide stip is attached to cabinet frame. Runner is attached to drawer. Back (dotted line) is removed to reveal parts. B—Cabinet drawers work freely with this metal and plastic glide system.

The corner guide may be formed in the cabinet by the side panel and frame. It may be necessary to add a spacer strip to hold the drawer square with the front facing. A single glide strip may be located in the center of the drawer opening with a runner placed on the bottom of the drawer, Fig. 49-20.

Special metal guides and load supports are manufactured for drawers. Many have rollers with nylon or ball bearings, as shown in view B of Fig. 49-20.

Drawers, like doors, are classified in two general types, the lip or offset and the flush mounted. Flush drawers must be carefully fitted and are commonly found in furniture construction. However, Fig. 49-21 shows flush drawers in kitchen cabinets. Lip or offset drawer fronts have a 3/8 x 3/8-in. (10 x 10 mm) rabbet along its edges which overlap the opening. The offset drawer allows looser fit.

Fig. 49-21. Flush mounted drawers may be used in kitchen cabinets. (Formica Corp.)

SHELVES

Shelves are used widely in cabinetwork. They are often found in wall units where drawers are not used.

In some designs it may be necessary to fit and glue the shelves into dadoes cut into the side pieces as in Fig. 49-22. The dado joint provides greater structural strength to the unit.

Whenever possible, it is a good idea to make a shelf adjustable. Such shelves can be adjusted for different sizes. See Fig. 49-23.

COUNTER TOPS

The work area of the cabinet is the counter top. This surface should be rugged, beautiful and waterproof while

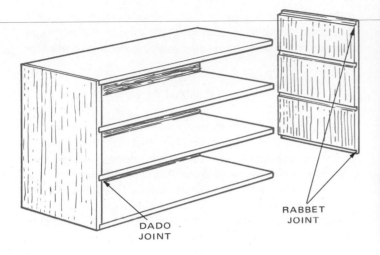

Fig. 49-22. Dado and rabbet joints are used for setting shelves into side panels. Such joints are very strong.

Fig. 49-24. Ceramic material used here for this counter top gives this kitchen a durable work area. Colors used in counter top materials go well with colors found in the walls or in flooring materials.
(American Olean Tile Co.)

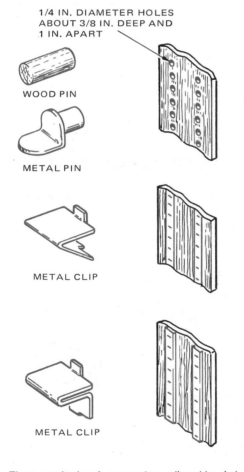

1/4 IN. DIAMETER HOLES ABOUT 3/8 IN. DEEP AND 1 IN. APART

WOOD PIN

METAL PIN

METAL CLIP

METAL CLIP

Fig. 49-23. Three methods of supporting adjustable shelves. These simple forms are easy and economical to install. Metal strips, metal insert pins and clips are available from hardware dealers.

resisting stain and heat. See Fig. 49-24. The counter top is often in a color that accents basic room color. Counter material should be easily cleaned because food is constantly being prepared there.

Counter tops are made up in many kinds of material. The plastic laminates are most popular. Plastic laminates have all the qualities needed. They are economical too. Laminate comes in many colors and decorator patterns.

Plastic laminate is bonded to a base of either 3/4-in. plywood or particle board. Edges of counter tops are built up to make the top look much thicker. Two methods of construction are shown in Fig. 49-25. When the laminate is applied on the job, the counter top has a straight-formed edge as shown in view A. The post-formed counter top, view B, is usually manufactured in a plant. Special equipment is needed to heat and form the plastic laminate. The counter top material generally is curved or carried up the back wall to a height of about 5 or 6 in. (13 or 15 cm). This provides a back

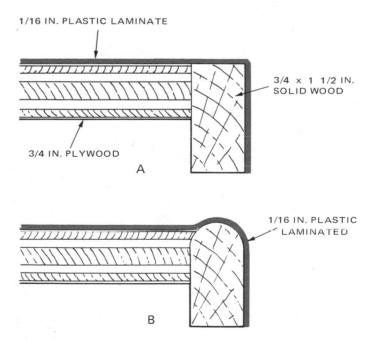

1/16 IN. PLASTIC LAMINATE

3/4 x 1 1/2 IN. SOLID WOOD

3/4 IN. PLYWOOD

A

1/16 IN. PLASTIC LAMINATED

B

Fig. 49-25. Edges for counter tops. A—Edge strip is separate piece of laminate. B—Edge forms single piece with top.

Fig. 49-26. Counter top material running up wall prevents water damage to wall covering.

splash to keep water off the wall covering material. Bathroom vanities, Fig. 49-26, sometimes use plastic laminate this way.

CABINET HARDWARE

Cabinet hardware includes knobs, pulls, hinges, catches and other metal fittings that are put on the cabinet. See Fig. 49-27. The hardware is prefitted before the finish is applied. Cabinetmakers use care in locating and drilling the holes for the hardware. They use sharp drill bits so that the wood will not splinter around the hole.

Drawer pulls usually look best when they are located slightly above the center line of the drawer front. However, they are centered horizontally.

Door pulls or knobs of the base unit doors are located somewhere in the top one-third of the door. For doors in the upper units the pulls are located in the lower one-third of the door.

SUMMARY

Cabinetry provides a structure with a finished appearance. More valuable however, is the fact that cabinets provide a special storage space for the small items. In some cases cabinets also provide work areas. In kitchens these work areas are called counter tops. Standard sizes are used for the convenience of the person who will use the work area.

Cabinets have drawers for smaller items and doors to close the shelf areas. Shelves of the upper units are spaced for the best use of the storage area. Adjustable shelves are convenient and easily built.

Joinery is the secret of good cabinet construction. The proper selection of joints will give the cabinet the greatest strength for the pieces used. With glued joints, the cabinet will become very rigid and durable.

CAREERS RELATED TO CONSTRUCTION

CABINETMAKER constructs and repairs wooden storage units and cabinets found in stores, offices and homes; cuts, fits and assembles wood pieces and attaches hardware. Mounts or attaches units to walls.

DISCUSSION TOPICS

1. What three methods are used to produce cabinets for construction?
2. What are cabinets used for? Where are they located?

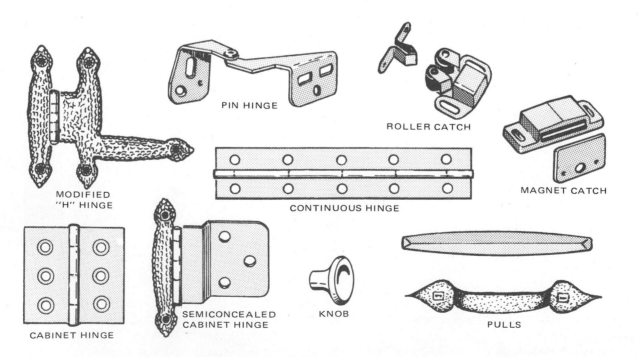

Fig. 49-27. Types of hardware for cabinet doors.

3. Why are cabinets built to standard sizes?
4. List the types of door construction.
5. List the parts of a drawer.
6. What is a dado joint best suited for in cabinet construction?
7. Why have a kicker for drawers?
8. Discuss the requirements for a good counter top.
9. What materials are generally used for cabinet construction?
10. Where are the pulls or knobs generally located on doors of base units?
11. Where are the pulls usually located on cabinet drawers?

Chapter 50

FINISHING THE PROJECT

Fig. 50-1. Decorating rooms and laying floor covering have added the final touch to office building construction project.
(The Verticel Co.)

Among the last steps in completing a construction project are painting, decorating, laying floor coverings, installing lighting fixtures and installing accessories. Fig. 50-1 shows a completed business office.

Painting and decorating put a protective and decorative coating on interior and exterior surfaces, Fig. 50-2. Decorative materials may be applied as a liquid or in the form of thin sheets, such as wall paper. The "wall paper" actually may be sheets of vinyl or wall fabric.

FLOOR COVERINGS

Flooring material varies. Carpet, wood and tiles of many materials are laid as floor coverings. Each has advantages and many need special surface preparation, Fig. 50-3.

Wood floors have been used in homes, offices, basketball courts and classrooms for many years. Oak, maple and birch are used for flooring planks today, Fig. 50-4.

Resilient flooring has grown in usefulness with advances in technology and development of material. Resilient flooring material usually is cemented to the subfloor. It comes in large rolls and is cut to fit the room with the least possible joints. Resilient materials include linoleum, asphalt, rubber, vinyl and cork, or combinations of these.

At present, there is increased interest in ceramic tile as a floor covering, Fig. 50-5.

APPLYING PAINT

Painted surfaces can be very serviceable, sanitary, clean and colorful. Before any paint can be applied, however, that surface must be thoroughly free from dirt, grease and other foreign material. For most paint, the wall and ceiling surfaces must be dry. Any loose paint or other flakiness of the surface must be removed.

Many methods of applying paint have been developed. Brushing is probably the oldest technique. Even today, more exterior paint is applied by brush than by any other method.

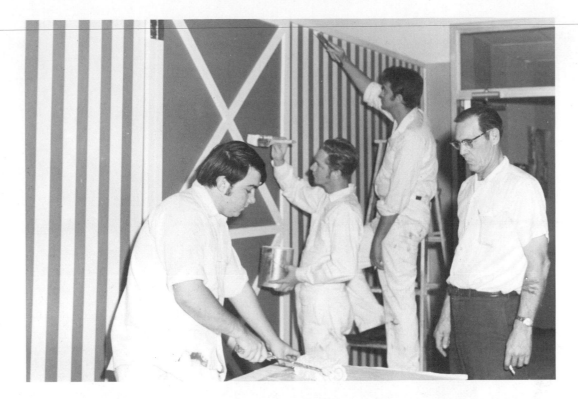

Fig. 50-2. Apprentice painters learn the many techniques of covering interior walls with a protective and decorative coat of paint. (Dallas Independent School District)

Fig. 50-3. Finished home has many features to enhance comfort and appearance of rooms. For example: Carpet deadens sound and establishes basic color scheme; vertical wall boards add warmth and beauty. (Western Wood Products Assoc.)

Fig. 50-4. Carpenter lays hardwood flooring over subflooring with a special nailing device. (National Oak Flooring Mfg. Assoc.)

Fig. 50-6. Spraying is the generally preferred method for applying paint over rough-textured surfaces. (Cook Paint Co.)

Fig. 50-5. Finishing the project includes setting and cleaning ceramic tile, installing plumbing fixtures, adding electrical switch plates and lighting fixtures. (American Olean Tile Co.)

are used for finished concrete and concrete block walls. There is a growing demand for masonry paints. Portland cement paint with water-resistant silicones are now being used to coat masonry walls for construction, Fig. 50-6.

Paint generally is applied in several layers or coats. The first is called a primer coat. *A primer coat is formulated for use over certain materials and under particular finish coats.*

Often, paint is used to provide colorful and contrasting trim for finishing certain areas, Fig. 50-7.

The rapid expansion of construction activities in the Twentieth Century has developed many new techniques for applying paint. Large surfaces can be rapidly painted by the use of paint rollers. Rolling paint is a popular method for home owners, because rollers are easy to handle and do the job quickly with very good results.

Paint spraying is used to gain professional results. Spraying requires a considerable amount of equipment. The spray gun and cup are attached to a compressed air system that propels the paint onto the surface. *Specially formulated spray paints*

Fig. 50-7. Painted trim, decorator panels and wall sconces provide the "finishing touches." (Western Wood Moulding & Millwork Producers)

Fig. 50-8. Lighting fixtures for exterior of structure usually are installed in the last stages of construction. (IDS Center)

Fig. 50-10. Lighting designed to accent ceiling provides contrasting light and shadows. (Aviation Div., Kansas City International Airport)

ELECTRICAL FIXTURES

The installation of electrical fixtures, Fig. 50-8, usually is the final task necessary to the completion of the construction project. Electrical connections are made with the fixtures in several different ways, Fig. 50-9. Light fixtures take many forms and provide lighting for different reasons. All fixture selections, however, are based on the illumination of space or areas, Fig. 50-10.

Switches and receptacles are wired and fastened to wall boxes, Fig. 50-11. The boxes, in turn, are finished by covering with plates of plastic or metal fastened by small machine screws, Fig. 50-12.

While the fixtures are being installed, workers are setting up

Fig. 50-11. Electrical connections in receptacles and switches are easy to make. Many simply push in a slot that grips the conductor. (Leviton Mfg. Co.)

other electrical appliances and accessories. Typically, these accessories include dishwashers, heating units, fans, garbage disposals and other built-in units.

Fig. 50-9. Electrical fixtures in this office area make up the entire ceiling. (Armstrong Architectural Ceiling Systems)

Fig. 50-12. Cover plate covers box and working parts of switch, making neat appearance and protecting persons from electrical hazard.

SUMMARY

Many finishing operations are needed to complete a project. Interior and exterior decoration are part of the finish work. Painting, wall papering and other wall coverings complete the interior design by adding pattern, texture and color to walls. The installation and fitting of electrical fixtures usually is the final stage of construction.

The last step in finishing the entire project is the removal of all construction debris from the site.

CAREERS RELATED TO CONSTRUCTION

PAPERHANGER applies sizing to seal wall surfaces; measures walls and cuts strips to fit from rolls of wallpaper or fabric; mixes paste and brushes it on back of strip; installs strip on wall, matching adjacent edges of figured paper; smooths joints with dry brush or felt roller.

ELECTRICIAN plans electrical layout and installs wiring, electrical fixtures, apparatus and control equipment; pulls wiring through conduit and connects to lighting fixtures and power equipment; installs grounding leads, tests continuity of circuits and observes function of installed equipment.

PLUMBER studies building plans to determine sequence of installation of plumbing units; assembles and installs pipes, fittings, valves and fixtures of heating, water and drainage systems, according to specifications and plumbing codes; fills system with water and checks for leaks.

PAINTER selects premixed paints or prepares paint to match specified colors; smooths and prepares surfaces for painting; applies coats of paint, varnish, stain, enamel or lacquer to decorate and protect interior or exterior surfaces, trimming and fixtures of buildings and other structures.

DISCUSSION TOPICS

1. Why are electrical fixtures installed last, instead of at the same time the major electrical work is completed?
2. What holds an electrical ceiling-mounted fixture to the ceiling?
3. How is a plumbing system tested?
4. What is the purpose of a primer coat in painting?
5. Why spray paint?
6. List as many ways as you can to finish a wall.
7. Is installing a telephone part of a construction activity?
8. Who is responsible for installing the telephone?
9. What is the advantage of carpeting for a floor covering?
10. How is wood strip flooring installed?
11. For what reason is paint used?

Electricians work on both new and old construction, installing and repairing electrical circuits. (Recruitment and Training Programs, Inc.)

Chapter 51

LANDSCAPING

Fig. 51-1. This shopping plaza landscaping plan calls for work on the inside as well as the outside of the structure.
(American Olean Tile Co.)

No construction project is completed until the lot is graded and landscaped. Landscaping starts with the original plan. A landscape architect prepares the landscape plan. In a large project, such as a shopping plaza, there is often indoor landscaping to consider. See Fig. 51-1. *Landscaping is not finished until carried out. Each tree, shrub, flower, rock, fountain or other focal point must be placed according to the landscape plan.*

The landscape plan may be as small as an apartment garden or as large as the planning of a complete town, park, highway or golf course. The first conditions to be considered include:
1. The use for which the site is intended.
2. Topography factors.
3. Climate and geological factors (wind, temperature, soil and moisture conditions).
4. Historical factors.
5. Economic factors which may affect all the others.

THE LANDSCAPE PLAN

A landscape plan puts down on paper or makes a model showing the location of the building and the other features

Fig. 51-2. Landscape planners sometimes prepare a model of the project to give customer clearer idea how finished landscaping will look.
(Hall's Crown Center)

357

which surround the structure. See Fig. 51-2. When the plan is for a residence, it will show the position of the house, garage, walks and drives on the building lot, as in Fig. 51-3. It then shows the location and species of the plantings. Other features of the plan are the spacing and shape of garden beds. Points of interest, such as stone outcroppings, will also be noted.

The landscape plan may be prepared by a nursery. Many nurseries offer free landscaping services and advice to their customers.

When the landscape plan is drawn by a landscape architect it can be used to get competition bids on the installation. A landscape contractor may be issued the contract to develop the landscape plan. The contractor will supply the plants, set them out, make beds and handle other landscaping arrangements.

LANDSCAPE DESIGN TIME

Time plays a greater role in landscaping than in other forms of construction planning. Plants and shrubs require time to achieve their intended beauty. Architectural landscaping planning may very well be worked around trees already on the property. Plantings selected for quick beauty may be worked into the plan as well as those which require more time.

LANDSCAPE TEXTURE

The importance of texture in a landscape increases as the pattern becomes more complete, Fig. 51-4. The "Japanese garden" gets contrast by varying plants, rock, sand, gravel and water, Fig. 51-5.

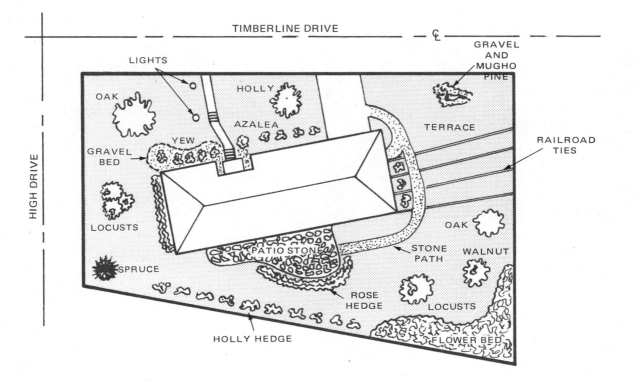

Fig. 51-3. Landscape plan for a home locates house, walks, driveway, all plants and other features of a yard.

METHODS OF LANDSCAPING

The landscape designer has a dual function:
1. To approach the problem first as an expert, examining and analyzing all the circumstances and conditions of the site.
2. To act as an artist with complete knowledge of materials available to create a landscape.

The most important principle of design is unity. To get unity, the designer plans for coherence and continuity of scale. This means the individual parts must seem to be a part of the whole. See Fig. 51-2. For large projects like shopping centers, the plan should have an underlying harmony between all its parts. Harmony might be the reoccurrence of fountains or like species of plantings.

LANDSCAPE COLOR

Color gives added contrast to the structure. Color can be put into the plan in several different ways. The easiest way is with blooming flowers, as in Fig. 51-6.

Large open areas are traditionally covered with clipped grass. Green predominates. Not all colorings go with it. For example purple-leaved trees should be used with restraint. Statues, steps, fountains and large stones of contrasting color provide the landscape plan with a special focal point as in Fig. 51-7. (A focal point is a main center of interest. It is the thing you will notice first.) Abstract forms have the same effect in a setting of contrasting foliage as do the natural stones of the Japanese gardens.

Fig. 51-4. Insert in the patio surface provides additional texture under foot while the arboretum provides texture overhead. (California Redwood Assoc.)

Fig. 51-5. Natural materials go well with plantings and existing trees. They give a very interesting texture and contrast to surrounding space. (Western Wood Products Assoc.)

Fig. 51-6. Small garden has blooming flowers to contrast with the large wood decking.

Fig. 51-7. A statue provides a center of attention (focal point) for this interior landscape area in a shopping center. (National Terrazzo and Mosaic Assoc.)

WATER IN LANDSCAPING

Water in any landscape scene focuses attention, Fig. 51-8. Moving water brings a feeling of life and involvement, Fig. 51-9. Expanses of still water, mirror pools, give the feeling of peace. When pools are designed to reflect the sky colors, they must be still and nearly level with the basin edge. When used, water should fit the general trend of the surrounding area. Water should always be placed in the lowest level of area. See Fig. 51-10.

TREES AND SHRUBS

Trees and shrubs have always been a part of the landscaping. Living plants give a feeling of life. Trees and shrubs give shelter, shade and protection. Certain plants provide texture, framing, screening, background color and form. See Fig. 51-11.

In large, open expanses, trees may be planted in clumps or small groves. A mound or small raised area provides a break in the monotony of the surroundings.

Fig. 51-8. Falling water calls attention to this four-story indoor tropical garden. It was carved out of a natural limestone outcropping to form a backdrop for a hotel lobby. Hundreds of plants, shrubs and full grown trees provide the landscape. (Hallmark's Crown Center)

Fig. 51-10. This pool is placed so that it reflects the sky and surrounding shrubbery when not in use. (California Redwood Assoc.)

Fig. 51-9. Fountains have been used for centuries in the landscape plan. Moving water gives the feeling of activity while refreshing the air in hot climates.

Fig. 51-11. Both broad leaf and evergreen shrubs provide a variation of color and texture as a background for these steps. (California Redwood Assoc.)

Fig. 51-12. The natural look. Care is taken not to disturb the surrounding shrubbery.

NATURAL LANDSCAPING

Natural landscaping adapts the structure to the site. The original condition of the site is preserved as much as possible. See Fig. 51-12. Natural landscaping is done mostly in residential construction. The site is generally carefully selected to have a desirable natural surrounding.

LAWNS

The lawn is an important part of the landscape plan. Lawns provide:
1. Ground cover.
2. Control of surface water.
3. A pleasing look as in Fig. 51-13.

Good lawns are achieved by one of two methods, seeding or sodding. Seeding is an inexpensive but slow method of getting a good ground cover. Sodding is quick but expensive. Sod is

Fig. 51-13. Lawns add color, help provide water control and help control blowing dust and debris in drier regions. (Hallmark's Crown Center)

usually grown on a special sod farm. Sod is cut just below the ground level, retaining the roots system. Rolled in small sections, it is transported to the landscaping site. It is unrolled, placed, tamped and watered to insure root growth.

SUMMARY

Landscaping is the final construction activity. Landscaping is a form of art, combining materials and plants into pleasing displays.

Landscaping is a planned activity done by a landscape architect or a nursery. The landscape architect is one of the consulting team of specialists who has worked on the project from the planning stage.

CAREERS RELATED TO CONSTRUCTION

LANDSCAPE ARCHITECT plans and designs surroundings for new structures or whole communities. Plans placement of trees, shrubs, decides contouring of site, studies soils, drainage, approves materials used in landscaping.

NURSERY MANAGER manages nursery where trees, shrubs and flowers are grown. Prepares and plants them on customers grounds. Selects and purchases seeds, fertilizer and gardening equipment. May provide landscaping service, advice or consultation. Hires and supervises NURSERY WORKERS.

DISCUSSION TOPICS

1. How do climate and weather affect the landscape plan?
2. What does a landscape plan include?
3. How does topography affect the landscape plan?
4. What are the duties of the landscape architect? Of the nursery person?
5. Name several species of trees used for ornamental landscaping.
6. Why would a landscape architect use both broadleaf and evergreen shrubs together?
7. Could lighting affect the landscape plan?
8. What are fountains? How do they fit into the landscape plan?
9. What methods would you use to add color to the landscape plan?
10. What is natural landscaping?

Chapter 52

SAFETY AND TECHNOLOGY

Construction is the process of building. It is the science of using proper tools, techniques and materials to erect structures in the most efficient manner.

From the beginning, construction has answered the human need for shelter and protection. Early civilizations built simple means of cover to shield families and belongings from adverse weather conditions. They also learned to construct towers, fortresses and castles to provide protection from invading hordes. Over the centuries, construction has served people in countless ways, either in fulfilling a need or in satisfying a desire to improve conditions.

Construction, however, can be a dangerous undertaking. In doing certain jobs, the construction worker is subjected to potentially harmful conditions; heights, heavy loads, massive equipment and handling of materials hazardous to the health.

Construction hazards can be reduced to a minimum if proper safety techniques are followed. Unfortunately, two problem areas exist:

1. Not all individual workers practice safety awareness.
2. Not all construction management officials see the need for providing tools and machines that are safe, nor do they provide workers with the necessary personal safety equipment.

The Williams-Steger Act, better known as the Occupational Safety and Health Act (OSHA), spells out safe working conditions. This law includes desired practice for construction workers. For example, no worker is expected to work in surroundings or under conditions that are unsanitary, hazardous or dangerous to one's health or safety. *OSHA requires that contractors provide a safe and healthful working environment for all workers on the job.*

OSHA places the safety responsibility on the employer, not only at the beginning or the start of the job, but at all times workers are on the job. It is the responsibility of the contractor's supervisor and/or foreman to carry out the safety function.

Accidents cost the contractor time and money, so a safe and secure working environment is good, sound business. It means greater efficiency and higher production from all workers and supervisors alike.

Every construction worker, regardless of the kind or nature of the construction work he does, uses tools and equipment. Regardless of who owns or supplies the tools or equipment, they must be safe to use on the job. Using an unsafe tool can cause serious injury to the user or to a fellow worker. Improperly maintained equipment also can be the source of accidents.

PERSONAL PROTECTION EQUIPMENT

Each construction worker is expected to have and use certain personal safety "protective wear." When a worker is properly protected, minor injuries can be eliminated and serious injuries prevented. At least, the severity of the injury will be reduced if the worker is wearing one of these protective devices.

Part of the overall responsibility of the construction supervisor is to make certain that each worker is properly equipped with and uses the correct protective equipment.

PROTECTIVE HARD HAT

The protective "hard hat," Fig. 52-1, has become almost universal in its application on the construction job. The hard

Fig. 52-1. The hard hat is a necessary piece of protective equipment worn by construction workers on any construction site.
(Corps of Engineers, Kansas City District)

hat has become a symbol of the occupations centered around construction. The hard hat protects the worker in two ways:

1. The hard shell resists penetration by falling objects.
2. The lacing holds the shell away from the skull and will cushion any heavy blow, preventing the force from being transmitted through the hat directly to the skull.

The hard hat does for the construction worker what the batting helmet does for the baseball batter.

PROTECTIVE FOOT WEAR

The construction worker continually handles heavy materials, such as concrete blocks, bricks, and reinforcement bars. Should one of these materials fall, it is very likely that the worker's foot will be in the path of its fall.

Special protective shoes can minimize injury from this type of accident. Workers in areas where foot and toe injuries are likely may be provided with steel caps that slip over the toe areas of the shoes to give extra protection.

PROTECTIVE HEARING DEVICES

The noise level of many pieces of construction equipment is hazardous to workers, especially if the noise continues for long periods of time. Ear protective devices are required in those situations. Ear protective devices are designed two different ways:

1. The ear plug.
2. The ear muff, Fig. 52-2.

The ear plug usually is designed of soft rubber or plastic. It fits into the ear and reduces the amount of sound waves reaching the inside ear. Muff type ear protectors are considered to be far better protection than plug type.

Sound level meters separate and record the various amounts of sound generated by a given piece of equipment. Results are measured in decibels. The decibel is a rating scale that indicates the magnitude of sound. Fig. 52-3 shows the relationship of decibels to some common construction noises.

TYPICAL SOUND LEVELS	
DECIBELS	
140	
130	F84 JET AT TAKEOFF
120	SAND BLASTING UNIT AT 4 FT. (122 cm)
110	CUTOFF SAW AT 2 FT. (61 cm)
100	
90	HEAVY TRUCK AT 20 FT. (6 m)
80	
70	
60	AVERAGE CONVERSATION AT 3 FT. (92 cm)
50	
40	
30	
20	RECORDING STUDIO
10	
0	THRESHOLD OF HEARING BY YOUNG MEN

Fig. 52-3. Sound levels for various voices and pieces of equipment are given in decibels.

Fig. 52-4. Wraparound safety glasses give side protection for construction worker's eyes. (Bostitch Div., Textron Inc.)

PROTECTIVE EYE SHIELDS

Eye protection is essential, especially if the worker is in an area where chips fly from the work. Eye protection is also necessary in welding and cutting operations. The smoke glass lens fitted in the welding helmet permits the welder to view his work. It also shields the head area.

Good eye protection for the general construction worker should be safety glasses with wraparound shields, Fig. 52-4. The full face shield is still better eye protection.

Fig. 52-2. "Ear muffs" protect worker from deafness due to continual exposure to above-normal noise levels.
(Western Wood Moulding and Millwork Producers)

Fig. 52-6. When arc welding, heavy leather gloves are used to prevent radiation burns. Welders also wear protective coats, knee pads, leggings and helmets.   (Exxon)

Fig. 52-5. Construction worker spraying urethane foam on a new roof is wearing a respirator. Otherwise, the gases emitted by urethane material could cause lung damage.   (General Electric Co.)

skin. Heavy leather gloves, Fig. 52-6, protect welders' hands from heat and radiation burns of the electric arc.

PROTECTIVE DUST RESPIRATORS

Dust respirators, Fig. 52-5, prevent the worker from inhaling fine dust particles. After a period of time, dust inhalation can be damaging to the lungs. Air laden with dust and fumes from welding and chemical reactions is very irritating unless protective equipment is made available and used.

PROTECTIVE WORK GLOVES

Work gloves protect construction workers' hands from rough materials being handled and to prevent blistering of the

PROTECTION AGAINST FALLING

Some phases of construction work must be performed high above the ground. When this kind of work is done, special safety devices and equipment must be used to insure the safety of the worker. Nets, Fig. 52-7, and safety straps are among the devices provided.

Actually, the net does not protect the worker against falling from a high structure. It prevents the person from falling to the surface below the net.

The safety strap is a harness of webbing fitted to the construction worker and tethered to a secure point above the area of work. In the event of a misstep, the worker does not fall, but simply swings at the end of the tether line. The safety strap minimizes the possibility of serious injury.

Safety does not end with the proper use of protective wearing apparel. Most accidents could have been prevented

Fig. 52-7. Construction net is used to "catch" worker who accidentally falls.   (Port of New York Authority)

SCAFFOLDS

Scaffolds and ladders fill an important need in the construction industry. In order to perform a job at heights beyond which a worker can reach, some kind of scaffolding is required.

Scaffolds are classified as either wood or metal. Wood scaffolds generally are made on the job. There is a limit to the height a wood scaffold can safely reach. Metal scaffolds usually are prefabricated in units that fit together to make a

Fig. 52-8. In building this structure, scaffolding and shoring procedures provide support for concrete forms. Prefabricated steel units simplify scaffold erection and disassembly. (Safeway Steel Products)

with proper care. *Accidents usually are the individual's own making, caused by the use of faulty equipment or improper techniques.*

Some accidents are the result of unsafe construction practices. It is the responsibility of the construction supervisor to see that these problems do not occur.

Fig. 52-10. Swing platform is supported by adjustable steel beams attached to wall parapets and anchored to roof. Generally, a "life line" for each worker is attached to roof.
(Dan Ebberts)

Fig. 52-9. Swing platform is raised and lowered by a series of cables and winches to allow work to be done at various levels. (Patent Scaffolding Co., Inc.)

strong work platform of the desired height. See Fig. 52-8. The prefabricated units offer the advantage of standardized size of units.

Swing or stage scaffolds are suspended from the roof for use by painters and other workers performing their tasks on the outside of the structure, Fig. 52-9. Swing scaffolds are suspended from the roof by a series of cantilevered beams anchored to the roof, Fig. 52-10.

CONSTRUCTION TECHNOLOGY

Construction technology is the carrying out of a well-planned building program that incorporates materials, processes, management and personnel in a well-balanced sequence. See Fig. 52-11. Note that the Table of Contents of CONSTRUCTION follows this same sequence, and the chapters make clear how the four segments of construction

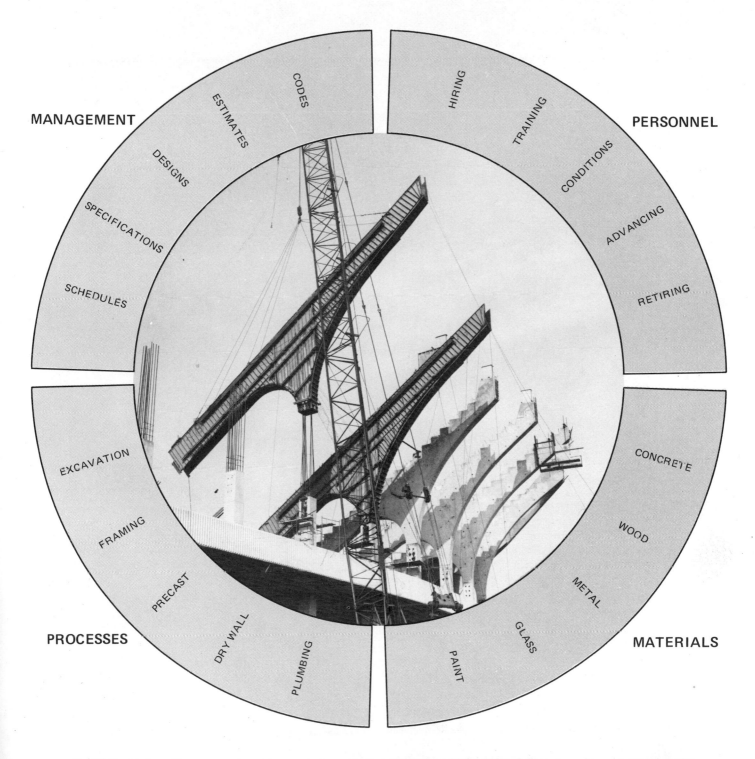

Fig. 52-11. Circle emblem represents total concept of construction terminology. Without any of these elements, construction would not be possible.

Fig. 52-12. Concrete ramp floats on teflon pads, which are unaffected by weather, chemical corrosion and extreme temperature change. (E. I. du Pont de Nemours & Co., Inc.)

Fig. 52-13. Skill in many trades is practiced by thousands of persons doing construction work. Cement masons are in demand and earn a good living. (Master Builders)

construction superintendent, the skilled carpenter, the expert electrician, the master plumber, the accomplished welder, the artistic stone mason and the strong yet nimble iron worker, to name a few. *Their skills are always in demand.*

Mastering a chosen trade gives the individual construction worker an "identity" that is unmatched in any other industry. Be sure to consider this "mark of the professional" when choosing *your* career.

technology fit together.

The total concept of construction technology, then, is the art and skill with which people utilize materials and processes, Figs. 52-12 and 52-13, and how well the builder organizes and manages labor (workers) and capital (money) needed to build a particular structure in a given length of time. See Fig. 52-14.

Some dedicated construction workers have spent their entire work lives studying and practicing specific building trades or skills. In this respect, recognition for consistently doing a good job goes to the fine architect, the knowledgeable

Fig. 52-14. Construction technology, the combination of materials, processes, management and personnel, produced this structure of great height and unusual design. (Canadian National)

USEFUL INFORMATION

The International System of Measurement is a serious effort by most nations of the world to create a common form of measurement. For the United States to get in step with other nations of the earth requires conversion of many of our present measurement standards into the metric standard. The base unit of linear measurement in SI metric is the metre. See Fig. 53-1.

MEASUREMENT IN THE UNITED STATES

The Constitution ratified by the United States gave the federal government the power to fix the standard of weights and measures. The authority was delegated (passed on) to Congress.

Secretary of State, Thomas Jefferson, submitted a report on weights and measures to Congress. One part of this report contained the proposal to establish a decimal system of weights and measures.

In 1866 use of the metric system in the United States became the one legal system. Yet, through common practice, the customary English system remained in use.

Congress passed a law in 1893 which set up a standard gauge thickness for many iron and steel products. Standard thicknesses and weights were given in both the metric and customary units. And most important, the yard and pound were identified on the basis of what fraction of the metre would make up a yard and what fraction of the kilogram would make up a pound. English length and mass measurements thus were defined by using the metric system.

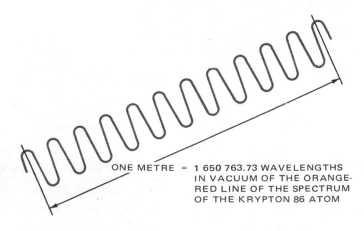

ONE METRE = 1 650 763.73 WAVELENGTHS IN VACUUM OF THE ORANGE-RED LINE OF THE SPECTRUM OF THE KRYPTON 86 ATOM

Fig. 53-1. The eleventh General Conference of Weights and Measures established the standard metre length.

The 1960 General Conference on Weights and Measures formally gave the revised metric system its new title "System International de Unite." The name has been shortened everywhere to "SI."

SI UNITS OF MEASURE

SI is by far the best system of measurement and calculation known. It is extremely convenient to use because of its base of 10. It provides greater speed in use. It is widely used by the scientific world. It is now also used by most of the world's population. A total measurement system, SI metric consists of the following:
1. Seven base units.
2. Two supplemental units.
3. Derived units which have special names and any number of combinations.
4. Non-SI units.

SEVEN BASE UNITS

QUANTITY	SI UNIT	SI SYMBOL
Length	metre	m
Mass (weight)	kilogram	kg
Time	second	s
Temperature	kelvin	K
Electric current	ampere	A
Luminous intensity	candela	cd
Amount of substance	mole	mol

Base units are standards based on a natural phenomenon. The standards can be reproduced and never have any change. Length, or one metre, is equal to 1 650 763.73 wavelengths of the orange-red lightwaves of krypton (kr) atom 86. The determination is accurate and can be reproduced accurately to 1 part in 100 000 000.

DERIVED UNITS

The SI derived units are formed by a simple mathematical multiplication and/or division of two or more SI base units. See Fig. 53-2. Therefore, the advantage of the SI system lies in the fact that the derived units are matched to the base units. This means the derived units are found by dividing or multiplying by base units. This is called a coherent system. Derived units are shown in Fig. 53-3.

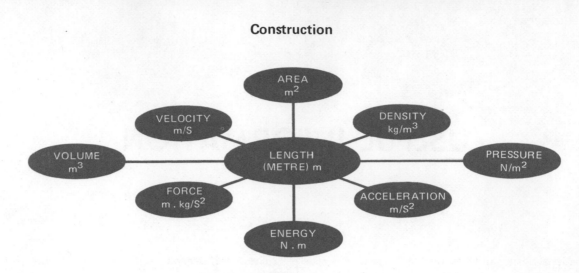

Fig. 53-2. The key to SI metrics is the metre.

QUANTITY	SI UNIT	SI SYMBOL
Energy (heat, work)	Joule	J
Force	newton	N
Power	watt	W
Electrical charge	coulomb	C
Electrical potential (voltage)	volt	V
Electrical resistance	ohm	Ω
Electrical conductance	siemers	S
Electrical capacitance	farad	F
Electrical inductance	henry	H
Frequency	hertz	Hz
Magnetic flux	weber	Wb
Magnetic flux density	tesla	T
Illumination	lux	Lx
Pressure	pascal	Pa
Luminous flux	lumen	Lm

Fig. 53-3. SI derived units.

There are 11 non-SI units shown in Fig. 53-4.

QUANTITY	SI NAME	SI UNIT	BASE UNIT
Liquid volume	litre	l	metre
Area	square metre	m^2	metre
Solid volume	cubic metre	m^3	metre
Temperature	celsius	$°C$	kelvin
Angle	degree	$...°$	radian
	minute	$...'$	radian
	second	$..."$	radian
Time	minute	min	second
	hour	h	second
	day	d	second
Mass	tonne	t	kilogram

Fig. 53-4. Non-SI units.

SUPPLEMENTAL UNITS

There are other units which are used so often that it has become convenient to give them names and symbols of their own. They are called supplemental units.

The supplemental unit is so called because the General Conference on Weights and Measures has not said whether they are base units or derived units. Supplemental units include:

QUANTITY	SI NAME	SI SYMBOL
Plane angle	Radian	rad
Solid angle	Steradian	sr

NON-SI UNITS

The General Conference on Weights and Measures has recognized that certain units are not part of the SI system. But these units are so widely used and accepted that they cannot be ignored or replaced. In most cases, the non-SI unit is:
1. An extension of a base unit.
2. A multiple or submultiple of a base unit.
3. A unit used with the SI in a specialized field.

PREFIXES

The multiples and sub-multiples of all SI units are formed by means of SI prefixes, Fig. 53-5. SI prefixes are based on the power of ten. Prefixes eliminate insignificant digits and decimals by the indicated order of magnitude. The powers of ten are, however, used in making calculations.

The prefix multiples are derived from Greek terms while the prefixes of sub-multiples are derived from Latin terms. All the SI prefixes may be used, the choice of prefix is governed only by its appropriateness in a particular circumstance. It is recommended that only one prefix be used in forming the value. For example:

Preferred: Mg (megagram) or (metric ton) = 1 000 000 grams
Not: kkg (1000 kilograms)   = 1 000 000 grams

USING THE PREFIXES

Numbers are added, subtracted, multiplied and divided in exactly the same manner as any decimal number. When using the SI metric system a step-by-step procedure will reduce the possibility of error when changing the prefixes. See the example shown in Fig. 53-6.

Prefix	Symbol Base Unit	Exponential Power	Decimal Value
tera	T	10^{12}	1 000 000 000 000
giga	G	10^{9}	1 000 000 000
mega	M	10^{6}	1 000 000
kilo	k	10^{3}	1 000
hecto	h	10^{2}	100
deka	da	10	10
SI base unit (no prefix)			1
deci	d	10^{-1}	0 . 1
centi	c	10^{-2}	0 . 01
milli	m	10^{-3}	0 . 001
micro	μ	10^{-6}	0 . 000 001
nano	n	10^{-9}	0 . 000 000 001
pico	p	10^{-12}	0 . 000 000 000 001
femto	f	10^{-15}	0 . 000 000 000 000 001
atto	a	10^{-18}	0 . 000 000 000 000 000 001

Note: The three prefixes underlined are the most commonly used.

Fig. 53-5. SI prefixes.

0.452	Prefix of kilo	kilometre
4.52	Prefix of hecto	hectometre
45.2	Prefix of deka	dekametre
452.0	Base unit of	metre
4520.0	Prefix of deci	decimetre
45200.0	Prefix of centi	centimetre
452000.0	Prefix of milli	millimetre

Fig. 53-6. Examples of using prefixes.

DRAWING TO SCALE

All construction drawings are drawn to scale except for schematics and tables. Drawing to scale refers to a drawing that has been reduced proportionally from actual size. Reduced size allows the total drawing to be placed on the drawing sheet. In some cases the drawing may be enlarged proportionally, especially a detail for clarity purposes.

RECOMMENDED SI SCALES FOR CONSTRUCTION AND MAP DRAWINGS

Recommended SI scales for construction and map drawings are listed in the table in Fig. 53-7.

TYPE OF DRAWING	SI SCALE	NEAREST ENGLISH EQUIVALENTS
Engineering Drawing	1:1	Full size
	1:2	6 in. to 1 ft.
	1:5	3 in. to 1 ft.
	1:10	1 in. to 1 ft.
	1:20	1/2 in. to 1 ft.
Architectural and Working Drawings	1:50	1/4 in. to 1 ft.
	1:100	1/8 in. to 1 ft.
	1:200	1/16 in. to 1 ft.
Site Plans	1:500	1/32 in. to 1 ft.
	1:1000	1 in. to 100 ft.
	1:2000	1 in. to 200 ft.
Surveys	1:5000	1 ft. to 1 mile
	1:10000	6 in. to 1 mile
	1:20000	3 in. to 1 mile
	1:50000	1 in. to 1 mile
Maps	1:100000	1/2 in. to 1 mile
	1:200000	1/4 in. to 1 mile
	1:500000	1/8 in. to 1 mile
	1:1000000	1/16 in. to 1 mile

Fig. 53-7. Recommended SI scales for construction and map drawing.

COMPARING THE COMMON MEASURING UNITS

To convert the many individual customary measurement units which we use each day, we simply multiply the unit and a common numerical factor. The resultant figure is the SI equivalent base unit and prefix. See Fig. 53-8.

USEFUL CONVERSIONS

The construction industry operates on standard sizes; quality control demands it. The conversion of American construction will, no doubt, bring about revision of these standard sizes. See Fig. 53-9. For example, the standard size of a sheet of plywood is 48 in. by 96 in. Direct conversion would be 1219.2 millimetres by 2438.4 millimetres. The American National Metric Council has recommended that when standards are revised they be changed to read to the nearest whole metric value. The above plywood size may be established at 1200 mm by 2400 mm.

In either case, think of the many other changes this one standard revision will make. In residential construction, a sheet size change will affect the stud spacing, joist spacing and ceiling heights.

The universal adoption of the SI metric system is underway, Fig. 53-10. It is essential that we learn the basics of this system of weights and measurements. All of the old standard sizes will eventually be changed or converted to metric sizes.

ENGLISH CUSTOMARY UNIT	MULTIPLIED BY	(equals)	METRIC EQUIVALENT
Length:			
inches	25.4		millimetres
feet	30.48		centimetres
yards	0.9		metres
miles	1.6		kilometres
Area:			
square inches	6.5		square centimetres
square feet	0.09		square metres
square yards	0.8		square metres
square miles	2.6		square kilometres
acres	0.4		square hectometres (hectares)
Mass:			
ounces	28.0		grams
pounds	0.45		kilograms
tons (short ton, 2000 lb.)	0.9		megagrams (metric ton)
Liquid Volume:			
ounces	30.0		millilitres
pints	0.47		litres
quarts	0.95		litres
gallons	3.8		litres

Fig. 53-8. Approximate conversion of customary to metric units.

Lumber	board feet	to	cubic metre or (board metre measure)
Plywood, particle board, hardboard, celotex	square feet	to	square metre
Molding and trim	lineal foot	to	metre
Steel beams	pounds/foot	to	kilograms/metre
length	feet	to	metres
height	inches	to	centimetres
width	inches	to	centimetres
thickness	inches	to	millimetres
Sheet metal, thickness	decimal inch	to	millimetre
sheet size	inches	to	centimetres
Nails	pound	to	kilogram
Water, paint, other liquids	gallon	to	litre
Cement (sack)	pound	to	kilogram
Sand	short ton	to	metric ton
Gravel	short ton	to	metric ton
Window units	inches	to	centimetres

Fig. 53-9. Useful conversions to be used in the construction industry.

Fig. 53-10. Workers will one day communicate spaces and length in metres. Excavated material will be measured in cubic metres. Loads to be supported will be calculated in kilograms. The relatively new language of measurement will require that each person be trained or retrained for its use. (Dickson-Basford, Inc.)

DICTIONARY OF CONSTRUCTION TERMS

The following definitions have been selected from a list compiled by the U.S. Department of Housing and Urban Development.

A

ABS: Acrylontrile Butadiene Styrene, a common type of plastic used for water and drainage pipe.

ABSTRACT: A document providing a history of all ownerships of a piece of property.

ACOUSTICAL TILE: Special tile for walls and ceilings made of mineral, wood, vegetable fibers, cork or metal. Its purpose is to control sound while providing cover.

ADHESIVES: Products used to bond several pieces of material together. Commonly referred to as glue.

AGGREGATES: Materials such as sand, rock, ground perlite and vermiculite used to make concrete.

AIR DUCT: Pipes that carry warm and cold air to rooms and back to furnace or air-conditioning system.

AMPERE: Rate of flow of electricity through electric wires.

APRON: A paved area such as the juncture of a driveway with the street or with a garage entrance.

ARCH: A curved lintel (top) used to span wide openings such as doorways.

ARCHITECT: Person trained to design residences and buildings.

ASPHALT: Mineral pitch used as a waterproofing material. It is extensively used in construction of roofs, exterior walls and in paints.

B

BACKFILL: The gravel or earth replaced in the space around a building wall after foundations are in place.

BALUSTERS: Upright supports of a hand rail.

BALUSTRADE: A row of balusters topped by a rail, edging a balcony or a staircase.

BASEBOARD: A board along the floor placed against walls and partitions to hide gaps and protect plaster or dry wall.

BATT: Insulation in the form of a blanket, rather than loose filling. Used in walls and ceilings.

BATTEN: Small thin strips covering joints between wider boards on exterior building surfaces.

BEAM: One of the principal horizontal wood or steel members of a building.

BEARING WALL: A wall that supports a floor or roof of a building.

BEDROCK: Solid formations of rock usually in layers under several thicknesses of soil, clay and rock; however, sometimes found exposed on the surface.

BENZENE: Solvent sometimes used in paints; derived from petroleum.

BIB OR BIBCOCK: A water faucet to which a hose may be attached. Also called a hose bib or sill cock.

BLEEDING: Seeping of resin or gum from lumber. Term also used in referring to the process of drawing air from water pipes.

BOARD FOOT MEASURE: A volume of wood which occupies the space of 1 ft. in width, 1 ft. in height, and 1 ft. in length (a cubic foot).

BOLTS: Metal fastening devices threaded to receive a nut.

BRACE: A piece of wood or other material used to form a triangle with some part of a structure to reinforce it or hold it up.

BRACED FRAMING: Construction technique using posts and crossbracing for greater stiffness.

BRICK: A building material made by burning and "firing" natural earth minerals, namely clay, into stone-like products. Produced in many colors, sizes, shapes and textures.

BRICK VENEER: Brick used as the outer surface of a framed wall.

BRIDGING: Small wood or metal pieces placed diagonally (crisscrossed) between floor joists.

BUILDING CODES: Written standards regulating and controlling design, construction methods and maintenance of all structures within a given political jurisdiction. Designed to protect private and public welfare.

BUILDING PAPER: Heavy paper used in walls or roofs as dampproofing.

BUILT-UP ROOF: A roofing material applied in sealed, waterproof layers where there is only a slight slope to the roof.

BUTT JOINT: Joining point of two pieces of wood or molding.

BX CABLE: Cable for conducting electricity. Consists of two or more wires wrapped in rubber with a flexible steel outer covering.

C

CANTILEVER: A projecting beam or joist, not supported at one end, used to support an extension of a structure.

CARRIAGE: The member which supports the steps or treads of a stair.

CASEMENT: A window sash that opens on hinges along a vertical edge.

CASING: Door and window framing.

CAVITY WALL: A hollow wall providing an insulating air space between firmly linked masonry walls.

CEMENT: A bonding material for masonry building materials.

CHAIR RAIL: Wooden molding on a wall around a room at the level of a chair back.

CHAMFERED EDGE: Molding with pared-off corners.

CHASE: A groove in a masonry wall or through a floor to accomodate pipes or ducts.

CHIMNEY CAP: Concrete capping around the top of chimney bricks and around the flues to protect masonry from the elements.

CHIMNEY BREAST: The horizontal projection, usually inside a building, of a chimney from the wall in which it is built.

CIRCUIT: The path or loop which electrical current follows to complete its movement.

CIRCUIT BREAKER: A safety device which opens (breaks) an electric circuit automatically when it becomes over-loaded.

CISTERN: A tank, often underground, to catch and store rain water.

CLAPBOARD: A long, thin board, thicker on one edge, overlapped and nailed on for exterior siding.

CLIENT: A person for whom a service is being provided. In construction, the customer for whom structure is being built.

COKE: Coal which has had the gases burnt off. Used in the production of iron and steel.

COLLAR BEAM: A horizontal beam fastened above the lower ends of rafters to add rigidity.

COLONADE: A row of columns.

COMPACTION: To pack down as earth; to increase density of road bed by rolling with drum-like apparatus.

CONCRETE: A building material made by mixing cement and various aggregates of sand and gravel.

CONDUCTIVITY: 1—The rate at which a material will transfer heat. 2—The rate at which a material will transfer electrical current.

CONDUIT: Small metal tubes, easily bent, to house the electrical conductors.

CONSTRUCTION: Those activities of man requiring skill and knowledge of materials, tools, processes and organized management to erect structures for shelter and other purposes.

CONSTRUCTION MATERIAL: The building product used to erect and finish a structure; includes wood, metals, concrete, gypsum, sand, gravel, plastics and glass.

CONSULTANT: A professional called in to advise in some phase of important projects.

CONTOUR MAP: A map drawn to show the changes in elevation of a piece of land. Lines connect like elevations.

CONTRACT: A legal document agreed upon by two or more competent persons to do or not to do something. In construction, an agreement between builder and owner to erect the structure in a certain specified manner and at a specified cost.

COPING: Tile or brick used to cap or cover the top of a masonry wall.

CORBEL: A horizontal projection from a wall, forming a ledge or supporting a structure above it.

CORNER BEAD: A strip of wood or metal for protecting the outside corners of plastered walls.

CORNICE: Horizontal projection at the top of a wall or under the overhanging part of the roof.

CORROSION: The accumulation of oxidized metal on the surface of that metal.

COURSE: A horizontal row of bricks, cinder blocks or other masonry materials.

COVE LIGHTING: Concealed light sources behind a cornice or horizontal recess which direct the light upon a reflecting ceiling.

CRAWL SPACE: A shallow, unfinished space beneath the first floor of a house which has no basement. Used for visual inspection and access to pipes and ducts. Also a shallow space in the attic, immediately under the roof.

CREOSOTE: A preservative for wood.

CRITICAL PATH METHOD: A plan or method of scheduling a construction project.

CRIPPLES: Short framing members above and below windows.

D

DAMPPROOFING: Application of sealants to prevent water seepage into the substructure.

DEED: A written document establishing the terms of transfer of property from one person to another.

DIMENSION LUMBER: Lumber two or more inches thick used in construction. Can be up to 12 inches wide.

DOME: A round-topped structure, simply thought of as a series of arches all having the same center.

DOOR BUCK: Rough frame of a door.

DORMER: A projection in a sloping roof to provide light and additional room to the interior space.

DOUBLE GLAZING: An insulating window pane formed of two thicknesses of glass with a sealed air space between them.

DOUBLE HUNG WINDOWS: Windows with an upper and lower sash, each supported by spring-tensioned counter-weights.

DOWNSPOUT: A spout or pipe to carry rain water from a roof or gutter.

DOWNSPOUT LEADER: A pipe for conducting rain water from the roof to a cistern or to the ground by way of a downspout.

DOWNSPOUT STRAP: A piece of metal attaching down-spouts to eaves or walls of a building.

DRIP: The projecting part of a cornice which sheds rain water.

DRY WALL: A wall surface of plasterboard or material other than plaster.

DUCTS: Sheet metal areas, may be round or square, for the channeling of air from one space to a second space.

E

EAVES: The extension of a roof beyond outside walls.

EFFLORESCENCE: White powder that forms on the surface of brick.

EFFLUENT: Treated sewage from a septic tank or sewage treatment plant.

ELEVATION: Simple rectangular drawing of front surface of an object.

EMINENT DOMAIN: A constitutional provision of a governmental agency to take or condemn private property for public purposes, even against the objection of the owner, at a fair and equitable price.

EMULSION: In paints, a preparation in which minute particles of one liquid are suspended in another.

EPICENTER: The part of the earth's surface directly above the focus of an earthquake.

EXCAVATION: Removal of earth and rock from a construction site usually to receive foundation.

F

FASCIA: Member used for outer face of a cornice, nailed to ends of rafters.

FELDSPAR: A reddish mineral found in natural stone used for building material.

FILL-TYPE INSULATION: Loose insulating material which is poured or blown into wall spaces or ceiling spaces mechanically.

FLASHING: Noncorrosive metal used around valleys or junctions in roofs and exterior walls to prevent leaks.

FLINT: Hard quartz rock which produces sparks when struck by steel.

FLITCH: A thin layer or slab of wood from which pieces of veneer are cut.

FLOOR JOISTS: Framing pieces which rest on outer foundation walls and interior beams or girders and support floorboards.

FLUE: A passageway in a chimney for conveying smoke, gases or fumes to the outside air.

FLUX: Material that prevents oxidizing of metals; used in soldering, welding and other metal-joining techniques or processes.

FOOTING: Concrete base on which a foundation sits.

FOUNDATION: Lower parts of walls on which the structure is built. Foundation walls of masonry or concrete are mainly below ground level.

FRAMING: The rough lumber of a house including joists, studs, rafters and beams.

FURRING: Thin wood or metal applied to a wall to level the surface for lathing, boarding, or plastering, to create an insulating air space, and to dampproof the wall.

FUSE: Safety device used in an electric panel box which opens (breaks) an electrical circuit when it becomes overloaded.

G

GABLE: Triangular part of a wall under the inverted "v" of the roof line.

GALVANIZING: The process of coating iron and steel with molten zinc to form a protective coating against oxidation (rust).

GAMBREL ROOF: A roof with two pitches, designed to provide more space on upper floors. The roof is steeper on its lower slope and flatter toward the ridge.

GEOLOGY: The study and science of earth's structure.

GIRDER: A main member in a framed floor supporting the joists which carry the flooring boards. It carries the weight of a floor or partition.

GLASS: A clear, transparent material produced by melting silica and other minerals. Formed into sheets, used as window panes, blocks and other items for construction.

GLAZING: Fitting glass into windows or doors.

GRADE LINE: The point at which the ground rests against the foundation wall.

GREEN LUMBER: Lumber which has been inadequately dried and which tends to warp or "bleed" resin.

GROUNDS: Pieces of wood embedded in plaster of walls to which skirtings are attached. Also wood pieces used to stop the plaster work around doors and windows.

GUSSET: A brace or bracket used to strengthen a structure.

GUTTER: A channel below and parallel to eaves for taking away rain water.

GYPSUM: A soft, chalky-white mineral found in the earth. Used in the production of wallboard for "dry wall" construction.

H

HARDBOARD: A thin sheet of construction material manufactured by heating and applying pressure to very small wood fibers forming them into a sheet.

HARDWOOD: The wood from broad leaved trees such as oak or maple.

HEADERS: Double wood pieces supporting joists in a floor or double wood members placed on edge over windows and doors to transfer the roof and floor weight to the studs.

HEEL: The end of a rafter that rests on the wall plate.

HINGE: The hardware which attaches a door to its frame but allows it to open and close.

HIP: The external angle formed by the juncture of two slopes of a roof.

HIP ROOF: A roof that slants upward on three or four sides.

HONEYCOMBING: In concrete, the result of trapped air in the pour, causing voids in the concrete.

I

INSPECTIONS: The process of evaluation of the techniques, materials, and workmanship being carried out by the contractor during construction.

INSULATION: 1—Resistance to heat transfer. 2—A material used to prevent heat or sound transfer.

INTEREST: Money paid by the borrower for the use of larger sums of money.

IRON: The basic ingredient of steel, usually has a high carbon content and other impurities.

J

JAMB: An upright surface that lines an opening for a door or window.

JOIST: A small, rectangular sectional member arranged parallel from wall to wall in a building, or resting on beams or girders. They support a floor or the laths and furring strips of a ceiling.

K

KILN-DRIED: Artificial drying of lumber, superior to most lumber that is air dried.

KING-POST: The middle post of a truss.

L

LAG-SCREWS OR COACH-SCREWS: Large, heavy screws, used where great strength is required, as in heavy framing or when attaching ironwork to wood.

LALLY COLUMN: A steel tube sometimes filled with concrete, used to support girders or other floor beams.

LAMINATION: The process of forming a product by adhering thin sheets together into a unit, such as veneer into plywood.

LAND DESCRIPTION: A formula by which a tract of land can be identified. It is based on relative known points or monuments.

LATH: One of a number of thin narrow strips of wood nailed to rafters, ceiling joists or wall studs. Makes a groundwork or key for slats, tiles or plastering.

LEACHING BED: Drainage tiles in trenches carrying treated liquid waste from septic tanks.

LEDGER: A piece of wood attached to a beam to help support joists.

LEHR: Glassmaking machine.

LETTER OF COMMITMENT: A written document stating to the borrower the intention of a lender to provide money to borrow for building purposes.

LEVER: A principle by which a relationship between force and counter force multiplies with distance.

LIEN: A legal claim against the property of another for the satisfaction of a debt.

LIME: A building material (calcium oxide) obtained by the action of heat on limestone. Used in making masonry mortar.

LIMESTONE: A sedimentary stone, high in calcium content, found commonly all over the earth's surface.

LINTEL: The top piece over a door or window which supports walls above the opening.

LIQUID SILICONE: Used to provide a completely transparent sealant over exterior masonry walls.

LOAD-BEARING WALL: In building construction a wall which supports weight of floors or structure above it.

LOCKSET: The mechanical device installed into a door which secures door against entry without a key.

LOUVER: An opening with horizontal slats to permit passage of air, but excluding rain, sunlight, and view.

LUMBER: Strips of wood sawed from logs to make boards, planks and timbers.

M

MANTEL: A ledge made of masonry material or wood located above a fireplace opening.

MASONRY: Art of constructing walls and floors using brick, stone, tile or similar materials.

METES AND BOUNDS: Identifying parcel of irregularly shaped tract of land by establishing boundaries by measuring angles and distances beginning at natural marker.

MOISTURE BARRIER: Treated paper or metal that retards or bars water vapor; used to keep moisture from passing into walls or floors.

MOLDING: A strip of decorative material having a plane or curved narrow surface prepared for ornamental application. These strips are often used to hide gaps at wall junctures.

MORTAR: A mixture of cement, sand and hydrated lime for bonding brick and other masonry units.

MORTGAGE: A written contract whereby real estate property is pledged to secure a loan.

MULLION: Slender framing which divides lights or panes of windows.

N

NAILS: Slender, round metal devices to pin pieces of material together.

NEWEL: The upright post or the upright formed by the inner or smaller ends of steps about which steps a circular staircase wind. In a straight flight staircase, the principal post at the foot or the secondary post at the landing.

NOSING: The rounded edge of a stair tread.

P

PARGET: A rough coat of mortar applied over a masonry wall as protection or finish; may also serve as a base for an asphaltic waterproofing compound below grade.

PARTICLE BOARD: Construction material in sheet form. Manufactured by gluing small chips of wood particles into sheets under pressure.

PARQUET: Type of floor made by assembling small blocks of wood into squares or rectangles.

PENNY: Length designation of nails.

PERLITE: A type of volcanic rock which expands under heat. Used to make lightweight concrete.

PIG IRON: The first metallic-like material produced in the making of iron. Contains many impurities and must be refined further.

PIGMENT: The body of paint or the solid portion. Also contains the color.

PILASTER: A projection on the foundation wall used to support a floor girder or to stiffen the wall.

PILING: Wood, steel or concrete columns driven into the earth to reach bedrock and provide a support for large structures.

PIPE: Hollow cylinder of metal, usually steel, used for transporting liquids and gases. Often used for structural building materials.

PITCH: The angle of slope of a roof.

PLASTERBOARD: (See dry wall.) Gypsum board, used instead of plaster.

PLATE GLASS: A thick, strong glass used for store fronts and other large glass areas.

PLATES: Lengths of 2 by 4 material forming upper and lower sides of walls. The bottom member of the wall is the sole plate and the top member is the rafter plate.

PLENUM: A chamber which can serve as a distribution area for heating or cooling systems.

PLYWOOD: A construction material in sheet form, manufactured by gluing an odd number of sheets of veneer together running in alternate directions.

POINTING: Treatment of joints in masonry by filling with mortar to improve appearance or protect against weather.

PORTLAND CEMENT: A bonding material used to produce concrete.

POST-AND-BEAM CONSTRUCTION: Wall construction in which beams are supported by heavy posts rather than many smaller studs.

POST-AND-LINTEL CONSTRUCTION: Simplest form of construction. Comprised of two upright columns and a third horizontal member laid across their top surfaces.

INDEX

Index

Index